International Union of Theoretical
and Applied Mechanics

Thermoinelasticity

Symposium East Kilbride
June 25—28, 1968

Edited by Bruno A. Boley
Ithaca, N.Y., U.S.A.

Springer-Verlag Wien · New York 1970

With 133 Figures

ISBN-13: 978-3-7091-8246-8 e-ISBN-13: 978-3-7091-8244-4
DOI: 10.1007/.978-3-7091-8244-4

Softcover reprint of the hardcover 1st edition 1970
Library of Congress Catalog Card Number 75-94050

Title No. 9252

Preface

The proposal for a Symposium on high temperature effects in solids originated in 1964, at which time a pilot committee, headed by Professor F. K. G. ODQVIST, was formed. Upon the recommendation of this committee, a site for the Symposium and a Scientific Organizing Committee were chosen, and some guidelines delineating the field to be covered were established.

The Symposium was held at the National Engineering Laboratory in East Kilbride (Glasgow), Scotland, from June 25 to 28, 1968. The Scientific Organizing Committee was composed of the following members:

Professor B. A. BOLEY, Chairman (U. S. A.)

Professor F. BUCKENS (Belgium)

Professor F. K. G. ODQVIST (Sweden)

Professor W. OLSZAK (Poland)

Professor H. PARKUS (Austria)

Dr. D. G. SOPWITH (U. K.)

The Local Organizing Committee was subsequently formed; it was directed by Dr. SOPWITH with Mr. R. BRODIE as Secretary, and Mr. C. E. PHILLIPS and Dr. J. GLEN as additional members. It deserves ample praise for excellent arrangements and hospitality.

In the discussions concerned with the goals of the Symposium, it was decided that its principal aim should be the presentation of a comprehensive view of the behavior of solids and structures at high temperature under any type of inelastic regime. It soon became apparent that this was a very wide field, since it would embrace questions ranging from those of basic thermodynamic to those of engineering analysis and design. Accordingly, an effort was made to have as many aspects of the field as possible represented in the program; this fulfilled the secondary aim of the Symposium, namely that of bringing together persons with different points of view, and of providing them with a forum for open discussion. The lively discussions which followed all papers testify to the fact that this aim was indeed accomplished.

Of course, it was not possible to treat all topics, and regretfully even some very important ones had to be omitted. For example, no paper discussing high-temperature behavior from the point of view of a lattice-type theory was included, no paper on the mechanical behavior of melting or solidifying bodies was presented, and many specialized applications (e.g. in the nuclear or aerospace field) were not treated. Nevertheless, a great deal of ground was covered, the principal topics discussed being the following: fatigue; fracture; the formulation of constitutive relations; thermomechanical coupling; stress analysis; experimental determination of material behavior. The papers themselves as well as the discussions indicated clearly that all of these topics must be considered at the present time as open, and that research is being vigorously pursued in all of them. The present Symposium thus represents an instantaneous picture of the field, and helped bring into focus the activities and interests of a variety of investigators.

Twenty papers were scheduled for presentation, and are included in this volume. Of these, two were not read at the Symposium since their authors were unable to be present.

The attendance of the Symposium comprised 72 persons, including 24 authors and co-authors. The program and the delegates represented 15 countries, namely: Austria (1), Belgium (1), Canada (2), Denmark (1), France (8), Germany (1), United Kingdom (34), Iran (1), Italy (3), Japan (2), Netherlands (1), Poland (2), Sweden (3), U.S.A. (12), U.S.S.R. (no delegates; 3 papers in the Proceedings).

A summary of the program of the Symposium, and a list of participants are included in this volume.

Ithaca, New York, U.S.A., September 1969

BRUNO A. BOLEY

Summary of Program
IUTAM Symposium on Thermoinelasticity

East Kilbride, June 25—28, 1968

Tuesday, June 25, 1968

Tour of National Engineering Laboratories
Informal Cocktail Party

Wednesday, June 26, 1968

Welcoming address by Mr. F. D. PENNY, Director of NEL

First Technical Session

Dr. D. G. SOPWITH, Chairman

W. OLSZAK and P. PERZYNA: "Thermal effects in viscoplasticity".
H. PARKUS and J. L. ZEMAN: "Some stochastic problems of thermo-visco-elasticity".
E. T. ONAT: "On the determination of the number and nature of state variables from phenomenological experiments".

Second Technical Session

Prof. J. F. BESSELING, Chairman

F. K. G. ODQVIST and N. G. OHLSON: "Thermal fatigue and thermal shock investigations".
A. I. SMITH, D. MURRAY and R. H. KING: "An applied mechanics approach to strain concentration effects at elevated temperature".
G. I. BARENBLATT, V. M. ENTOV and R. L. SALGANIK: "The effect of small vibrations on the deformations and fracture of viscoelastic materials". [Read by Dr. J. R. DIXON of NEL.]
R. MAZET: "Sur la normalité de la fonction de fluage lors d'essais de traction d'un dural à 200 °C".

Thursday, June 27, 1968

Third Technical Session

Prof. F. K. G. Odqvist, Chairman

R. A. Schapery: "On the application of a thermodynamic constitutive equation to various non-linear materials".

E. H. Lee: "Viscoelastic stress analysis with thermo-mechanical coupling".

M. J. Crochet and P. M. Naghdi: "On a class of material with fading memory".

Fourth Technical Session

Prof. H. Parkus, Chairman

J. D. Achenbach and G. Herrmann: "Propagation of second-order thermomechanical disturbances in viscoelastic solids".

Jan A. H. Hult: "Cyclic thermal loading in viscoelastic pressure vessels".

W. Johnson and R. Slater: "The dynamic blanking, forging and indenting of hot metals".

A. M. Freudenthal: "Creep acceleration in some metals under rapidly increasing temperature".

Evening: Banquet at Forest Hills Hotel, Aberfoyle, Perthshire.

Friday, June 28, 1968

Fifth Technical Session

Prof. B. A. Boley, Chairman

A. Phillips: "Yield surfaces of pure aluminum at elevated temperatures".

T. R. Tauchert: "Heat generation in viscoelastic solids with emphasis on experimental results".

P. Thirion and R. Chasset: "Influence de la température sur les réactions viscoélastiques de vulcanisats de caoutchouc susceptibles de cristalliser au cours d'essais de traction monoaxiale".

S. Taira and R. Ohtani: "Some problems on creep fracture under combined stress system".

List of Participants

ACHENBACH, Prof. J. D., Department of Civil Engineering, Northwestern University, Evanston, Ill. 60201, U. S. A.

AUSANGÉE, R., Électricité de France, Dèpartement de Mécanique et Technologie Nuclèaire, Les Renardières, Route de Sens, 77 Écuelles, près Moret-sur-Loing, France.

BARROWMAN, E. M., Department of Mechanical Engineering, University of Strathclyde, Glasgow, C. 1, Scotland.

BESSELING, Prof. J. F., Technische Hogeschool Delft, Mekelweg 2, Delft, Netherlands.

BEVERIDGE, Mrs. A. A., National Engineering Laboratory, East Kilbride, Scotland.

BLACKBURN, Dr. W. S., International Research and Development Company Limited, Fossway, Newcastle-upon-Tyne, 6, England.

BLAND, Prof. D. R., Department of Mathematics, The University of Manchester, Manchester 13, England.

BOISSERIE, J. M., Électricité de France, S. E. R. Nu. T. H., 6 Quai Watier, 78 Chatou, Paris, France.

BOLEY, Prof. B. A., Department of Theoretical and Applied Mechanics, Cornell University, Ithaca, N. Y. 14850, U. S. A.

BOLLANI, Dr. G., Laboratorio Ricerche e Controlli Auto-Avia, F. I. A. T., C. Agnelli, 200, Torino, Italy.

BRADLEY, D. M., National Engineering Laboratory, East Kilbride, Scotland.

BUTLER, S., Bristol College of Science and Technology, Ashley Down, Bristol 7, England.

BYCHAWSKI, Dr. Z., Krowoderska 39, m. 11, Kraków, Poland.

CARTER, G. A., Engineering Department, Cambridge University, Trumpington Street, Cambridge, England.

CORBEL, J., Électricité de France, S. E. R. Nu. T. H., 6 Quai Watier, 78 Chatou, Paris, France.

CRAIK, I. C., National Engineering Laboratory, East Kilbride, Scotland.

CROCHET, Dr. M. J., Institut de Mécanique et Mathématiques Appliquées, Université Catholique de Louvain, Heverlee, Belgium.

den Hartog, Prof. J., Department of Mechanical Engineering, Massachusetts Institute of Technology, Cambridge, Mass. 02139, U. S. A.

Dixon, Dr. J. R., National Engineering Laboratory, East Kilbride, Scotland.

Estruch, Dr. B., Imperial Chemical Industries Limited, Wilton Works, PO Box 54, Middlesbrough, Tees-side, England.

Ferguson, F. R., National Engineering Laboratory, East Kilbride, Scotland.

Freudenthal, Prof. A. M., Department of Civil Engineering and Engineering Mechanics, Columbia University, New York, N. Y. 10027, U. S. A.

Glenn, Dr. J., Central Research Department, Colvilles Limited, Meadow Road, Motherwell, Lanarkshire, Scotland.

Graham, Prof. G. A. C., Department of Mathematics, Simon-Fraser University, Burnaby 2, British Columbia, Canada.

Guelin, P., Attaché de Récherches, Laboratoire de Mécanique des Fluides, 44 Avenue Félix Viallet, 38 Grenoble, France.

Henderson, J., National Engineering Laboratory, East Kilbride, Scotland.

Herrmann, Prof. G., Department of Civil Engineering, Northwestern University, Evanston, Ill. 60201, U. S. A.

Hicks, E., United Kingdom Atomic Energy Authority, Reactor Fuel Element Laboratories, Springfields, Salwick, Preston, Lancs, England.

Hopkins, Prof. H., Department of Mathematics, The University of Manchester, Manchester 1, England.

Hult, Prof. J. A. H., Chalmers Tekniska Högskola, Göteborg, Sweden.

Johnson, Prof. W., Institute of Science and Technology, The University of Manchester, Manchester 1, England.

Kennedy, P. A., National Engineering Laboratory, East Kilbride, Scotland.

King, Dr. R. H., National Engineering Laboratory, East Kilbride, Scotland.

Leckie, Prof. F. A., Department of Engineering, University of Leicester, Leicester, England.

Lee, Prof. E. H., Division of Engineering Mechanics, Stanford University, Stanford, Calif. 94305, U. S. A.

Lemaitre, Dr. J., Office National d'Études et de Récherches Aérospatiales (O. N. E. R. A.), 29 Avenue de la Division Leclerc, 92 Châtillon (Seine), France.

Libove, Dr. C., Mechanical and Aerospace Engineering Department, Syracuse University, Syracuse, N. Y. 13210, U. S. A.

Louden, J. S., National Engineering Laboratory, East Kilbride, Scotland.

MACKENZIE, Dr. A. C., Department of Mechanical Engineering, University of Glasgow, Glasgow W. 2, Scotland.

MCPAKE, P. J., National Engineering Laboratory, East Kilbride, Scotland.

MARCONI, Dr. M., Centro Consultivo, Studi e Ricerche, Via dei Pontefici, Roma, Italy.

MAZET, Prof. R., Office National d'Études et de Récherches Aérospatiales (O. N. E. R. A.), 29 Avenue de la Division Leclerc, 92 Châtillon (Seine), France.

MEIKLE, G. G., National Engineering Laboratory, East Kilbride, Scotland.

MIRZAD, Dr. S. H., Ministry of Water and Power, Teheran, Iran.

MURRAY, D., National Engineering Laboratory, East Kilbride, Scotland.

NAGHDI, Prof. P. M., Department of Mechanical Engineering, University of California, Berkeley, Calif. 94720, U. S. A.

ODQVIST, Prof. F. K. G., The Royal Institute of Technology, Stockholm 70, Sweden.

OHLSON, Prof. N. G., The Royal Institute of Technology, Stockholm 70, Sweden.

OHTANI, Prof. R., Department of Mechanical Engineering, Kyoto University, Kyoto, Japan.

OLSZAK, Prof. W., Polish Academy of Sciences, ul. Świętokrzyska 21, Warszawa, Poland.

ONAT, Prof. E. T., Department of Engineering and Applied Science, Yale University, New Haven, Conn. 06520, U. S. A.

OWEN, D. G., Engineering Department, Cambridge University, Cambridge, England.

PARKUS, Prof. H., Institut für Mechanik, Technische Hochschule Wien, Karlsplatz 13, 1040 Wien, Austria.

PERZYNA, Dr. P., Polish Academy of Sciences, ul. Świętokrzyska 21, Warszawa, Poland.

PHILLIPS, Prof. A., Department of Engineering and Applied Science, Yale University, New Haven, Conn. 06520, U. S. A.

PHILLIPS, C. E., National Engineering Laboratory, East Kilbride, Scotland.

QUOZZO, Dr. G., Scuola di Ingegneria Aerospaziale, Università degli Studi, Città Universitaria, Roma, Italy.

RIEDER, Prof. G., Technische Hochschule, 51 Aachen, Federal Republic of Germany.

ROBSON, K., National Engineering Laboratory, East Kilbride, Scotland.

ROGUET, R., Commissariat à l'Énergie Atomique, B. P. 5—10, 75 — Paris XVe, France.

SCHAPERY, Prof. R. A., School of Aeronautics, Astronautics and Engineering Sciences, Purdue University, Lafayette, Ind. 47907, U. S. A.

SLATER, Dr. R. A. C., Institute of Science and Technology, The University of Manchester, Manchester 1, England.

SMITH, A. I., National Engineering Laboratory, East Kilbride, Scotland.

SMITH, I. G., National Engineering Laboratory, East Kilbride, Scotland.

SOPWITH, Dr. D. G., National Engineering Laboratory, East Kilbride, Scotland.

TANAKA, Prof. K., Department of Aeronautical Engineering, Kyoto University, Kyoto, Japan.

TAUCHERT, Prof. T. R., Department of Aerospace and Mechanical Sciences, Princeton University, Princeton, N. J. 08540, U. S. A.

THIRION, Dr. P., Institut Français du Caoutchouc, 42, Rue Scheffer, Paris XVIe, France.

TOWNLEY, Dr. C., Central Electricity Generating Board, Berkeley Nuclear Laboratories, Berkeley, Gloucestershire, England.

VALLIAPPAN, S., School of Engineering, University College of Swansea, Singleton Park, Swansea, Wales.

VAUGHAN, Dr. H., Department of Mechanical Engineering, University of British Columbia, Vancouver 8, British Columbia, Canada.

WOLFE, Dr. M. O. W. Ministry of Technology, Room 228, St. Giles Court, 1—13 St. Giles High Street, London, W. C. 2, England.

Contents

Propagation of Second-Order Thermomechanical Disturbances in Viscoelastic Solids

By

J. D. Achenbach and **G. Herrmann**

Evanston, Ill. (U.S.A.)

Summary

The change of magnitude of a strain discontinuity is studied as it propagates in a heat conducting viscoelastic medium. The paper includes an investigation of the effects of thermomechanical coupling, viscous flow, second-order strains, and temperature-dependent material properties.

Introduction

If a continuous body is suddenly subjected to a mechanical or a thermal disturbance on its bounding surface, a signal will propagate into the interior with a finite velocity, provided that the material shows initial elastic behavior. In this paper a surface traction varying in time as a step function is applied, and interest is focused on the change of magnitude of the propagating discontinuity in the strain field, particularly as caused by the coupling of mechanical and thermal disturbances (anelasticity) and as caused by viscous flow of the solid (inelasticity). The wave propagation problem is formulated in one dimension, and in addition to thermomechanical coupling and viscoelasticity we also include second-order strains and temperature-dependent material properties, which are of importance for certain viscoelastic solids, particularly polymers.

As the point of departure we take a Fréchet-type expansion of the free energy in Stieltjes convolutions, in which terms of the third order of strains and temperature increments are included. By established methods the free energy leads to expressions relating the stress and the specific entropy to the strain and the temperature increment. The system of constitutive equations is completed by a relation between the heat flux and the temperature, for which we choose Fourier's law. By employing the balance laws of momentum and energy, in conjunction with the kinematic condition of compatibility, several conclusions can be drawn with regard to the jumps of field quantities at the wave front. It can be shown that for a conducting solid the temperature is continuous, but the specific entropy suffers a jump of the same order as the strain (as opposed to a nonconductor where this jump is of the third order of the strain). An ordinary differential equation relating the jump in strain to the deformation just behind the wave front and the temperature distribution ahead of the wave front, is established. The temperature distribution ahead of the wave front enters into this equation through the thermal precursor effect due to thermomechanical coupling, and the dependence of the jump on the deformation just behind the wave front is typical for wave propagation problems that include large deformations. If the deformation due to temperature increments ahead of the wave front is neglected, and if the viscoelastic solid is semi-linear, i.e., if it exhibits linear instantaneous response and nonlinear memory response, the differential equation governing the jump in stress reduces to a nonlinear ordinary differential equation with constant coefficients, which is integrable. If the assumption of semi-linearity is dropped, an integrable nonlinear ordinary differential equation governing the jump in stress at small times can be obtained.

Balance Laws and Constitutive Relations

The objective of this paper is to present a discussion of the effects on the propagation of stress discontinuities in viscoelastic solids of such thermodynamic phenomena as heat conduction, thermomechanical coupling, and temperature-dependence of the material properties. Because of this deliberate emphasis on the interaction of mechanical and thermal behavior, geometrical complications are excluded, and the kinematical and thermodynamical variables are all taken as one-dimensional. A motion of a particle is then defined by a scalar function $x(X, t)$, which gives the position at time t of the material particle which occupied the position X in some fixed reference configuration with mass density ϱ_0. The deformation gradient F at X at time t is given by

$$F(X, t) = \partial x(X, t)/\partial X. \tag{1}$$

The particle velocity and the acceleration are, respectively,

$$\dot{x}(X, t) = \partial x(X, t)/\partial t \quad \text{and} \quad \ddot{x}(X, t) = \partial^2 x(X, t)/\partial t^2. \tag{2a, b}$$

In the present one-dimensional treatment, the strain $E(X, t)$ is defined as the gradient of the displacement with respect to X, i.e.,

$$E(X, t) = F(X, t) - 1. \tag{3}$$

Other field quantities that frequently appear in the sequel are: the stress $\sigma(X, t)$, the absolute temperature $T(X, t)$, the specific internal energy $\varepsilon(X, t)$, the specific entropy $\eta(X, t)$, and the heat flux $q(X, t)$.

In the absence of body forces and external heat sources, the laws of balance of momentum and balance of energy for an element $X_1 \leq \leq X \leq X_2$ take the form [1]:

$$\frac{d}{dt} \int_{X_1}^{X_2} \varrho_0 \dot{x} \, dX = \sigma(X_2, t) - \sigma(X_1, t), \tag{4}$$

and

$$\frac{d}{dt} \int_{X_1}^{X_2} \varrho_0 \left(\frac{1}{2} \dot{x}^2 + \varphi + T\eta\right) dX = \sigma(X_2, t) \, \dot{x}(X_2, t) - \\ - \sigma(X_1, t) \, \dot{x}(X_1, t) - q(X_2, t) + q(X_1, t), \tag{5}$$

respectively, where $\varphi(X, t)$ is the specific free energy, defined as

$$\varphi(X, t) = \varepsilon(X, t) - T(X, t) \, \eta(X, t). \tag{6}$$

The balance laws are supplemented by the principle of irreversibility (the Clausius-Duhem inequality), which may be written as (see [1]):

$$\frac{d}{dt} \int_{X_1}^{X_2} \varrho_0 \eta \, dX + \frac{q(X_2, t)}{T(X_2, t)} - \frac{q(X_1, t)}{T(X_1, t)} \geq 0. \tag{7}$$

The mechanical behavior of the continuum model is principally determined by the expression that is chosen for the specific free energy φ. Here we consider a viscoelastic material for which φ depends on the temporal histories of both the strain and the temperature increment $\theta(X, t)$, where the latter is defined as

$$\theta(X, t) = T(X, t) - T_0. \tag{8}$$

In eq. (8), T_0 is the reference temperature. Restricting attention to a homogeneous medium which is initially of constant temperature T_0, and at rest in an unstressed state, we can express $\varrho_0 \varphi$ in terms of second- and third-order functionals as

$$\varrho_0 \varphi = I^{(2)}(t; G_2, E, E) + I^{(2)}(t; K_2, E, \theta) + I^{(2)}(t; M_2, \theta, \theta) + \\ + I^{(3)}(t; G_3, E, E, E) + I^{(3)}(t; L_3, E, E, \theta) + I^{(3)}(t; N_3, E, \theta, \theta) + \\ + I^{(3)}(t; M_3, \theta, \theta, \theta) + O(E^4), \tag{9}$$

1*

where the Stieltjes integrals $I^{(2)}$ and $I^{(3)}$ are defined as

$$I^{(2)}(t; F_2, f, g) = \frac{1}{2!}\int\limits_{0^-}^{t}\int\limits_{0^-}^{t} F_2(t-\tau_1, t-\tau_2)\, df\, dg, \tag{10}$$

and

$$I^{(3)}(t; F_3, f, g, h) = \frac{1}{3!}\int\limits_{0^-}^{t}\int\limits_{0^-}^{t}\int\limits_{0^-}^{t} F_3(t-\tau_1, t-\tau_2, t-\tau_3)\, df\, dg\, dh. \tag{11}$$

The kernels in $I^{(n)}$ are symmetric functions of their arguments.

If the kernels are constants, eq. (9) reduces to the free energy for an elastic solid, where second-order strains and linear temperature-dependence of the material properties, as well as thermomechanical coupling, are taken into account. Denoting by $\overline{G}_2$, $\overline{N}_3$, etc., the initial values of the relaxation functions, the specific free energy of the corresponding elastic solid is thus obtained as

$$\varrho_0\varphi_{ie} = \frac{1}{2}\,\overline{G}_2E^2 + \frac{1}{2}\,\overline{K}_2E\theta + \frac{1}{2}\,\overline{M}_2\theta^2 + \frac{1}{6}\,\overline{G}_3E^3 + $$
$$+ \frac{1}{6}\,\overline{L}_3E^2\theta + \frac{1}{6}\,\overline{N}_3E\theta^2 + \frac{1}{6}\,\overline{M}_3\theta^3 + O(E^4). \tag{12}$$

A specific free energy of the type eq. (12) was studied by HERRMANN [2], who also identified the physical significance of the constants.

If in eq. (9) we set $G_3(\tau_1, \tau_2, \tau_3) \equiv 0$, an expression for the free energy, which was studied earlier by ACHENBACH, VOGEL and HERRMANN [1], is obtained. For $G_3(\ ) \equiv 0$, $L_3(\ ) \equiv 0$, $N_3(\ ) \equiv 0$, $M_3(\ ) \equiv$ $\equiv 0$, the free energy reduces to a form which yields a linear theory of viscoelasticity with thermomechanical coupling, as discussed by CHRISTENSEN and NAGHDI [3]. Here we shall consider the full form of eq. (9), i.e., a second-order theory of viscoelasticity with thermomechanical coupling and temperature-dependent material properties. We shall consider only materials with initial elasticity, i.e., $\overline{G}_2 \neq 0$.

The main result of studies on the thermodynamics of materials with memory is a theorem relating the stress and specific entropy functionals to the specific free energy functional. This theorem states that [4]

$$\sigma(X, t) = \varrho_0 D_E\varphi, \tag{13}$$

$$\eta(X, t) = -D_\theta\varphi. \tag{14}$$

Thus, once φ is postulated, one can determine the functional representing σ and η by taking the partial derivatives of φ with respect to the present values of the strain and the temperature increment, respectively, while keeping the past histories fixed. To actually carry out these operations on the form of φ as defined by eq. (9), it is convenient to rewrite eq. (9) by carrying out several integrations by parts of the type:

$$\begin{aligned} I^{(2)}(F_2, f, g) = {} & \frac{1}{2}\,\overline{F}_2 f g + \frac{1}{2}\, f J\{F_{2,2}(0, t-\tau_2); g\} + \\ & + \frac{1}{2}\, g J\{F_{2,1}(t-\tau_1, 0); f\} + \\ & + \frac{1}{2}\, J\{F_{2,12}(t-\tau_1, t-\tau_2); f, g\}, \end{aligned} \tag{15}$$

$$\begin{aligned} I^{(3)}(F_3, f, f, g) = {} & \frac{1}{6}\,\overline{F}_3 f^2 g + \frac{1}{6}\, f^2 J\{F_{3,3}(0, 0, t-\tau_3); g\} + \\ & + \frac{1}{3}\, f g J\{F_{3,1}(t-\tau_1, 0, 0); f\} + \\ & + \frac{1}{3}\, f J\{F_{3,13}(t-\tau_1, 0, t-\tau_3); f, g\} + \\ & + \frac{1}{6}\, g J\{F_{3,12}(t-\tau_1, t-\tau_2, 0); f, f\} + \\ & + \frac{1}{6}\, J\{F_{3,123}(t-\tau_1, t-\tau_2, t-\tau_3); f, f, g\}. \end{aligned} \tag{16}$$

In eqs. (15) and (16), a bar-superscript denotes, as before, a value of a function for vanishing argument, thus

$$\overline{F}_3 = F(0, 0, 0). \tag{17}$$

A subscript comma denotes a differentiation with respect to the appropriate argument, i.e.,

$$F_{2,1} = \frac{\partial F(s_1, s_2)}{\partial s_1}; \qquad F_{3,13} = \frac{\partial^2 F(s_1, s_2, s_3)}{\partial s_1\, \partial s_3}; \text{ etc.} \tag{18a, b}$$

The functionals $J\{\ \}$ are defined as

$$J\{F_{2,12}(t-\tau_1, t-\tau_2); f, g\} = \int_0^t \int_0^t \frac{\partial^2 F_2(t-\tau_1, t-\tau_2)}{\partial(t-\tau_1)\,\partial(t-\tau_2)}\, f(\tau_1)\, g(\tau_2)\, d\tau_1\, d\tau_2 \tag{19}$$

and

$$J\{F_{3,12}(t-\tau_1, t-\tau_2, 0\}; f, g\} = \int_0^t \int_0^t \frac{\partial F_3(t-\tau_1, t-\tau_2, 0)}{\partial(t-\tau_1)\,\partial(t-\tau_2)}\, f(\tau_1)\, g(\tau_2)\, d\tau_1\, d\tau_2. \tag{20}$$

Other forms of the type $J\{\ \}$ appearing in eqs. (15) and (16) are easily constructed if cognizance is taken of eqs. (19) and (20). After some manipulation, the free energy may then be rewritten as

$$\varrho_0 \varphi = a_{00} + (a_{10} + a_{11}\theta + a_{12}\theta^2)E + (a_{20} + a_{21}\theta)E^2 + \\ + a_{30}E^3 + a_{01}\theta + a_{02}\theta^2 + a_{03}\theta^3. \tag{21}$$

In eq. (21),

$$a_{00} = \frac{1}{2} J\{G_{2,12}(t-\tau_1, t-\tau_2); E, E\} + \frac{1}{2} J\{K_{2,12}(t-\tau_1, t-\tau_2); E, \theta\} + \\ + \frac{1}{2} J\{M_{2,12}(t-\tau_1, t-\tau_2; \theta, \theta\} + \\ + \frac{1}{6} J\{G_{3,123}(t-\tau_1, t-\tau_2, t-\tau_3; E, E, E\} + \\ + \frac{1}{6} J\{L_{3,123}(t-\tau_1, t-\tau_2, t-\tau_3); E, E, \theta\} + \\ + \frac{1}{6} J\{N_{3,123}(t-\tau_1, t-\tau_2, t-\tau_3); E, \theta, \theta\} + \\ + \frac{1}{6} J\{M_{3,123}(t-\tau_1, t-\tau_2, t-\tau_3); \theta, \theta, \theta\} \tag{22}$$

$$a_{10} = J\{G_{2,1}(t-\tau_1, 0); E\} + \frac{1}{2} J\{K_{2,1}(t-\tau_1, 0); \theta\} + \\ + \frac{1}{2} J\{G_{3,12}(t-\tau_1, t-\tau_2, 0); E, E\} + \\ + \frac{1}{3} J\{L_{3,12}(t-\tau_1, t-\tau_2, 0); E, \theta\} + \\ + \frac{1}{6} J\{N_{3,12}(t-\tau_1, t-\tau_2, 0); \theta, \theta\} \tag{23}$$

$$a_{11} = \frac{1}{2} \bar{K}_2 + \frac{1}{3} J\{L_{3,1}(t-\tau_1, 0, 0); E\} + \frac{1}{3} J\{N_{3,1}(t-\tau_1, 0, 0); \theta\} \tag{24}$$

$$a_{12} = \frac{1}{6} \bar{N}_3 \tag{25}$$

$$a_{20} = \frac{1}{2} \bar{G}_2 + \frac{1}{2} J\{G_{3,1}(t-\tau_1, 0, 0); E\} + \frac{1}{6} J\{L_{3,1}(t-\tau_1, 0, 0); \theta\} \tag{26}$$

$$a_{21} = \frac{1}{6} \bar{L}_3 \tag{27}$$

$$a_{30} = \frac{1}{6} \bar{G}_3 \tag{28}$$

$$a_{01} = J\{M_{2,1}(t-\tau_1, 0); \theta\} + \frac{1}{6} J\{L_{3,12}(t-\tau_1, t-\tau_2, 0); E, E\} + \\ + \frac{1}{2} J\{K_{2,1}(t-\tau_1, 0); E\} + \frac{1}{3} J\{N_{3,12}(t-\tau_1, t-\tau_2, 0); E, \theta\} + \\ + \frac{1}{2} J\{M_{3,12}(t-\tau_1, t-\tau_2, 0); \theta, \theta\} \tag{29}$$

$$a_{02} = \frac{1}{2} \overline{M}_2 + \frac{1}{6} J\{N_{3,1}(t-\tau_1, 0, 0); E\} + \frac{1}{2} J\{M_{3,1}(t-\tau_1, 0, 0); \theta\} \tag{30}$$

$$a_{03} = \frac{1}{6} \overline{M}_3. \tag{31}$$

Substitution of eq. (21) into eqs. (13) and (14) yields

$$\sigma(X, t) = a_{10} + a_{11}\theta + a_{12}\theta^2 + 2(a_{20} + a_{21}\theta)E + 3a_{30}E^2 \tag{32}$$

$$-\varrho_0 \eta(X, t) = (a_{11} + 2a_{12}\theta)E + a_{21}E^2 + a_{01} + 2a_{02}\theta + 3a_{03}\theta^2. \tag{33}$$

The system of constitutive relations is completed by a relation between the heat flux q and the absolute temperature T. Here we assume that Fourier's law is valid, i.e.,

$$q = -k \frac{\partial T}{\partial X}, \tag{34}$$

where k is the thermal conductivity, which is considered constant. A material is termed a non-conductor for $k \equiv 0$, and a definite conductor for $k \neq 0$. A modification of Fourier's law to a relation of the type $q + \tau\, \partial q/\partial t = -k\, \partial T/\partial X$ was recently discussed by ACHENBACH [5]. A more general functional relation between heat flux and temperature distribution was postulated by CHRISTENSEN and NAGHDI [3].

The Propagation of Discontinuities

We consider a propagating discontinuity whose position in the reference configuration at time t is denoted by

$$X = Y(t). \tag{35}$$

It is assumed in this paper that the integrity of the material is not affected, i.e., $x(X, t)$ is continuous for all X and t. The position in space, $y(t)$, of the wave front is then given by $y(t) = x(Y(t), t)$, and the velocity of the wave front, as seen by an observer at rest, is defined by

$dy(t)/dt = dx(Y(t), t)/dt$. The intrinsic velocity U, which, crudely speaking, is the velocity of the wave relative to the material, is given by

$$U = \frac{dY(t)}{dt}. \tag{36}$$

A discontinuity of the function $f(X, t)$ at $X = Y(t)$ is denoted, in the usual bracket notation, by

$$[f](t) = f^-(t) - f^+(t) = \lim_{X\to Y(t)^-} f(X, t) - \lim_{X\to Y(t)^+} f(X, t). \tag{37}$$

An indispensable relation for the study of propagating discontinuities is the kinematical condition of compatibility, which for a discontinuous function $f(X, t)$, with discontinuous first-order derivatives, where the discontinuities propagate with velocity U, takes the form

$$\frac{d}{dt}[f](t) = \left[\frac{\partial f}{\partial t}\right] + U\left[\frac{\partial f}{\partial X}\right]. \tag{38}$$

For a function $f(X, t)$ continuous in all X, with jump discontinuities in $\partial f/\partial t$ and $\partial f/\partial X$ at $X = Y(t)$, eq. (38) reduces to

$$\left[\frac{\partial f}{\partial t}\right] = -U\left[\frac{\partial f}{\partial X}\right]. \tag{39}$$

A propagating discontinuity is termed a shock wave if the particle velocity at X changes discontinuously as X is traversed by the front, but the displacement remains continuous. Thus, for a shock wave we have, according to eqs. (1), (2a), (3) and (39):

$$[\dot{x}] = -U[E]. \tag{40}$$

By application of a limiting process to the balance law of linear momentum, eq. (4), it is found that jumps in the particle velocity and the stress must be related by

$$[\sigma] = -\varrho_0 U[\dot{x}]. \tag{41}$$

At the wave front, the balance law of energy becomes

$$-\varrho_0 U\left(\frac{1}{2}[\dot{x}^2] + [\varphi] + [T\eta]\right) = [\sigma\dot{x}] - [q]. \tag{42}$$

The Clausius-Duhem inequality yields

$$\varrho_0 U[\eta] - \left[\frac{q}{T}\right] \geq 0. \tag{43}$$

It can be shown that for shock waves in one-dimensional geometry propagating in a medium satisfying Fourier's law of heat conduction, the temperature remains continuous at the wave front [6]. Thus, one obtains from eqs. (32) and (33):

$$[\sigma] = 2(a_{20} + a_{21}\theta)[E] + 3a_{30}[E^2] \tag{44}$$

and

$$-\varrho_0[\eta] = (a_{11} + 2a_{12}\theta)[E] + a_{21}[E^2]. \tag{45}$$

It is noted once again, see also Ref. [1], that in a definite conductor the jump in specific entropy is of the order of the jump in strain, in contrast to a non-conductor, where the jump in specific entropy is of the order of the third power of $[E]$, see Ref. [7].

We now return to eq. (42), which is rewritten as

$$[\sigma\dot{x}] + \varrho_0 U\left(\frac{1}{2}[\dot{x}^2] + [\varphi]\right) = [q] - \varrho_0 U T[\eta]. \tag{46}$$

After some manipulation we can write

$$[\sigma\dot{x}] + \frac{1}{2}\varrho_0 U[\dot{x}^2] = -\frac{1}{2}U(\sigma^- + \sigma^+)[E]. \tag{47}$$

By employing the condition that the temperature is continuous, we obtain from (21)

$$\varrho_0[\varphi] = (a_{10} + a_{11}\theta + a_{12}\theta^2)[E] + (a_{20} + a_{21}\theta)[E^2] + a_{30}[E^3]. \tag{48}$$

Similarly, eq. (32) yields the result

$$\frac{1}{2}(\sigma^- + \sigma^+)[E] = (a_{10} + a_{11}\theta + a_{12}\theta^2)[E] + (a_{20} + a_{21}\theta)[E^2] + \\ + \frac{3}{2}a_{30}([E]^3 + 2E^+E^-[E]). \tag{49}$$

It is now also easily shown that

$$[E^3] = [E]^3 + 3E^+E^-[E]. \tag{50}$$

Now substituting eq. (49) into eq. (47) and eq. (50) into eq. (48), and subsequently substituting the resulting expressions into eq. (46), we obtain

$$\varrho_0 U T[\eta] - [q] = \frac{1}{2}a_{30}U[E]^3. \tag{51}$$

Returning to eq. (43), we note that as an implication of eq. (51) and of the continuity of the temperature, the principle of irreversibility reduces at the wave front to the statement

$$a_{30}[E]^3 \geq 0. \tag{52}$$

If the mass density of the reference configuration is known, the mass density ϱ at time t is determined through the relation

$$\varrho(X, t) = \varrho_0/F(X, t), \tag{53}$$

where F is the deformation gradient defined by eq. (1). A stress wave is said to be compressive if $[\varrho] > 0$, and expansive if $[\varrho] < 0$. It follows from eqs. (53) and (3) that

$$\operatorname{sgn}[\varrho] = -\operatorname{sgn}[F] = -\operatorname{sgn}[E]. \tag{54}$$

The implications of the principle of irreversibility, eq. (52), and eq. (54), are that a compressive stress wave can exist only if $a_{30} < 0$, and an expansive stress wave can exist only if $a_{30} > 0$. In other words, a requirement for the existence of a stress wave is that $\bar{G}_3$ and $[E]$ have the same sign. This condition may also be viewed as requiring that the curve of stress versus initial elastic strain be concave if the wave is compressive, and convex if the wave is expansive. The material parameter $\bar{G}_3$ is the instantaneous second-order stress-strain modulus at fixed present temperature. If $\bar{G}_3 \equiv 0$, the principle of irreversibility places no restriction on the propagation of stress waves, and one has returned to the semi-linear case.

By combining eqs. (40) and (41), and subsequently substituting eq. (44), the intrinsic velocity U is readily obtained as

$$\varrho_0 U^2 = [\sigma]/[E] = 2(a_{20} + a_{21}\theta) + 3a_{30}[E^2]/[E]. \tag{55}$$

The last term of eq. (55) can be simplified, and we obtain

$$\varrho_0 U^2 = 2(a_{20} + a_{21}\theta) + 6a_{30}E^+ + 3a_{30}[E]. \tag{56}$$

Equation (56) reveals the dependence of the intrinsic velocity on the state of strain and the temperature ahead of the wave front, as shown by the first three terms, and on the magnitude of the jump in strain, as shown by the last term. The dependence on the temperature and strain fields, as manifested by the first two terms, enters through the thermo-mechanical coupling and is also present if second-order strains are neglected. The dependence on E^+ and $[E]$ is typical for the propagation of shock waves if second-order strain effects are included. It is noted that this dependence vanishes for the semi-linear case, i.e., when $\bar{G}_3 \equiv 0$. It can be shown that the intrinsic velocity is supersonic at the front side of a shock wave, and subsonic at the back side.

The transport equation for $[E]$ can be derived in a similar manner, as discussed in Ref. [1]. We start out with the kinematical conditions of compatibility, eq. (38), for E and $\dot{x}$, and then eliminate $[\dot{E}]$ from the two equations to yield

$$\frac{d[\dot{x}]}{dt} - [\ddot{x}] = U \frac{d[E]}{dt} - U^2 \left[\frac{\partial E}{\partial X}\right]. \tag{57}$$

By differentiating the compatibility relation $[\dot{x}] = -U[E]$, eq. (40), with respect to time, we obtain

$$\frac{d[\dot{x}]}{dt} = -U \frac{d[E]}{dt} - \frac{dU}{dt}[E]. \tag{58}$$

Substitution of (58) into (57) yields

$$2U \frac{d[E]}{dt} + \frac{dU}{dt}[E] = -[\ddot{x}] + U^2 \left[\frac{\partial E}{\partial X}\right]. \tag{59}$$

The law of balance of momentum, eq. (4), implies that for $X \neq Y(t)$ we have $\partial\sigma/\partial X = \varrho_0 \ddot{x}$, while for $X = Y(t)$ we have

$$\left[\frac{\partial \sigma}{\partial X}\right] = \varrho_0 [\ddot{x}]. \tag{60}$$

Thus, in view of eq. (60), the transport equation for $[E]$, eq. (59), may be rewritten as

$$2\frac{d[E]}{dt} + \frac{1}{U}\frac{dU}{dt}[E] = -\frac{1}{\varrho_0 U}\left[\frac{\partial \sigma}{\partial X}\right] + U\left[\frac{\partial E}{\partial X}\right]. \tag{61}$$

The evaluation of $[\partial\sigma/\partial X]$ from eq. (32) involves a number of lengthy but straightforward manipulations. The result is

$$\begin{aligned}
\left[\frac{\partial \sigma}{\partial X}\right] = &-\frac{1}{U}\Big\{\bar{G}_{2,1} + J\{G_{3,12}(t-\tau_1, 0, 0); E\} + \frac{1}{3} J\{L_{3,12}(t-\tau_1, 0, 0); \theta\} + \\
&+ \frac{1}{3}\bar{L}_{3,1}\theta + \bar{G}_{3,1}E^+\Big\}[E] + 2\left(\frac{\partial a_{20}}{\partial X}\right)^+[E] + \frac{1}{3}\bar{L}_3\left(\frac{\partial \theta}{\partial X}\right)^+[E] + \\
&+ \bar{G}_3\left(\frac{\partial E}{\partial X}\right)^+[E] + \Big\{\frac{1}{2}\bar{K}_2 + \frac{1}{3}J\{L_{3,1}(t-\tau_1, 0, 0); E\} + \\
&+ \frac{1}{3}J\{N_{3,1}(t-\tau_1, 0, 0); \theta\} + \frac{1}{3}\bar{N}_3\theta + \frac{1}{3}\bar{L}_3 E^+\Big\}\left[\frac{\partial \theta}{\partial X}\right] + \\
&+ \Big\{\bar{G}_2 + J\{G_{3,1}(t-\tau_1, 0, 0); E\} + \frac{1}{3}J\{L_{3,1}(t-\tau_1, 0, 0); \theta\} + \\
&+ \frac{1}{3}\bar{L}_3\theta + \bar{G}_3E^+\Big\}\left[\frac{\partial E}{\partial X}\right] + \frac{1}{3}\bar{L}_3[E]\left[\frac{\partial \theta}{\partial X}\right] - \frac{1}{U}\bar{G}_{3,1}[E]^2 + \\
&+ \bar{G}_3[E]\left[\frac{\partial E}{\partial X}\right].
\end{aligned} \tag{62}$$

As before, a superscript bar denotes the value of the function for vanishing argument (s).

Equation (61) is a non-linear ordinary differential equation for the jump in the strain. The initial condition on $[E]$ is the strain jump applied at $t=0$ at $X=0$. Usually the external disturbance is applied as a traction. If the stress is discontinuous at $X=0$, $t=0$, the strain jump can be determined, however, from eq. (32) at $X=0$, $t=0$, i.e.,

$$[\sigma]_{t=0} = \bar{G}_2[E]_{t=0} + \frac{1}{2}\bar{G}_3[E]^2_{t=0} \tag{63a}$$

or

$$\sigma_0 = \bar{G}_2 E_0 + \frac{1}{2}\bar{G}_3 E_0^2. \tag{63b}$$

It is rather apparent that the equation resulting from the substitution of eq. (62) into eq. (61) poses insurmountable problems for its analytical solution. In physical terms the reason for the difficulties is that heat generated at and behind the wave front flows to positions ahead of the wave front, thus causing the wave front to enter a region that is already in a state of inhomogeneous deformation. The simplest way to make the transport equation for $[E]$ amenable to analytical solution is based on the physical argument that the amount of heat that precurses the wave front is very small, giving rise to very small temperature changes and correspondingly to very small deformations before the wave front arrives. Taking this view, we assume that θ and E vanish ahead of the wave front. It should be mentioned here that this in no way means that the effect of thermomechanical coupling is herewith neglected. The effect is still there, but the assumption implies that it is felt only behind the wave front. It is clear that the assumption is best justifiable for small times.

If the simplifications discussed in the previous paragraph are introduced in eq. (61), we obtain drastic simplifications, and $[\partial\sigma/\partial X]$ reduces to

$$\left[\frac{\partial\sigma}{\partial X}\right] = -\frac{1}{U}\bar{G}_{2,1}[E] + \frac{1}{2}\bar{K}_2\left[\frac{\partial\theta}{\partial X}\right] + \bar{G}_2\left[\frac{\partial E}{\partial X}\right] + \\ + \frac{1}{3}\bar{L}_3[E]\left[\frac{\partial\theta}{\partial X}\right] - \frac{1}{U}\bar{G}_{3,1}[E]^2 + \bar{G}_3[E]\left[\frac{\partial E}{\partial X}\right]. \tag{64}$$

The discontinuity in the gradient of the temperature distribution can be computed from eqs. (34) and (51) as

$$\left[\frac{\partial\theta}{\partial X}\right] = -\frac{1}{k}[q] = \frac{1}{k}\left\{\frac{1}{2}a_{30}U[E]^3 - \varrho_0 U T[\eta]\right\}. \tag{65}$$

By introducing eq. (45), and again assuming that the strains and temperature increments ahead of the wave front are negligible, we can reduce eq. (65) to

$$\left[\frac{\partial \theta}{\partial X}\right] = \frac{U}{k}\left\{\frac{1}{12}\,\bar{G}_3[E]^3 + \frac{1}{6}\,T_0\bar{L}_3[E]^2 + \frac{1}{2}\,T_0\bar{K}_2[E]\right\}. \tag{66}$$

For a medium which is assumed at rest ahead of the wave front, the expression for the intrinsic velocity U, eq. (56), reduces to

$$\varrho_0 U^2 = \bar{G}_2 + \frac{1}{2}\,\bar{G}_3[E]. \tag{67}$$

Substituting eq. (64) into eq. (61), we write

$$2\,\frac{d[E]}{dt} + \frac{1}{2U^2}\,\frac{d(U^2)}{dt}\,[E] = \frac{1}{\varrho_0 U^2}\left\{\bar{G}_{3,1}[E]^2 + \bar{G}_{2,1}[E]\right\} - $$
$$- \frac{1}{\varrho_0 U}\left\{\frac{1}{2}\,\bar{K}_2 + \frac{1}{3}\,\bar{L}_3[E]\right\}\left[\frac{\partial \theta}{\partial X}\right] - \frac{1}{2\varrho_0 U}\,\bar{G}_3[E]\left[\frac{\partial E}{\partial X}\right]. \tag{68}$$

In eq. (68), $[\partial\theta/\partial X]$ and U are defined by eqs. (66) and (67), respectively. If the deformation of the medium ahead of the wave front is neglected, the jumps in stress and strain at the wave front are related by

$$[\sigma] = \bar{G}_2[E] + \frac{1}{2}\,\bar{G}_3[E]^2. \tag{69}$$

Thus, once $[E]$ has been determined from eqs. (68) and (63), we can compute $[\sigma]$ from eq. (69).

It is noted that if viscoelastic and thermal effects are excluded, eq. (68) reduces to

$$2\,\frac{d[E]}{dt} + \frac{1}{2U^2}\,\frac{d(U^2)}{dt}\,[E] = -\frac{1}{2\varrho_0 U}\,[E]\left[\frac{\partial E}{\partial X}\right]. \tag{70}$$

Obviously, $[E] = E_0$ and $[\partial E/\partial X] = 0$ are solutions to eq. (70), which are valid if a step strain (or a step stress] is applied at $X = 0$, where the wave front separates a region of homogeneous deformation from the undisturbed region.

The ordinary differential equation (68), as it stands is, however, still not amenable to analytic solution because of the appearance of $[\partial E/\partial X]$. If we consider, however, the case $\bar{G}_3 \equiv 0$, but $\bar{G}_{3,1} \neq 0$, i.e., a solid with linear initial elastic response, but non-linear viscoelastic behavior, eq. (68) reduces to such an extent that analytic solution is possible. The equation then becomes

$$\frac{d[E]}{dt} = C_1[E] + C_2[E]^2 + C_3[E]^3, \tag{71}$$

wherein

$$C_1 = \frac{1}{2}\,\bar{G}_{2,1}/\bar{G}_2 - \frac{1}{8}\,(\bar{K}_2)^2 T_0/\varrho_0 k \tag{72}$$

$$C_2 = \frac{1}{2}\,\bar{G}_{3,1}/\bar{G}_2 - \frac{1}{8}\,\bar{K}_2 L_3 T_0/\varrho_0 k \tag{73}$$

$$C_3 = -\frac{1}{36}\,(\bar{L}_3)^2 T_0/\varrho_0 k. \tag{74}$$

Equation (69) now reduces to

$$[\sigma] = \bar{G}_2[E], \tag{75}$$

and eq. (75) may thus be substituted in eq. (71) to yield an equation for $[\sigma]$.

Further simplifications of eq. (71) are easily obtained. If the initial elastic behavior is independent of the temperature, i.e., if $\bar{L}_3 \equiv 0$, then eq. (71) can be solved to yield

$$[E] = \left(\frac{C_1}{C_2}\right)\left\{\left(\frac{C_1}{C_2}\frac{1}{E_0} + 1\right) e^{-C_1 t} - 1\right\}^{-1}, \tag{76}$$

where E_0 is the jump in stress at $t = 0$, applied at $X = 0$, which can be computed from σ_0 by the use of eq. (75). If, in addition, the material behavior is elastic, i.e., $\bar{G}_{2,1} = \bar{G}_{3,1} \equiv 0$, we recover the well-known result of coupled linear thermoelasticity. If second-order effects in strain, as well as temperature-dependence of the material properties, are disregarded, i.e., $\bar{G}_{3,1} \equiv 0$ and $\bar{L}_3 \equiv 0$, we have

$$[E] = E_0 \exp\left\{\frac{\bar{G}_{2,1}}{2\bar{G}_2}\,t - \frac{1}{8}\,\frac{(\bar{K}_2)^2 T_0}{\varrho_0 k}\,t\right\}. \tag{77}$$

The result for first-order uncoupled viscoelasticity follows from eq. (77) by setting $\bar{K}_2 \equiv 0$. The well-known result for coupled linear thermoelasticity, see [8], follows from eq. (77) by setting $G_{2,1} \equiv 0$.

The assumptions that have entered the analysis after eq. (62) and prior to eq. (71) are that the material may be assumed undisturbed prior to arrival of the wave front, and additionally, that $\bar{G}_3 \equiv 0$. The restrictions imposed by the latter assumption can be reduced and an approximation for small times can be constructed for the case $\bar{G}_3 \neq 0$. We thus return to eq. (68) and we observe that the really troublesome term is the one containing $[\partial E/\partial X]$. Writing the kinematical condition of compatibility, eq. (38), for E, we obtain

$$\frac{d}{dt}[E] = \left[\frac{\partial E}{\partial t}\right] + U\left[\frac{\partial E}{\partial X}\right]. \tag{78}$$

The term $[\partial E/\partial X]$ can, of course, not be eliminated from eq. (78), because $[\partial E/\partial t]$ is also unknown. By employing eqs. (68) and (78) we may, however, obtain a relation between $[\partial E/\partial X]$ and $[\partial E/\partial t]$ for $t = 0$, which may then be employed as an approximation for $t > 0$ to eliminate $[\partial E/\partial t]$ from eq. (78), and write $[\partial E/\partial X]$ in terms of $(d/dt)\,[E]$, where the latter expression can then be substituted into eq. (68) to yield an approximate relation between $(d/dt)\,[E]$ and $[E]$. Thus, we first eliminate $(d/dt)[E]$ from eqs. (68) and (78) to obtain

$$\left\{2 + \frac{\bar{G}_3}{4\varrho_0 U^2}\,[E]\right\}\left[\frac{\partial E}{\partial t}\right] + \left\{2U + \frac{3}{4}\,\frac{\bar{G}_3}{\varrho_0 U}\,[E]\right\}\left[\frac{\partial E}{\partial X}\right]$$
$$= \frac{1}{\varrho_0 U^2}\left\{\bar{G}_{3,1}[E]^2 + \bar{G}_{2,1}[E]\right\} - \frac{1}{\varrho_0 U}\left\{\frac{1}{2}\,\bar{K}_2 + \frac{1}{3}\,\bar{L}_3[E]\right\}\left[\frac{\partial \theta}{\partial X}\right]. \tag{79}$$

Equation (79) relates the three unknown quantities $[E]$, $[\delta E/\partial X]$, and $[\partial E/\partial t]$, where $[E] = E_0$ for $X = 0$, $t = 0$. For small times, $[E]$ may be approximated by its initial value, with the result

$$\left[\frac{\partial E}{\partial t}\right] \simeq \alpha + \beta U\left[\frac{\partial E}{\partial X}\right], \tag{80}$$

wherein

$$\alpha = \frac{4\bar{G}_{3,1}E_0^2 + 4\bar{G}_{2,1}E_0}{8G_2 + 5G_3E_0} - \frac{4}{\varrho_0 k} \times$$
$$\times \frac{\left(\bar{G}_2 + \frac{1}{2}\,\bar{G}_3E_0\right)\left(\frac{1}{2}\bar{K}_2 + \frac{1}{3}\,\bar{L}_3E_0\right)\left(\frac{1}{12}\,\bar{G}_3E_0^3 + \frac{1}{6}\,T_0\bar{L}_3E_0^2 + \frac{1}{12}\,T_0\bar{K}_2E_0\right)}{8G_2 + 5\bar{G}_3E_0} \tag{81}$$

$$\beta = -\frac{8\bar{G}_2 + 7\bar{G}_3E_0}{8G_2 + 5G_3E_0}. \tag{82}$$

Returning to eq. (78), and substituting eq. (80), we obtain

$$\left[\frac{\partial E}{\partial X}\right] \simeq \left\{\frac{d}{dt}\,[E] - \alpha\right\} \Big/ U(1 + \beta). \tag{83}$$

Substituting eq. (83) into eq. (68), we obtain after some manipulation

$$\{1 + p[E]\}\,\frac{d[E]}{dt} = \sum_{n=1}^{4} q_n[E]^n, \tag{84}$$

wherein

$$p = \left\{\frac{5}{8} + \frac{1}{4(1+\beta)}\right\}\frac{\bar{G}_3}{G_2} \tag{85}$$

$$q_1 = C_1 + \frac{\alpha}{4(1+\beta)}\,\frac{\bar{G}_3}{\bar{G}_2} \tag{86}$$

$$q_2 = C_2 + \frac{1}{2} \frac{\bar{G}_3}{\bar{G}_2} C_1 \tag{87}$$

$$q_3 = C_3 + \frac{1}{2} \frac{\bar{G}_3}{\bar{G}_2} C_2 \tag{88}$$

$$q_4 = \frac{1}{2} \frac{\bar{G}_3}{G_2} C_3 . \tag{89}$$

Equation (84) can be solved by standard methods.

Acknowledgments

This work was supported by the Office of Naval Research under Contract ONR Nonr. 1228(34) with Northwestern University, and by the Air Force Office of Scientific Research under Grant AF-AFOSR-100-67 with Northwestern University.

References

[1] ACHENBACH, J. D., S. M. VOGEL and G. HERRMANN: On Stress Waves in Viscoelastic Media Conducting Heat. Irreversible Aspects of Continuum Mechanics — Transfer of Physical Characteristics in Moving Fluids. Ed. by H. PARKUS and L. I. SEDOV. Wien-New York: Springer. 1968.

[2] HERRMANN, G.: On Second-Order Thermoelastic Effects. ZAMP **15**, 253—262 (1964).

[3] CHRISTENSEN, R. M., and P. M. NAGHDI: Linear Non-Isothermal Viscoelastic Solids. Acta Mechanica **3**, 1—12 (1967).

[4] COLEMAN, B. D.: Thermodynamics of Material with Memory. Arch. Rational Mech. Anal. **17**, 230—254 (1964).

[5] ACHENBACH, J. D.: The Influence of Heat Conduction on Propagating Stress Jumps. J. Mech. Phys. Solids **16**, 273—282 (1968).

[6] VOGEL, S. M.: The Influence of Thermodynamical Effects on Stress Waves in Viscoelastic Solids. Ph. D. Dissertation, Northwestern University, 1968.

[7] COLEMAN, B. D., and M. E. GURTIN: Thermodynamics and One-Dimensional Shock Waves in Materials with Memory. Proc. Roy. Soc., A, **292**, 562—574 (1966).

[8] ACHENBACH, J. D.: The Propagation of Stress Discontinuities According to the Coupled Equations of Thermoelasticity. Acta Mechanica **3**, 342—351 (1967).

Equations of the Theory of Thermal Stresses in Double-Modulus Materials

By

S. A. Ambartsumian

Yerevan (U.S.S.R.)

Summary

The study deals with the development of the theory of thermo-inelasticity for the "elastic" material whose elastic moduli are different in tension (E^+) and in compression (E^-).

The proposed theory is based upon the mathematical theory of elasticity of double-modulus materials and upon the familiar Neuman's hypothesis.

Proceeding from accepted principles, the general theory of thermo-inelasticity for the double-modulus material is developed and the solutions of some problems are presented.

A vital characteristic of the nonlinear equations and formulae derived is the fact that all the nonlinear terms representations depend upon the parameter $(E^+ - E^-)/(E^+ + E^-)$, which is assumed small.

It is shown that neglect of heteroresistance may introduce significant errors.

Proceeding from the general principles of the heteromodulus theory of elasticity [1, 2] and from the fundamental hypotheses of the mathematical theory of thermoelasticity [3, 4], a general stress-strain theory of heteromodulus materials, in a field of thermal effects, has been developed.

1. Basic Relations

Let the material under consideration be such that in pure tension in any direction its elastic modulus is E^+, and in pure compression in any direction it is E^-. Let the corresponding values of Poisson ratio be: ν^+ (indicating a transverse contraction in tension) and ν^- (indicating a transverse expansion in compression).

On simultaneous expansion and compression in different mutually perpendicular principal directions the elastic moduli and Poisson ratios are assumed to remain E^+, ν^+ and E^-, ν^- respectively. The material in question under any stressed state is considered to undergo only small strains and to obey the general regularity conditions of a continuous elastic medium [5, 6].

It is also assumed that the change of the body temperature T, reckoned from some initial state T_0, is in general a function of the body point coordinates $(x_1 = x,\ x_2 = y,\ x_3 = z)$ and the time t. $T = T(x_1, x_2, x_3, t)$ satisfies the equation of heat conduction and suitable initial and boundary conditions [3, 4]. It is assumed that the temperature change is sufficiently small that physical and mechanical properties of the material remain unaltered.

It is finally assumed that due to heating or cooling from the arbitrary initial state T_0, additional strains in the form of an isotropic thermal expansion or contraction with a constant coefficient of linear thermal expansion α develop in the body. The tensor of thermal strains is spherical [3—6].

In virtue of the above assumptions, the stress-strain relations of elasticity are written as follows (only one of the equivalent variants is given [2]):

$$e_{ij} = A_1\sigma_{ij} + a_{12}\sigma\delta_{ij} + B_3 m_i m_j \bar{\sigma}_2 + B_2 n_i n_j \bar{\sigma}_3 + \alpha T \delta_{ij} \tag{1.1}$$

where δ_{ij} is the Kronecker delta $(\delta_{ij} = 0,\ i \neq j;\ \delta_{ij} = 1,\ i = j,\ j$ not summed), and

$$\sigma = \sigma_{11} + \sigma_{22} + \sigma_{33} = \bar{\sigma}_1 + \bar{\sigma}_2 + \bar{\sigma}_3$$

$$\sigma_{11} = \sigma_x,\ \sigma_{22} = \sigma_y,\ \sigma_{33} = \sigma_z$$

$$\sigma_{13} = \sigma_{31} = \tau_{xz},\ \sigma_{32} = \sigma_{23} = \tau_{yz},\ \sigma_{12} = \sigma_{21} = \tau_{xy}$$

$$A_1 = a_{11} - a_{12},\ B_2 = a_{33} - a_{11},\ B_3 = a_{22} - a_{11} \tag{1.2}$$

In these formulae a_{ik} are the coefficients of elasticity which, depending on the signs of the principal stresses $\bar{\sigma}_1, \bar{\sigma}_2, \bar{\sigma}_3$, assume the values of either E^+ or E^-. For example:

1. $\bar{\sigma}_1 > 0,\ \bar{\sigma}_2 > 0,\ \bar{\sigma}_3 > 0$ (point or domain of the first kind [1, 2]), then $a_{11} = a_{22} = a_{33} = (1/E^+),\ a_{ik} = -(\nu^+/E^+)$,

2. $\bar{\sigma}_1 < 0, \bar{\sigma}_2 < 0, \bar{\sigma}_3 < 0$ (point or domain of the first kind [1, 2], then $a_{11} = a_{22} = a_{33} = (1/E^-),\ a_{ik} = -(\nu^-/E^-)$;

3. $\bar{\sigma}_1 > 0,\ \bar{\sigma}_2 < 0,\ \bar{\sigma}_3 > 0$ (point or domain of the second kind [1, 2]), then $a_{11} = a_{33} = (1/E^+),\ a_{22} = (1/E^-),\ a_{ik} = -(\nu^+/E^+) = -(\nu^-/E^-)$;

4. $\bar{\sigma}_1 > 0$, $\bar{\sigma}_2 < 0$, $\bar{\sigma}_3 < 0$ (point or domain of the second kind [1, 2]), then $a_{11} = (1/E^+)$, $a_{22} = a_{33} = (1/E^-)$, $a_{ik} = -(\nu^+/E^+) = -(\nu^-/E^-)$ and so forth.

The principal stresses $\sigma_1, \sigma_2, \sigma_3$ are represented by known relations [5, 6] in terms of the unknown stresses $\sigma_x, \ldots, \tau_{xy}$, namely[1]

$$\bar{\sigma}_1 = l_i l_j \sigma_{ij} \qquad (l, m, n) \tag{1.3}$$

The direction cosines l_i, m_i, n_i (see Table 1), which are related by the familiar equations of the following form:

$$\begin{aligned} l_1^2 + l_2^2 + l_3^2 &= 1, & l_1 l_2 + m_1 m_2 + n_1 n_2 &= 0, \\ l_1^2 + m_1^2 + n_1^2 &= 1, & l_1 m_1 + l_2 m_2 + l_3 m_3 &= 0, \end{aligned} \tag{1.4}$$

Table 1

	1	2	3
$x = x_1$	l_1	m_1	n_1
$y = x_2$	l_2	m_2	n_2
$z = x_3$	l_3	m_3	n_3

indicate the direction of the principal stresses (1, 2, 3) with respect to the initial coordinate system (x, y, z).

The direction cosines are found from the conditions $\bar{\tau}_{23} = 0$, $\bar{\tau}_{31} = 0$, $\bar{\tau}_{12} = 0$ ($\bar{\tau}_{ik}$ are the principal stresses), which in explicit form are as follows:

$$m_i n_j \sigma_{ij} = 0, \qquad l_i n_j \sigma_{ij} = 0, \qquad l_i m_j \sigma_{ij} = 0. \tag{1.5}$$

The direction cosines may also be found from the conditions $\bar{e}_{23} = 0$, $\bar{e}_{31} = 0$, $\bar{e}_{12} = 0$ (e_{ik} are the principal strains), namely

$$m_i n_j e_{ij} = 0, \qquad l_i n_j e_{ij} = 0, \qquad l_i m_j e_{ij} = 0 \tag{1.6}$$

When solving systems (1.5) or (1.6), relations (1.4) are to be used.

In a particular case, when the coordinate directions (x, y, z) are the principal ones, i.e. when, for example

$$\sigma_x = \bar{\sigma}_1,\ \sigma_y = \bar{\sigma}_2,\ \sigma_z = \bar{\sigma}_3,\ \tau_{yz} = \bar{\tau}_{23} = 0,$$
$$\tau_{x2} = \bar{\tau}_{13} = 0,\ \tau_{xy} = \bar{\tau}_{12} = 0,\ e_x = \bar{e}_1,\ e_y = \bar{e}_2,$$
$$e_z = \bar{e}_3,\ e_{yz} = \bar{e}_{23} = 0,\ e_{xz} = \bar{e}_{13} = 0,\ e_{xy} = \bar{e}_{12} = 0$$

[1] The notation (l, m, n) indicates that the values of the other two principal stresses can be obtained by cyclic permutation of indices.

so that $l_1 = 1$, $m_2 = 1$, $n_3 = 1$, and the other direction cosines are equal to zero, the relations of elasticity and (1.1) are significantly simplified and take the form

$$\bar{e}_i = A_i \bar{\sigma}_i + a_{12}\sigma + \alpha T \quad (i \text{ not summed})$$
$$A_i = a_{ii} - a_{12} \quad (i \text{ not summed}) \tag{1.7}$$

To complete the formulation of a problem of thermoelasticity, the above constitutive relations of elasticity must be supplemented [5, 6] by the equations of euqilibrium:

$$\sigma_{ij,j} + \varrho X_i = 0 \tag{1.8}$$

and the conditions on the surface:

$$X_i^* = \sigma_{ij}\Gamma^j, \tag{1.9}$$

where

$$X_1^* = X^*,\ X_2^* = Y^*,\ X_3^* = Z^* \tag{1.9a}$$

are the projections on the coordinate axes of the surface traction, acting on an area with normal Γ; Γ^j are the components of the unit normal Γ. Furthermore, the relationships between the strain and displacement components are

$$e_x = u_{,x},\ e_y = v_{,y},\ e_z = w_{,z}$$
$$e_{xy} = v_{,x} + u_{,y},\ e_{yz} = w_{,y} + v_{,z},\ e_{zx} = u_{,z} + w_{,x}, \tag{1.10}$$

where commas indicate differentiation. The equations of compatibility of strains are

$$e_{x,yy} + e_{y,xx} = e_{xy,xy} \qquad (x, y, z)$$
$$(e_{yz,x} + e_{zx,y} - e_{xy,z})_z = 2e_{zz,xy} \qquad (x, y, z) \tag{1.11}$$

Note that in problems of thermoelasticity of double modulus materials, at those points and domains of the body where simultaneously $\bar{\sigma}_1 > 0$, $\bar{\sigma}_2 > 0$, $\bar{\sigma}_3 > 0$ or $\bar{\sigma}_1 < 0$, $\bar{\sigma}_2 < 0$, $\bar{\sigma}_3 < 0$ (i.e., in the case of the domains of the first kind), the generalized law of elasticity (1.1) is simplified and coincides with Hooke's generalized law, with either E^+, ν^+ or E^-, ν^-. Thus the classical equations of thermoelasticity [3, 4] remain unchanged for domains of the first kind.

For domains of the second kind, i.e. at the points and domains of the body where one of the principal stresses has the sign different from two others (for example, $\bar{\sigma}_1 > 0$, $\bar{\sigma}_2 < 0$, $\bar{\sigma}_3 < 0$), relations (1.1) must be employed. In this case the equations of thermoelasticity of double modulus materials differ fundamentally from the corresponding ones of the classical theory. It is for these domains, therefore, that the equations of thermoelasticity are developed below.

2. Equations for Domains of the Second Kind

The equations of thermoelasticity for domains of the second kind will be represented first in terms of displacement components. Solving (1.1) for the stresses, one obtains

$$\sigma_{ij} = \frac{1}{A_1} e_{ij} + d_{12} e \delta_{ij} - d_{12}(B_3 \bar{\sigma}_2 + B_2 \bar{\sigma}_3) \delta_{ij} - \\ - \frac{B_3}{A_1} m_i m_j \bar{\sigma}_2 - \frac{B_2}{A_1} n_i n_j \bar{\sigma}_3 - d_* \alpha T \delta_{ij}, \tag{2.1}$$

where

$$d_{11} = \frac{a_{11} + a_{12}}{\Omega}, \qquad d_* = d_{11} + 2 d_{12} = \frac{a_{11} - a_{12}}{\Omega}$$

$$d_{12} = -\frac{a_{12}}{\Omega}, \qquad \Omega = (a_{11} - a_{12})(a_{11} + 2a_{12}) \tag{2.2}$$

and where

$$e = e_x + e_y + e_z = e_{11} + e_{22} + e_{33}. \tag{2.3}$$

Relationships (2.1) include the principal stresses $\bar{\sigma}_2$ and $\bar{\sigma}_3$ which, with (1.7), are as follows in terms of the principal strains $\bar{e}_1$, $\bar{e}_2$, $\bar{e}_3$:

$$\bar{\sigma}_i = c_{ij} \bar{e}_i - C_i \alpha T \equiv \sigma_i^0 - C_i \alpha T \tag{2.4}$$

where

$$c_{11} = \frac{a_{22} a_{33} - a_{12}^2}{\Delta}, \qquad c_{22} = \frac{a_{11} a_{33} - a_{12}^2}{\Delta}, \qquad c_{33} = \frac{a_{11} a_{22} - a_{12}^2}{\Delta}$$

$$c_{12} = -\frac{a_{12}(a_{33} - a_{12})}{\Delta}, \quad c_{13} = -\frac{a_{12}(a_{22} - a_{12})}{\Delta}, \quad c_{23} = -\frac{a_{12}(a_{11} - a_{12})}{\Delta}$$

$$C_1 = c_{11} + c_{12} + c_{13} = \frac{1}{\Delta}\left[a_{22} a_{33} - a_{12}(a_{22} + a_{33}) + a_{12}^2\right]$$

$$C_2 = c_{22} + c_{12} + c_{23} = \frac{1}{\Delta}\left[a_{11} a_{33} - a_{12}(a_{11} + a_{33}) + a_{12}^2\right]$$

$$C_3 = c_{33} + c_{13} + c_{23} = \frac{1}{\Delta}\left[a_{11} a_{22} - a_{12}(a_{11} + a_{22}) + a_{12}^2\right]$$

$$\Delta = a_{11} a_{22} a_{33} - a_{12}^2 (a_{11} + a_{22} + a_{33}) + 2 a_{12}^3.$$

The values of the principal stresses (that is σ_i^0) in terms of the displacements $u_1 = u(x, y, z)$, $u_2 = v(x, y, z)$, $u_3 = w(x, y, z)$ will be required below. For this reason, the values of the principal strains, through the strains e_x, e_y, e_z and thus through the displacements, should be substituted into (2.4), with the aid of the familiar relationships

$$\bar{e}_1 = l_i l_j e_{ij} \qquad (l,\ m,\ n). \tag{2.5}$$

The direction cosines, appearing in formula (2.1)—(2.5) are to be found from the system of eqs. (1.4), (1.6), taking into account (1.10).

Substituting the stresses from (2.1), considering (1.10) and (2.4), into the equilibrium eq. (1.8), one obtains

$$\begin{aligned} &\Delta u_i + M_{11} e_{,j} - 2B_3[M_{12}^0 \sigma_{2,i} + Q_{ij,j}(\sigma_2^0)] - 2B_2[M_{12}^0 \sigma_{3,i} + S_{ij,j}(\sigma_3^0)] \\ &= 2A_1 \varrho X_i + 2[M_{11} + M_{12}(1 - B_3 C_2 - B_2 C_3)]\alpha T_{,i} - 2B_3 C_2 \alpha Q_{ij,j}(T) - \\ &\quad - 2B_2 C_3 \alpha S_{ij,j}(T), \end{aligned} \tag{2.6}$$

where

$$\begin{aligned} Q_{ij}(k) &= m_i m_j k, \qquad S_{ij}(k) = n_i n_j k \\ M_{11} &= \frac{a_{11}}{a_{11} + 2a_{12}}, \qquad M_{12} = \frac{a_{12}}{a_{11} + 2a_{12}} \\ e &= e_{jj} = u_{,x} + v_{,y} + w_{,z} \\ u_{i,jj} &= u_{i,11} + u_{i,22} + u_{i,33} = \Delta u_i. \end{aligned} \tag{2.7}$$

The system of non-linear eqs. (2.6) represents equations of equilibrium in terms of displacements for domains of the second kind for double modulus materials in a thermal field. The values of σ_2^0 and σ_3^0 are not given explicitly here in terms of the unknown displacements because of the cumbersome structure of eqs. (2.6).

As in the case of the classical theory, if the temperature is assumed to be a given function of the coordinates, then the consideration of thermal effects is formally reduced to introduction of an additional body force into the equations of equilibrium in terms of displacements. However, unlike the classical theory, the additional fictitious force will be proportional both to the temperature gradient and to the temperature itself at a given point.

Similarly to the classical theory [4], the thermal problem of the double modulus theory of elasticity may be formulated in terms of stresses, as will now be done. The equations of equilibrium in stresses are represented in a conventional manner (1.8), but the equations of compatibility of strains (1.11) must be written in terms of stresses as follows:

$$\begin{aligned} &a_{11}\Delta\sigma - A_1 \Delta\sigma_{ii} - a_{11}\sigma_{,ii} + \frac{1}{2} B_3 M_{mm}(\bar{\sigma}_2) \,|\, e^{imn} \,| + \frac{1}{2} B_2 N_{mn}(\bar{\sigma}_3) \,|\, e^{imn} \,| \\ &= A_1 \varrho X_{k,k} - \alpha T_{,kk} + \alpha T_{,ii} \qquad (i \text{ not summed}) \end{aligned} \tag{2.8}$$

$$\begin{aligned} &A_1 \Delta \sigma_{ij} + a_{11} \sigma_{,ij} + B_3 L_{ijk}(\bar{\sigma}_2) \,|\, e^{ijk} \,| + B_2 K_{ijk}(\bar{\sigma}_3) \,|\, e^{ijk} \,| \\ &= -A_1 \varrho (X_{i,j} + X_{j,i}) - \alpha T_{,ij} \qquad (i \neq j\ i \text{ and } j \text{ not summed}), \end{aligned} \tag{2.9}$$

where

$$M_{ik}(\beta) = (m_i^2\beta)_{,kk} + (m_k^2\beta)_{,ii} - 2(m_i m_k \beta)_{,ik}$$
$$N_{ik}(\beta) = (n_i^2\beta)_{,kk} + (n_k^2\beta)_{,ii} - 2(n_i n_k \beta)_{,ik}$$
$$L_{ikj}(\beta) = (m_j^2\beta)_{,ik} + (m_i m_k \beta)_{,jj} - (m_j m_k \beta)_{,ij} - (m_j m_i \beta)_{,kj}$$
$$K_{ikj}(\beta) = (n_j^2\beta)_{,ik} + (n_i n_k \beta)_{,jj} - (n_j n_k \beta)_{,ij} - (n_j n_i \beta)_{,kj}$$
$$(i,\ j,\ k \text{ not summed}) \qquad (2.10)$$

$e^{ijk} = 0$ when any two indices are equal

$e^{ijk} = 1$ for even permutations of indices

$e^{ijk} = -1$ for odd permutations of indices.

The first set of the stress equations of compatibility (2.8) may be somewhat transformed. By adding all the three eqs. (2.8), one obtains

$$\Delta\sigma = -\frac{2\alpha}{a_{11}+a_{12}}\Delta T - B_3F_3(\bar\sigma_2) - B_2F_2(\bar\sigma_3) - \frac{a_{11}-a_{12}}{a_{11}+a_{12}}\varrho X_{i,i}$$
$$F_3(\bar\sigma_2) = \frac{1}{a_{11}+a_{12}}[M_{12}(\bar\sigma_2) + M_{23}(\bar\sigma_2) + M_{31}(\bar\sigma_2)]$$
$$F_2(\bar\sigma_3) = \frac{1}{a_{11}+a_{12}}[N_{12}(\bar\sigma_3) + N_{23}(\bar\sigma_3) + N_{31}(\bar\sigma_3)]. \qquad (2.11)$$

Substituting the values of $\Delta\sigma$ from (2.11) into (2.8) one obtains

$$A_1\Delta\sigma_{ii} + a_{11}\sigma_{,ii} + B_3\Big[a_{11}F_3(\bar\sigma_2) - \frac{1}{2}M_{mn}(\bar\sigma_2)|e^{imn}|\Big] + B_2\Big[a_{11}F_2(\bar\sigma_3) - $$
$$- \frac{1}{2}N_{mn}(\bar\sigma_3)|e^{imn}|\Big] = -2A_1\varrho X_{i,i} + a_{12}\frac{a_{11}-a_{12}}{a_{11}+a_{12}}\varrho X_{k,k} -$$
$$- \alpha\left(T_{,ii} + \frac{a_{11}-a_{12}}{a_{11}+a_{12}}\Delta T\right) \qquad (i \text{ not summed}) \qquad (2.12)$$

As before, for simplicity, the values of the principal stresses $\bar\sigma_2$ and $\bar\sigma_3$, represen ted by unknown stresses, are not substituted into corresponding relationships. If required, this may be performed with the aid of formula (1.3). The direction cosines may be found through the stresses from eqs. (1.4) and (1.5). Finally, to complete the formulation of problems of thermal stress and strain, boundary conditions formulated in the same manner as in the classical theory [3—6], must be added to eqs. (2.6), (2.8), (2.9) and (2.12).

The non-linear equations and relationships obtained here coincide with the familiar linear equations and relationships of the classical theory of thermoelasticity [3—6] for the particular case of a classical unimodulus material, or in the domains of the first kind of a double modulus material.

Finally, it should be noted that the considerations, equations and formulas presented here are sufficient to solve various problems of the determination of thermal stresses and displacements in bodies of double modulus materials. In what follows, the problems of plane strain and plane stress for double modulus materials in a thermal field are discussed.

3. Plane Strain

Consider a prismatic body and let the direction of generatrices Z be the principal one and coincide with direction "3". The strain is assumed to be plane, i.e., the component w of displacement in direction A is equal to zero, and the other two components of displacement u and v are functions of the independent variables x and y alone [2—6]. In this case thus, according to (1.1), (1.4) and (1.10), one may write

$$\begin{gathered} u = u(x, y), \quad v = v(x, y), \quad w = 0, \quad e_z = 0, \quad e_{xz} = 0, \\ e_{yz} = 0, \quad l_3 = m_3 = n_1 = n_2 = 0, \quad n_3 = 1, \quad \sigma_z = \bar{\sigma}_3, \\ \tau_{xz} = \bar{\tau}_{13} = 0, \quad \tau_{yz} = \bar{\tau}_{23} = 0. \end{gathered} \tag{3.1}$$

From the third of eqs. (1.1), one obtains

$$\sigma_2 = -\frac{a_{12}}{a_{33}} (\sigma_x + \sigma_y) - \frac{\alpha}{a_{33}} T. \tag{3.2}$$

The value of the factor a_{33} may be found only after the sign of the principal stress $\sigma_z = \bar{\sigma}_3$ is determined, since if $\sigma_z > 0$ for a_{33} one has $1/E^+$, and if $\sigma_z < 0$, then $a_{33} = 1/E^-$.

Substituting the value of σ_z from (3.2) into the non-trivially vanishing ones of the set (1.1), the following equations of the generalized law of elasticity are obtained for plane strain:

$$\begin{aligned} e_x &= b_{11}\sigma_x + b_{12}\sigma_y + B_3 m_1^2 \bar{\sigma}_2 + \frac{A_3}{a_{33}} \alpha T \\ e_y &= b_{11}\sigma_y + b_{12}\sigma_x + B_3 m_2^2 \bar{\sigma}_2 + \frac{A_3}{a_{33}} \alpha T \\ e_{xy} &= 2A_1 \tau_{xy} + 2B_3 m_1 m_2 \bar{\sigma}_2, \end{aligned} \tag{3.3}$$

where and later on

$$b_{ik} = a_{ik} - \frac{a_{12}^2}{a_{33}}, \qquad A_3 = a_{33} - a_{12}, \qquad A_2 = a_{22} - a_{12}. \tag{3.4}$$

Solving eqs. (3.3) for the stresses, one obtains:

$$\sigma_x = \frac{1}{A_1} e_x - \frac{\mu}{A_1} e + B_3 \left(\frac{\mu}{A_1} - \frac{m_1^2}{A_1}\right) \bar{\sigma}_2 - \left(\frac{1}{A_1} - 2\frac{\mu}{A_1}\right) \frac{A_3}{a_{33}} \alpha T$$

$$\sigma_y = \frac{1}{A_1} e_y - \frac{\mu}{A_1} e + B_3 \left(\frac{\mu}{A_1} - \frac{m_2^2}{A_1}\right) \bar{\sigma}_2 - \left(\frac{1}{A_1} - 2\frac{\mu}{A_1}\right) \frac{A_3}{a_{33}} \alpha T$$

$$\tau_{xy} = \frac{1}{2A_1} e_{xy} - \frac{B_3}{A_1} m_1 m_2 \bar{\sigma}_2, \tag{3.5}$$

where

$$\mu = \frac{b_{12}}{b_{11} + b_{12}} = \frac{a_{12}(a_{33} - a_{12})}{a_{33}(a_{11} + a_{12}) - 2a_{12}^2}$$

$$e = e_x + e_y. \tag{3.6}$$

The principal stress $\bar{\sigma}_2$ contained in (3.3) and (3.5), in virtue of (3.1), from (1.3), (2.4) may be written as follows:

$$\bar{\sigma}_2 = m_1^2 \sigma_x + m_2^2 \sigma_y + 2 m_1 m_2 \tau_{xy} \tag{3.7}$$

or, in terms of strains,

$$\bar{\sigma}_2 = \frac{a_{11}a_{33} - a_{12}^2}{a_{33}\omega} e_2 - \frac{a_{12}}{a_{33}} \frac{A_3}{\omega} e_1 - \frac{A_1}{\omega} \frac{A_3}{a_{33}} \alpha T$$

$$\omega = a_{11}a_{22} - a_{12}^2 - (a_{11} + a_{22} - 2a_{12}) \frac{a_{12}^2}{a_{33}} \tag{3.8}$$

while for the principal strains one has

$$\bar{e}_1 = l_1^2 e_x + l_2^2 e_y + l_1 l_2 e_{xy}. \tag{3.9}$$

The direction cosines, contained in the above equations and relationships, are found from the conditions

$$\bar{\tau}_{12} = l_1 m_1 \sigma_x + l_2 m_2 \sigma_y + (l_1 m_2 + l_2 m_1) \tau_{xy} = 0$$

or

$$\bar{e}_{12} = 2(l_1 m_1 e_x + l_2 m_2 e_y) + (l_1 m_2 + l_2 m_1) e_{xy} = 0$$

i.e. from the condition that directions "1" and "2" are the principal ones. Without going into details [2], the final values of the direction cosines are given by

$$m_1^2 = \frac{1}{1 + k^2}, \qquad m_2^2 = \frac{k^2}{1 + k^2}, \qquad m_1 m_2 = \frac{k}{1 + k^2}$$

$$k = \frac{m_2}{m_1} = -\frac{(\sigma_x - \sigma_y) + \sqrt{(\sigma_x - \sigma_y)^2 + 4\tau_{xy}^2}}{2\tau_{xy}} =$$

$$= -\frac{(e_x - e_y) + \sqrt{(e_x - e_y)^2 + e_{xy}^2}}{e_{xy}}. \tag{3.10}$$

We now consider the more particular case of $\bar{\sigma}_1 > 0$, $\bar{\sigma}_2 < 0$, for which, with (1.3), (3.1), (3.7), (3.10), one obtains the familiar formulas for the principal stresses:

$$\bar{\sigma}_1 = \frac{\sigma_x + \sigma_y}{2} + \sqrt{\tau_{xy}^2 + \frac{(\sigma_x - \sigma_y)^2}{4}}$$

$$\bar{\sigma}_2 = \frac{\sigma_x + \sigma_y}{2} - \sqrt{\tau_{xy}^2 + \frac{(\sigma_x - \sigma_y)^2}{4}}. \tag{3.11}$$

Proceeding in the same manner, from corresponding relationships for the principal strains one obtains

$$\bar{e}_1 = \frac{e_x + e_y}{2} + \frac{1}{2}\sqrt{e_{xy}^2 + (e_x - e_y)^2}$$

$$\bar{e}_2 = \frac{e_x + e_y}{2} - \frac{1}{2}\sqrt{e_{xy}^2 + (e_x - e_y)^2}. \tag{3.12}$$

The principal stresses, in terms of the strains, are derived taking into account the fact that $e_2 = \bar{e}_3 = 0$, and solving the set (17) with respect to stresses, and remembering (3.12); thus

$$\bar{\sigma}_1 = \frac{A_2}{2\omega} e + \frac{b_{22} + b_{12}}{2\omega}\sqrt{e_{xy}^2 + (e_x - e_y)^2} - \frac{A_2 A_3}{\omega a_{33}} \alpha T$$

$$\bar{\sigma}_2 = \frac{A_1}{2\omega} e - \frac{b_{11} + b_{12}}{2\omega}\sqrt{e_{xy}^2 + (e_x - e_y)^2} - \frac{A_1 A_3}{\omega a_{33}} \alpha T. \tag{3.13}$$

All the equations and relationships required for finding stresses and displacements have now been treated, but nothing has so far been said about the temperature function $T = T(x, y, z)$, and about the conditions to be satisfied by it so that a state of plane strain results. This will now be done.

Substitute the values of σ_x and σ_y from (3.2) into (3.2), and take into account (1.10), (3.1) and (3.13), to obtain:

$$\sigma_z = f(x, y) - \left[1 + \frac{a_{12} A_3 (1 - 2\mu)}{a_{33}}\left(\frac{B_3}{\omega} - \frac{2}{A_1}\right)\right]\frac{\alpha}{a_{33}} T.$$

The third equation of equilibrium (1.8) for the case of the homogeneous plane problem is of the form $\sigma_{z,z} = 0$. Substituting the value of σ_z into the above equation, one obtains $T_{,z} = 0$, i.e., to achieve a state of plane strain, as in the case of the classical problem, it is sufficient to have $T = T(x, y)$, i.e. the temperature function should depend upon the two coordinates x and y only.

The equations of equilibrium for the case of the plane problem are rewritten as follows:

$$\sigma_{x,x} + \tau_{xy,y} + \varrho X = 0, \qquad \tau_{xy,x} + \sigma_{y,y} + \varrho Y = 0, \tag{3.14}$$

where the components of the body forces are also functions of only two coordinates, i. e. $X = X(x, y)$, $Y = Y(x, y)$.

Substituting the values of stresses from (3.5) into (3.14), taking into account (1.10), one obtains the following set of non-linear differential equations in the unknown displacements:

$$\Delta_1 u + (1 - 2\mu) e_{,x} + 2 B_3 [\mu \sigma^0_{2,x} - (m_1^2 \sigma_2^0)_{,x} - (m_1 m_2 \sigma_2^0)_{,y}] = -2 A_1 \varrho X +$$
$$+ 2\left[1 - 2\mu + B_3 \frac{A_1}{\omega}\right]\mu \frac{A_3}{a_{33}} \alpha T_{,x} - 2 B_3 \frac{A_1 A_3}{\omega a_{33}} \alpha [(m_1^2 T)_{,x} + (m_1 m_2 T)_{,y}]$$

$$\Delta_1 v + (1 - 2\mu) e_{,y} + 2 B_3 [\mu \sigma^0_{2,y} - (m_2^2 \sigma_2^0)_{,y} - (m_1 m_2 \sigma_2^0)_{,x}] = -2 A_1 \varrho Y +$$
$$+ 2\left[1 - 2\mu + B_3 \frac{A_1}{\omega} \mu\right] \frac{A_3}{a_{33}} \alpha T_{,y} - 2 B_3 \frac{A_1 A_3}{\omega a_{33}} \alpha [(m_2^2 T)_{,y} + (m_1 m_2 T)_{,x}], \tag{3.15}$$

where along with the preceding notation one also has

$$\Delta_1(\ ) = (\ )_{,xx} + (\ )_{,yy}, \qquad e = e_x + e_y = u_{,x} + v_{,y}$$

$$\sigma_2^0 = \frac{A_1}{2\omega}(u_{,x} + v_{,y}) - \frac{b_{11} + b_{12}}{2\omega} \sqrt{(u_{,y} + v_{,x})^2 + (u_{,x} - v_{,y})^2}.$$

For the direction cosines eq. (3.10) still holds, where however, by virtue of (1.10),

$$k = -\frac{(u_{,x} - v_{,y}) + \sqrt{(u_{,x} - v_{,y})^2 + (u_{,y} + v_{,x})^2}}{u_{,y} + v_{,x}}.$$

The boundary conditions, if they are expressed in terms of displacements, remain unchanged. If the boundary conditions are given in terms of stresses, the values of stresses from (3.5) should be substituted into (1.9), written for the plane problem, considering the value of the principal stress $\bar{\sigma}_2$ in the displacements (3.13).

If the body forces have a potential, then their components are [7]:

$$\varrho X = -U_{,x}, \qquad \varrho Y = -U_{,y}, \tag{3.16}$$

where $U = U(x, y)$ is the potential function.

With (3.16) and (3.1), the equations of equilibrium (1.8) for the plane problem become

$$(\sigma_x - U)_{,x} + \tau_{xy,y} = 0, \quad (\sigma_y - U)_{,y} + \tau_{xy,x} = 0 \tag{3.17}$$

and introduction of the stress function $\varphi = \varphi(x, y)$ gives

$$\sigma_x - U = \varphi_{,yy}, \quad \sigma_y - U = \varphi_{,xx}, \quad \tau_{xy} = -\varphi_{,xy} \tag{3.18}$$

which identically satisfy the equations of equilibrium (3.17). The strains e_x, e_y, e_{xy} follow from (3.3) as:

$$e_x = A_1\varphi_{,yy} + b_{12}(\varphi_{,yy} + \varphi_{,xx}) + B_3 m_1^2 \bar{\sigma}_2 - (A_1 + 2b_{12})\, U + \frac{A_3}{a_{33}} \alpha T$$

$$e_y = A_1\varphi_{,xx} + b_{12}(\varphi_{,yy} + \varphi_{,xx}) + B_3 m_2^2 \bar{\sigma}_2 - (A_1 + 2b_{12})\, U + \frac{A_3}{a_{33}} \alpha T$$

$$e_{xy} = -2A_1\varphi_{,xy} + 2B_3 m_1 m_2 \bar{\sigma}_2 \tag{3.19}$$

where for $\bar{\sigma}_2$ and k (which appears in m_i) one has respectively:

$$\bar{\sigma}_2 = \frac{\varphi_{,yy} + \varphi_{,xx}}{2} - \sqrt{\varphi_{,xy}^2 + \frac{(\varphi_{,yy} - \varphi_{,xx})^2}{4}} \tag{3.20}$$

$$k = \frac{\varphi_{,yy} - \varphi_{,xx} + \sqrt{(\varphi_{,yy} - \varphi_{,xx})^2 + 4\varphi_{,xy}^2}}{2\varphi_{,xy}}.$$

Only one of the six compatibility equations (1.11) is not identically zero. Substituting the values of the strains e_x, e_y, e_{xy} into this equation, the following non-linear equation is obtained for the unknown stress function $\varphi(x, y)$:

$$\Delta_1\Delta_1\varphi + \frac{B_3}{b_{11}} [(m_1^2\bar{\sigma}_2)_{,yy} - 2(m_1 m_2 \bar{\sigma}_2)_{,xy} + (m_2^2\bar{\sigma}_2)_{,xx}] =$$

$$= -\frac{A_3}{a_{33}b_{11}} \alpha \Delta_1 T - \frac{b_{11} + b_{12}}{b_{11}} \Delta_1 U. \tag{3.21}$$

From the above considerations the boundary conditions for the plane problem can be written in terms of the stress function in the conventional manner [3—7]. The plane strain formulation is thus complete.

4. State of Plane Stress

As in the case of plane strain, the direction Z is assumed to be the principal one and to coincide with direction "3". The stressed state is assumed to be plane, when the stresses acting on sections $Z =$ const. are equal to zero. Then one can write

$$l_3 = m_3 = n_1 = n_2 = 0, \quad n_3 = 1, \quad \sigma_z = \bar{\sigma}_3 = 0$$

$$\tau_{xz} = \bar{\tau}_{13} = 0, \quad \tau_{yz} = \bar{\tau}_{23} = 0 \tag{4.1}$$

and the generalized stress-strain law (1.1), becomes, with (4.1):

$$\begin{aligned} e_x &= a_{11}\sigma_x + a_{12}\sigma_y + B_3 m_1^2 \bar{\sigma}_2 + \alpha T \\ e_y &= a_{11}\sigma_y + a_{12}\sigma_x + B_3 m_2^2 \bar{\sigma}_2 + \alpha T \\ e_z &= a_{12}(\sigma_x + \sigma_y) + \alpha T, \qquad e_{yz} = 0 \\ e_{xy} &= 2A_1 \tau_{xy} + 2B_3 m_1 m_2 \bar{\sigma}_2, \qquad e_{zx} = 0. \end{aligned} \tag{4.2}$$

Solving eqs. (4.2) explicitly for the stresses, one obtains

$$\begin{aligned} \sigma_x &= \frac{1}{A_1} e_x - \frac{\lambda}{A_1} e + B_3\left(\frac{\lambda}{A_1} - \frac{m_1^2}{A_1}\right)\bar{\sigma}_2 - \left(\frac{1}{A_1} - 2\frac{\lambda}{A_1}\right)\alpha T \\ \sigma_y &= \frac{1}{A_1} e_y - \frac{\lambda}{A_1} e + B_3\left(\frac{\lambda}{A_1} - \frac{m_2^2}{A_1}\right)\bar{\sigma}_2 - \left(\frac{1}{A_1} - 2\frac{\lambda}{A_1}\right)\alpha T \\ \tau_{xy} &= \frac{1}{2A_1} e_{xy} - \frac{B_3}{A_1} m_1 m_2 \bar{\sigma}_2, \end{aligned} \tag{4.3}$$

where

$$\lambda = \frac{a_{12}}{a_{11} + a_{12}}, \qquad e = e_x + e_y. \tag{4.4}$$

Comparing the expressions for the stresses in the case of plane strain (3.5) with those for plane stress (4.3), it is noted that they differ only by the replacement of the coefficient μ by λ, and by the presence of the factor A_3/a_{33} in the temperature term for the plane strain case alone.

The main difference lies in strain e_z, since in the case of plane strain $e_z = 0$, while here

$$e_z = \lambda(e - B_3\bar{\sigma}_2) + (1 - 2\lambda)\alpha T. \tag{4.5}$$

In virtue of (4.1), the equations of equilibrium for plane stress are of the form (3.4). Substituting the values of stresses from (4.3) into (3.14), taking into account (1.10), one obtains the following set of non-linear differential equations for the unknown displacements $u = u(x, y, z)$ and $v = v(x, y, z)$:

$$\begin{aligned} \Delta_1 u + (1 - 2\lambda) e_{,x} + 2B_3[\lambda\sigma^0_{2,x} - (m_1^2\sigma_2^0)_{,x} - (m_1 m_2 \sigma_2^0)_{,y}] = -2A_1\varrho X + \\ + 2[1 - 2\lambda + B_3\frac{A_1}{\omega_a}\lambda)\alpha T_{,x} - 2B_3\frac{A_1}{\omega_a}\alpha[(m_1^2 T)_{,x} + (m_1 m_2 T)_{,y}] \end{aligned} \tag{4.6}$$

$$\begin{aligned} \Delta_1 v + (1 - 2\lambda) e_{,y} + 2B_3[\lambda\sigma^0_{2,y} - (m_2^2\sigma_2^0)_{,y} - (m_1 m_2 \sigma_2^0)_{,x}] = -2A_1\varrho Y + \\ + 2\left[1 - 2\lambda + B_3\frac{A_1}{\omega_a}\lambda\right]\alpha T_{,y} - 2B_3\frac{A_1}{\omega_a}\alpha[(m_1^2 T)_{,y} + (m_1 m_2 T)_{,x}]. \end{aligned}$$

The principal stress $\bar{\sigma}_2$ and its basic portion σ_2^0 are found from the first two equations of the set (1.7). Assuming $\sigma_z = \bar{\sigma}_3 = 0$ and solving (1.7) with respect to the principal stresses $\bar{\sigma}_1$ and $\bar{\sigma}_2$, and considering (3.12), (1.10), one obtains:

$$\bar{\sigma}_2 = \sigma_2^0 - \frac{A_1}{\omega_a}\alpha T = \frac{A_1}{2\omega_a} e - \frac{a_{11}+a_{12}}{2\omega_a}\sqrt{e_{xy}^2 + (e_x - e_y)^2} - \frac{A_1}{\omega_a}\alpha T =$$

$$= \frac{A_1}{2\omega_a}(u_{,x} + v_{,y}) - \frac{a_{11}+a_{12}}{2\omega_a}\sqrt{(u_{,y} + v_{,x})^2 + (u_{,x} - v_{,y})^2} - \frac{A_1}{\omega_a}\alpha T$$

$$\omega_a = a_{11}a_{22} - a_{12}^2.$$

The direction cosines are found from (3.10), where, with the use of (1.10), for k one has.

$$k = -\frac{(u_{,x} - v_{,y}) + \sqrt{(u_{,x} - v_{,y})^2 + (u_{,y} + v_{,x})^2}}{u_{,y} + v_{,x}}.$$

Comparing eqs. (4.6) with the analogous equations for plane strain, (3.15), it should be noted that they also differ only by the replacement of the coefficient μ by λ (actually by the coefficients a_{ik} and b_{ik}), and by the factor A_3/a_{33} in the temperature terms. However, as in the classical theory, the apparent similarity only masks the fundamental difference that in the problem of plane strain all the variables are functions of only x and y, while in the problem of plane stress the variables, as a rule, depend also on z [3—6].

The condition to be satisfied by the temperature function, to achieve a state plane stress, is rather rigorous. As in the classical theory [4], it can be found from the relationships of compatibility (2.8)—(2.12), by reducing a three-dimensional problem to a "two-dimensional" problem of plane stress. In this case, by virtue of (3.18), the equations of equilibrium (3.17) are completely satisfied.

Plane stress may be achieved with sufficient accuracy [3, 5, 8] in thin plates, assuming that stresses act only in the plane of the plate and that they are uniformly distributed throughout its thickness. The assumption adopted, of course, does not fully agree with the requirements of the three-dimensional problem; in spite of this fact, however, it is acceptable in the case of thin plates when the plate surfaces are free from stresses, and the temperature depends on two coordinates only. In this case, the direction Z is assumed to be normal to the plane of the plate (x, y), principal and coincides with direction "3". Then for the problem of plane stress one can write

$$l_3 = m_3 = n_1 = n_2 = 0, \quad n_3 = 1, \quad \sigma_z = \overline{\sigma}_3 = 0,$$
$$\tau_{xz} = \overline{\tau}_{13} = 0, \quad \tau_{yz} = \overline{\tau}_{23} = 0, \quad \sigma_x = \sigma_x(x, y),$$
$$\sigma_y = \sigma_y(x, y), \quad \tau_{xy} = \tau_{xy}(x, y), \quad u = u(x, y),$$
$$v = v(x, y), \quad T = T(x, y). \tag{4.7}$$

In virtue of (4.7), the set of eqs. (4.6), whose appearance does not change, acquires a new significance, since, similarly to the problem of plane strain, all the variables will be functions of only two variables, x and y.

Furthermore, with these assumptions, all the equations of strain compatibility, except one, are satisfied to the required degree of accuracy. Unsatisfied remains only the first equation of the form

$$e_{x,yy} + e_{y,xx} = e_{xy,xy}. \tag{4.8}$$

With (3.18), the equations of equilibrium (3.17) are identically satisfied and the strains are, from (4.2),

$$e_x = a_{11}\varphi_{,yy} + a_{12}\varphi_{,xx} + B_3 m_1^2 \overline{\sigma}_2 + (a_{11} + a_{12})\, U + \alpha T$$
$$e_y = a_{11}\varphi_{,xx} + a_{12}\varphi_{,yy} + B_3 m_2^2 \overline{\sigma}_2 + (a_{11} + a_{12})\, U + \alpha T$$
$$e_{xy} = -2A_1\varphi_{,xy} + 2m_1 m_2 \overline{\sigma}_2, \tag{4.9}$$

where for m_i and $\overline{\sigma}_2$ the conventional formulae of the plane problem are used (see plane strain). Upon substitution of these strains into (4.8), the following non-linear equation is obtained for the stress function φ in the case of plane stress:

$$\Delta_1\Delta_1\varphi + \frac{B_3}{a_{11}}\left[(m_1^2\overline{\sigma}_2)_{,yy} - 2(m_1 m_2 \overline{\sigma}_2)_{,xy} + (m_2^2\overline{\sigma}_2)_{,xx}\right] =$$
$$= -\frac{\alpha}{a_{11}}\Delta_1 T - \frac{a_{11} + a_{12}}{a_{11}}\Delta_1 U, \tag{4.10}$$

where m_i and $\overline{\sigma}_2$ are calculated from (3.20).

Thus the solution of the thermal problem of plane stress for double modulus materials is also reduced to the consideration one non-linear equation. It is evident that eq. (4.10) differs from the similar equation of plane strain for double modulus materials only in the value of the coefficient a_{ik} (in eq. (3.21) b_{ik} appears instead of a_{ik}) and by the factor in the temperature term.

References

[1] Ambartsumian, S. A., and A. A. Khachaturian: Basic Equations of the Theory of Elasticity for Materials Which Resist Tension and Compression in a Different Manner. Mechanics of Solids (Mekhanika Tverdogo Tela), No. 2 (1966).
[2] Ambartsumian, S. A.: Equations of the Plane Problem of the Elastic Theory of Materials Which Resist Differently Tension and Compression or of Different Modulus in Tension and Compression. News of the Academy of Sciences of the Armenian Socialist Republic, Mechanics, **19**, No. 2 (1966).
[3] Nowacki, W.: Thermoelasticity. London: Pergamon Press. 1962.
[4] Boley, B. A., and J. H. Weiner: Theory of Thermal Stresses. New York: J. Wiley and Sons. 1960, and MIR. 1964.
[5] Libenson, L. S.: Course in the Theory of Elasticity. Gostekhizdat. 1947.
[6] Novozhilov, V. V.: Theory of Elasticity. Sudpromgiz. 1958.
[7] Timoshenko, S. P.: Theory of Elasticity. New York: Mc Graw-Hill. 1934.
[8] Melan, E., and H. Parkus: Wärmespannungen infolge stationärer Temperaturfelder. Wien: Springer. 1953.

On the Influence of Vibrational Heating on the Fracture Propagation in Polymeric Materials

By

G. I. Barenblatt, V. M. Entov and **R. L. Salganik**

Moscow (U.S.S.R.)

Introduction

According to the basic concept of S. N. Zhurkov [1, 2], the fracture of solids is a process which takes place in time under any stress and is controlled by certain kinetics due to thermal fluctuations and is strongly affected by temperature and stress. Let us denote the average sample lifetime at temperature T and tensile stress σ by $t_0 = t_0(\sigma, \mathrm{T})$. For the evaluation of the time of failure t_0^* for the case of an alternating stress $\sigma(t)$, Bailey's rule of damage summation will be used [3], i.e.

$$\int_0^{t_0^*} \frac{dt}{t_0(\sigma(t), T)} = 1 \tag{0.1}$$

following from the assumption that at each moment the average rate of the damage process is the same as if the acting stress were constant in time (quasi-stationarity).

It is evident that rule (0.1) is valid only under conditions of rather slow change of $\sigma(t)$, when the specific time of this change exceeds considerably the relaxation time. Equation (0.1) was confirmed many times by experiments at low rate of stress variations. However, experiments performed with samples of polymeric materials exposed to cyclic stress, have shown that the actutally observed failure time was in a number of cases far less than the time determined from eq. (0.1). Within the framework of the concept of the fluctuational nature of failure these discrepancies might be explained by the influence of relaxation processes and by the heating of the material due to the mechanical losses in a cyclic deformation. The fact that the latter effect might play an important role is confirmed by rather significant experiments performed by V. R. REGEL and A. M. LEKSOVSKY [4], in which they succeeded in reducing and in some cases in eliminating completely the difference between the estimated and observed failure time by blowing intensively on the samples.

Estimates show that the temperature rise necessary for an explanation of the observed change of lifetime is much higher than the actual change of the sample bulk temperature. Hence, it follows that the temperature rise should be essentially localized and concentrated in the very places where the failure process takes place. The failure process for rigid materials is localized at the crack tips where stress concentration takes place. Therefore the intensity of heat generation, which is proportional to square of the stress amplitude, is low far from the crack tips, but might be considerable and involve a substantial temperature rise near the crack tips. Thus under the influence of vibrational stress a non-uniform temperature distribution in the sample arises, activating the failure process (rupture of bonds) just in the places where it is localized; in the remaining part of the sample the temperature rise might be negligible. The localization of heat generation makes the influence of vibrations on material strength essentially different from the influence of vibrations on the deformation: the acceleration of deformation processes is connected with an almost uniform temperature rise in the whole sample. Hence it follows, in particular, that the vibration effect on the strength might be observed as well under conditions in which the vibration effect on deformation does not take place. It is natural that the effect of the strength reduction under the vibrational load is most prominent with polymers; since relative high mechanical losses and low thermal conductivity are characteristic for these materials.

The concept of the local heating effect on the rate of crack propagation and on lifetime reduction was apparently introduced for the first time by G. M. BARTENEV et al. [5], but there is now no complete quantitative theory of this effect. The present report is an attempt to develop such a theory.

1. Vibrational Heating Near the Crack Tips

The heat release rate per unit volume of viscoelastic medium is equal [6]

$$\Phi = 1/2\,\omega I'' \left[(\sigma^a_{\alpha\beta} - 1/3\,\delta_{\alpha\beta}\sigma^a_{\gamma\gamma})(\sigma^a_{\alpha\beta} - 1/3\,\delta_{\alpha\beta}\sigma^a_{\gamma\gamma}) + 1/3\,\sigma^a_{\gamma\gamma}\sigma^a_{\gamma\gamma}\right], \quad (1.1)$$

where I'' is loss compliance [7], a material property which is generally speaking dependent on the temperature and to a limited extent on the frequency ω; $\sigma^a_{\alpha\beta}$ are the amplitudes of the vibrational stress tensor components. Assuming that the material temperature is not too close to the softening point it is possible to neglect in first approximation the dependence of loss compliance on the temperature and consider I'' as constant.

Consider the body to be a plane plate with an average temperature T across its thickness. The temperature distribution $T(x, y, t)$ in this conditions will satisfy the following equation:

$$\varrho c \frac{\partial T}{\partial t} = \lambda\left(\frac{\partial^2 T}{\partial x^2} + \frac{\partial^2 T}{\partial y^2}\right) - a(T - T_0) + \Phi, \tag{1.2}$$

where ϱ is the density, c the specific heat, λ the thermal conductivity of the material, a the coefficient of heat exchange, and T_0 the ambient temperature.

Let us suppose further that the period of vibrations is sufficiently small compared with the characteristic time of the temperature change; eq. (1.2) in this case is already considered as averaged in a time interval sufficiently large as compared with the time of substantial change of temperature.

The failure process takes place mainly in a region near the crack edge with linear dimensions of the order of d, which is assumed small in comparison with the characteristic length l (crack length). In addition, there exists another dimension, s, specifying the distance from the heat source, at which the temperature rise caused by this source disappears as a result of surface cooling. The length s depend on the heat loss value, decreasing as the latter grows. In a certain intermediate range of values of the heat exchange coefficient a the inequalities hold

$$d \ll s \ll l, \tag{1.3}$$

which will here be considered as fulfilled. The left-hand inequality means that it is possible to neglect the distortion of stress field due to the forces present in the edge region and to consider its temperature as constant and equal to the value at the crack tip if the crack is considered as an infinitely thin cut. The right-hand inequality renders unimportant the heat generation distribution far from the crack tip, so that in calculations of heat release it is possible to use only the main terms of the asymptotic expansions for the stress tensor components near the crack tips.

It is possible to demonstrate in a very simple way that in a linear viscoelastic body the stress distribution around the moving crack is the same as that in the static theory of elasticity (if the dynamic effects may be neglected) [8]. Then, using the known asymptotic expansion of the stress components close to crack tip (see [9—11]), we get the equations for the amplitudes of the components of the vibrational stress tensor as

$$\sigma_{xx}^a = N_0^a r^{-1/2} \cos\frac{\theta}{2}\left[1 - \sin\frac{\theta}{2}\sin\frac{3\theta}{2}\right]$$

$$\sigma_{yy}^a = N_0^a r^{-1/2} \cos\frac{\theta}{2}\left[1 + \sin\frac{\theta}{2}\sin\frac{3\theta}{2}\right], \qquad \sigma_{zz}^a = 2\nu N_0^a r^{-1/2}\cos\frac{\theta}{2}$$

$$\sigma_{xy}^a = N_0^a r^{-1/2} \cos\frac{\theta}{2}\sin\frac{\theta}{2}\cos\frac{3\theta}{2}, \qquad \sigma_{xz}^a = \sigma_{yz}^a = 0, \tag{1.4}$$

where ν is Poisson's ratio, r is the distance from the crack end to point in question, θ is the polar angle, and N_0^a the amplitude of the stress-intensity factor, whose value is based on the amplitudes of added vibrational stress and may be calculated in exactly the same way as the stress intensity factor in static problems (see for example [12]).

Substituting (1.4) in (1.1), we get the simplified expression for the heat generation function:

$$\Phi = \frac{I''\omega}{2}\frac{(N_0^a)^2}{r}\Psi(\theta)$$

$$\Psi(\theta) = 11/8 + 2\nu^2 + (11/8 + 2\nu^2)\cos\theta - 3/8\cos^2\theta - - 7/8\cos^3\theta + 1/2\cos^5\theta. \tag{1.5}$$

From (1.3) it follows that the crack propagates at a distance considerably exceeding s without significant change of the outside conditions. This means, in particular, that the amplitudes of the stress-intensity factor N_0^a in the solution of the heat conduction problem may be considered as constant and the temperature distribution as quasi-steady, or

$$T(x, y, t) = T(\xi, y), \qquad \xi = x - ut, \tag{1.6}$$

where u is the velocity with which the crack tip moves along the x-axis to the right. The velocity u in the solution of the heat conduction problem must be considered as constant and sufficiently low to neglect the formation time of the heating zone, $s^2/\varkappa$, as compared with the time of passage of the heating zone by the crack tip, s/u. Thus, short-time failure is omitted from consideration. Substituting in (1.2) the expressions (1.5) and (1.6) we get:

$$u\frac{\partial T}{\partial \xi} + \varkappa\left(\frac{\partial^2 T}{\partial \xi^2} + \frac{\partial^2 T}{\partial y^2}\right) - \varkappa\alpha^2(T - T_0) + \frac{I''\omega}{2}\frac{(N_0^a)^2}{\varrho c r}\Psi(\theta) = 0, \tag{1.7}$$

where $\varkappa = \lambda/\varrho c$ is the thermal diffusivity, and $\alpha^2 = a/\lambda$.

The boundary conditions for eq. (1.7) should be prescribed at the crack surface and at infinity. Since the crack is very thin and air motion practically does not occur inside the crack, one may consider the heat flux as vanishing over the whole crack surface as well, i.e.

$$\partial T/\partial y = 0 \qquad (x \leq 0, \quad y = 0).$$

As a consequence of the symmetry of the whole problem with respect to axis x, eq. (1.7) may be solved in the whole space: the result will automatically satisfy the boundary condition of zero heat flux over the crack surface. The condition of boundedness of temperature at infinity gives

$$T(x, y, t) \to T_0 \text{ as } \sqrt{x^2 + y^2} \to \infty. \tag{1.8}$$

2. Temperature Distribution Close to the Crack Tip

The solution of the above-formulated problem, eqs. (1.7) to (1.8), may be expressed as follows:

$$T(r, \theta) = T_0 + \frac{\omega I''(N_0^a)^2}{4\pi\lambda} \int_0^{2\pi} \int_0^{\infty} \frac{K_0(\eta\zeta)\, \Psi(\theta'')\, e^{k\zeta\cos\theta}\, \zeta\, d\zeta\, d\theta'}{\sqrt{r^2 + \zeta^2 - 2r\zeta\cos(\theta' - \theta)}} \tag{2.1}$$

where K_0 stands for the Mc Donald function, $k = u/2\varkappa$, $\eta = (k^2 + \alpha^2)^{1/2}$ and the other symbols are evident from Fig. 1. In particular, the value of the temperature rise $\Delta T = T - T_0$ at the crack tip, $r = 0$, is equal to

$$\Delta T_0 = \frac{I''\omega(N_0^a)^2}{2\pi\lambda} \int_0^{\pi} \Psi(\theta)\, d\theta \int_0^{\infty} e^{k\zeta\cos\theta} K_0(\eta\zeta)\, d\zeta. \tag{2.2}$$

Taking into consideration the fact that the function $\Psi(\theta)$, as follows from (1.5), can be expressed as:

$$\Psi(\theta) = \sum_{m=0}^{5} B_m \cos m\theta$$

$B_0 = 19/16 + 2\nu^2$, $B_1 = 33/32 + 2\nu^2$,

$B_2 = -3/16$, $B_3 = -1/16$,

$B_4 = 0$, $B_5 = 1/32$

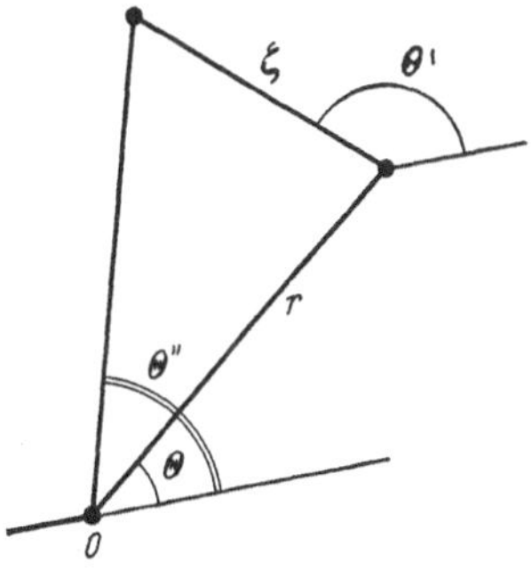

Fig. 1

and performing the integration at first with respect to θ, and then with respect to ζ, by the use of known formulas of integral representation for Bessel functions and for the integrals of a product of Bessel functions

[13], we reduce the expression for ΔT_0 to the following form:

$$\Delta T_0 = \frac{I''\omega(N_0^a)^2}{4\lambda k} \sum_{m=0}^{5} B_m \frac{k^{m+1}\Gamma^2(1/2(m+1))}{\eta^{m+1}\Gamma(m+1)} F\left(\frac{m+1}{2}, \frac{m+1}{2}, m+1, \frac{k^2}{\eta^2}\right)$$

$$\equiv \frac{I''\omega(N_0^a)^2}{2\alpha\lambda}\Theta(z); \quad z = \frac{u}{2\varkappa\alpha} = \frac{k}{\alpha}, \tag{2.3}$$

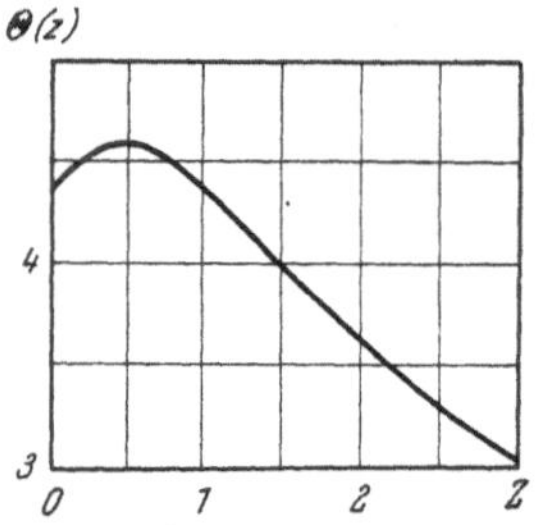

Fig. 2

where the symbols Γ and F indicate the gamma and hypergeometric functions respectively. The function $\Theta(z)$ is represented in Fig. 2, for Poisson ratio $\nu = 0.3$. It has simple asymptotic expansions for small and large values of the argument. With small z it gives the following equation:

$$\Theta(z) = \pi(19/16 + 2\nu^2) + O(z) \tag{2.4}$$

With large z, which corresponds to a high speed of crack propagation and slow cooling, use of eq. (2.1) and of the formula [13] gives the result:

$$\int_0^\infty e^{k\zeta\cos\theta} K(\zeta\eta)\,d\zeta = \frac{1}{\eta}\frac{\arccos(-k/\eta)\cos\theta}{\sqrt{1-(k^2/\eta^2)\cos^2\theta}} \tag{2.5}$$

$$\Delta T_0 = \frac{I''\omega(N_0^a)^2}{\alpha\lambda z}\left[(1+2\nu^2)\ln z + \frac{25}{32} + o(1)\right], \tag{2.6}$$

The range of applicability of the above scheme will now be examined. With large values of the argument the following asymptotic formula for the Mc.Donald function holds [13]:

$$K_0(x) \sim (\pi/2x)^{1/2}e^{-x}. \tag{2.7}$$

Substituting (2.7) into integral (2.1), and bearing in mind that $(k^2+\alpha^2)^{1/2} > k$, we come to the conclusion that the heat generation function occurs in the integral multiplied by a function decreasing at infinity, at least as fast as $\exp[-(\eta - k)\zeta]$. Since the heat generation function is integrable in any finite domain, it follows that in the evaluation of the integral the values of this function are essential only at a distance of the order of $s = 1/(\eta - k)$ and smaller from the point where the temperature is determined. Here s is the distance mentioned in inequalities (1.3) with $z \gg 1$ we have $s \simeq u/\alpha^2\varkappa$.

As the cooling intensity, i.e. α, is reduced, the characteristic size s grows, the lefthand inequality (1.3) starts to be violated and in the calculation of the heat generation not only the first terms of asymptotic expansion near the crack tip but consideration of the complete pattern of stress field becomes essential. The scheme under discussion here is no longer acceptable.

The temperature is decreasing at distances r greater than the dimension s specifying the extent of the heating zone at the crack end, but small as compared with l.

It is easy to define the principal asymptotic term which determines the rate of temperature decrease at such distances. As the circle of radius s with the point (r, θ) as center gives the main contribution in the value of the integral, we may replace in first approximation $\Psi(\theta'')$ by $\Psi(\theta)$ under the integral sign in (2.1) and disregard all the terms under the square root with the exception of the first. Thus we get the first term of the asymptotic representation for the temperature distribution as:

$$\Delta T(r, \theta) = \frac{\omega I''(N_0^a)^2}{2\lambda r} \Psi(\theta) \int\limits_0^\infty K_0(\eta\zeta) I_0(k\zeta) \zeta d\zeta = \frac{\omega I''(N_0^a)^2 \Psi(\theta)}{2\lambda r \alpha^2}. \quad (2.8)$$

According to (2.8), the temperature decreases as r^{-1} with a coefficient highest at $\theta = 0$; at $\theta = \pi$ the coefficient equals to zero; between those values it changes monotonically.

3. Kinetics of the Crack Propagation Under Vibrational Heating Conditions

Up to now the vibrational stress amplitudes were limited only by the demands on the accuracy of expression (1.1) for the heat release and of expressions of elasticity theory for stress tensor components, which should be valid at distances r, such that $d \ll r \ll l$; in other respects those amplitudes are arbitrary.

Generally speaking, the kinetics of crack propagation will be influenced by relaxation processes caused both by vibrations and by vibrational heating. However, there exists the possibility of singling out in some special case the vibrational heating effect. Such is the case in which the sample is under a constant tensile load and a superimposed vibrational load of small amplitude, the latter being of high frequency so that it contributes only to the heat generation at the crack tip but may be neglected in all other respects. The crack propagation will be influenced only by the stress field, due to the principal loads; the vibrational stress will have only the indirect influence caused by heating the region around the crack tip. In the paper [14] this general approach is applied to the problems in which the small vibrational components are superimposed on the main field and it is shown that the amplitude distribution of these vibrational components satisfies the equations of elasticity theory. After calculation of these amplitudes the heat generation may be determined by formula (1.1) and then the temperature distribution is determined as above. As to the main field, inasmuch as rigid materials are considered, it might be taken as elastic at a sufficient distance from the crack tip (at distances greatly exceeding d). In this case, if the applied main loads are changing sufficiently slowly, it can be shown [15] that at slow quasi-static propagation of a crack the stress intensity factor N_0 based on the main loads is equal to a universal (i.e., independent of geometry and applied loads) function of the crack propagation velocity u and temperature T, or

$$N_0 = \pi^{-1} K, \tag{3.1}$$

where $K = K(u, T)$ is a cohesion modulus. If this function (obtained theoretically or measured experimentally) and the relation between the stress intensity factor N_0 and the crack length (which may be calculated on the basis of elasticity theory) are known, then (3.1) is an ordinary differential equation from which the dependence of the crack length on time can be established.

Once the latter relation has been determined in a range of temperature and static loads, it is possible to check quantitatively the theory developed here by finding the same relation in the presence of the vibrational by-load.

In the case in which the elastic bonds are broken under the influence of thermal fluctuations only (purely fluctuational failure), the dependence $K(u, T)$ may be established theoretically [16] as

$$u = \frac{U_* K_*}{K} \exp \frac{K}{K_*} \tag{3.2}$$

$$U_* = U_*^0 \exp\left(\frac{U}{kT_0} - \frac{U}{kT}\right), \quad K = K_*^0 T/T_0, \tag{3.3}$$

where T_0 is some initial temperature, k is Boltzmann's constant, and U^0, K_*^0 are constants, which can be considered as temperature independent (K_*^0 is proportional to the reference temperature T_0).

In the most important case of Griffith's fracture (isolated rectilinear crack in an infinite plate under unifrom tension at infinity, perpendicular to the direction of crack propagation) the mean values of the stress intensity factor N_0 and of its amplitude N_0^a are expressed by following relations:

$$N_0 = \sigma(1/2l)^{1/2}, \quad N_0^a = \sigma_*(1/2l)^{1/2}, \tag{3.4}$$

where σ and σ_* are the mean stress and the amplitude of stress vibration respectively.

Substituting (3.3), (3.1) in (3.2) we obtain

$$u = \frac{dl}{dt} = \frac{U_*^0 K_*^0}{\pi\sigma(1/2l)^{1/2}} \exp \frac{U}{kT_0}\left(1 + \frac{\varDelta T_0}{T_0}\right) \times$$

$$\times \exp\left\{-\frac{U}{k(T_0 + \varDelta T_0)} + \frac{\pi\sigma(1/2l)^{1/2}\, T_0}{K_*(T_0 + \varDelta T_0)}\right\}. \qquad (3.5)$$

The temperature increase $\varDelta T_0$ is determined by eq. (2.3), with the above expression for N_0^a substituted, to get

$$\frac{\varDelta T_0}{T_0} = \frac{l}{\tilde{\tau}\alpha\varkappa}\,\Theta(z), \qquad \tilde{\tau} = \frac{\varrho c T_0}{\omega\sigma_*^2 I''} \qquad (3.6)$$

In the absence of the vibrational load ($\sigma_* = 0$, $\varDelta T_0 = 0$, $T = T_0$) we have for the velocity ($u = u_0$):

$$u_0 = \frac{U_*^0 K_*^0}{\pi\sigma\sqrt{1/2l}} \exp\left(\frac{\pi\sigma}{K_*^0}\sqrt{\frac{l}{2}}\right). \qquad (3.7)$$

It is possible to neglect the temperature dependence of the coefficient of the exponential factor in (3.5) because of the strong dependence of the exponential factor itself on the temperature. With this assumption we get:

$$\ln\frac{u}{u_0} = \left(\frac{U}{kT_0} - \frac{\pi\sigma}{K_*^0}\sqrt{\frac{l}{2}}\right)\frac{\varDelta T_0/T_0}{1 + \varDelta T_0/T_0} \qquad (3.8)$$

Bearing in mind that the relative temperature increase $\varDelta T_0/T_0$ is small, we may disregard it as compared with 1 in the denominator of (3.8) (such neglection is similar to the D. A. Frank-Kamenetsky's transformation [17] used in chemical kinetics). As a result, taking into account (3.6), we get the following transcendental equation for z:

$$\ln\frac{z}{z_0} = A\Theta(z), \qquad z_0 = \frac{u_0}{2\alpha\varkappa} \qquad (3.9)$$

$$A = \left(\frac{U}{kT_0} - \frac{\pi\sigma}{K_*^0}\sqrt{\frac{l}{2}}\right)\frac{l}{2\alpha\varkappa\tilde{\tau}}. \qquad (3.10)$$

The behavior of the function $\Theta(z)$ (see Fig. 2) shows that the eq. (3.10) has a unique solution. The solution gives the crack propagation velocity $u = dl/dt$ as a function of crack length l involved in the expressions for u_0 and A.

The failure time t_0^v with vibrations present is given by the relation:

$$t_0^v = \int_{l_0}^{\infty} \frac{dl}{u(l)}. \qquad (3.11)$$

In (3.11) the expression for u resulting from the solution of eq. (3.9) should be substituted. It is important to note that, because of the assumed smallness of $\Delta T_0/T_0$ and the strong stress-dependence on the velocity u, evaluation of the integral in (3.11) can be drastically simplified and thus the result obtained in a rather simple form.

It is possible to present the expression for the velocity u as a function of l as an approximation for small $\Delta T_0/T_0$ as:

$$u = \frac{U_*^0}{\beta}\left(\frac{l_0}{l}\right)^{1/2} \exp\left\{\left[\frac{U}{kT_0} - \beta\left(\frac{l}{l_0}\right)^{1/2}\right]\frac{\Delta T_0}{T_0}\right\} \exp\left[\beta\left(\frac{l}{l_0}\right)^{1/2}\right]$$
$$= u_0(l)\exp\left\{\left[\frac{U}{kT} - \beta\left(\frac{l}{l_0}\right)^{1/2}\right]\frac{\Delta T_0}{T_0}\right\}, \qquad \beta = \frac{\pi\sigma}{K_*^0}\left(\frac{l_0}{2}\right)^{1/2}. \quad (3.12)$$

In expression (3.12) the factor $\exp\left(\beta\sqrt{l/l_0}\right)$ is changing extremely rapidly; the second exponential factor is changing more slowly as it contains the small ratio $\Delta T_0/T_0$. Therefore substituting (3.12) into (3.11) and integrating by parts we obtain

$$t_0^v = \frac{2l_0}{U_*^0}e^{-\beta}\exp\left[\left(\beta - \frac{U}{kT_0}\right)\left(\frac{\Delta T_0}{T_0}\right)_{l=l_0}\right]\left\{1 + O\left(\frac{1}{\beta}\right) + O\left(\frac{\Delta T_0}{T_0}\right)\right\} \quad (3.13)$$

or, since the factor $2l_0e^{-\beta}/U_*^0$ is simply equal to t_0, i.e. the failure time when vibration is absent, we get, disregarding the small terms, the result:

$$\frac{t_0^v}{t_0} = \exp\left[-\left(\frac{U}{kT_0} - \beta\right)\left(\frac{\Delta T_0}{T_0}\right)_{l=l_0}\right] = \frac{u_0(l_0)}{u(l_0)} = \frac{z_0(l_0)}{z(l_0)}. \quad (3.14)$$

In other words, the failure times with or without vibration are inversely proportional to the corresponding initial velocities of crack propagation.

4. Analysis of Results

Let us examine formula (3.14). When the velocity of crack propagation is such that the value of $z = u/2\varkappa\alpha$ is small, we obtain from this formula:

$$\frac{t_0^v}{t_0} = \frac{z_0(l_0)}{z(l_0)} = e^{-A\Theta(0)} = e^{-4.3A}(\nu = 0.3). \quad (4.1)$$

From the graph of the function $\Theta(z)$ we can observe that the approximation considered holds when $z \leq 1$; this means that the temperature increase at the crack tips does not depend on the velocity of its propagation and consequently when solving the heat transfer problem it is possible to disregard the convective part of heat transfer.

The value of A decreases in a manner inversely proportional to T_0^2, as follows from formulas (3.6) and (3.10), and linearly with average stress σ, and increases proportionally to the square of vibrational stress amplitude σ_*^2. As a result, the effect of vibrational heating, characterized by the ratio t_0/t_0^v, is decreasing with temperature and mean stress growth if the vibrational component is constant; if the proportional increase of both stress compoments takes place without alteration of the cycle form, then the vibrational effect increases (the ratio t_0/t_0^v increases).

Let us consider now the opposite case (large z). Then, in accordance with (2.6)

$$\Theta(z) = z^{-1}[(1 + \nu^2)\ln z + 25/32] \tag{4.2}$$

and the value of ΔT_0 strongly depends on the crack propagation velocity u, decreasing with its growth. As a result, any factor causing a velocity increase will decrease the direct effect of vibration. In particular, with a growth in the test temperature and a growth in the mean stress the velocity u strongly (exponentially) rises, with the result that the failure times approach each other:

$$\frac{t_0^v}{t_0} = \frac{z_0(l_0)}{z(l_0)} = \exp\left\{-\frac{A}{z}[(1 + 2\nu^2)\ln z + 25/32]\right\}. \tag{4.3}$$

Even in the case in which the vibration component increases proportionally to the mean stress (i.e. the cycle form is preserved), the effect of vibration should be reduced (in view of the compartively rapid velocity growth with mean stress increase).

The above conclusions are in agreement with the known experimental fact that the estimated and experimental life-times approach each other as the test temperature or stress level increase, if it is assumed [2] that for stiff polymers the principal contribution in the observed time of failure gives the crack development stage.

It is possible to assume that this stage starts with a crack length whose size is of the order of the inherent flaw in Berry's meaning [18], naturally formed in a body under the stress action. For stiff amorphous polymers this size is 10^{-2}—10^{-1} cm. [18] and is hardly affected by the temperature [19].

Let us take for the purpose of estimating the effects discussed above, $\varrho c = 0.4$ cal/cm² degree; $I'' = 2.10^{-13}$ cm²/dyn, $l_0 = 10^{-1}$ cm, $t_0 = 10^5$ sec, $U/kT_0 = 60$, $\varkappa = 10^{-3}$ cm²/sec, $\beta = 20$, $T_0 = 300$°K. These data refer to polystyrene; the value I'' is taken from [20] and that of l_0 from [19]. For the initial velocity with no vibrational load we have $u_0 = 2l_0/t_0\beta = 10^{-7}$ cm/sec. Accordingly $z_0(l_0) = u_0/2\varkappa\alpha = 5.10^{-5}$. Further, $\tilde{\tau} = \varrho c T_0(\sigma_*^2 \omega I'')^{-1} = 0{,}8\ 10^4$ sec, whereupon $\omega = 300$ sec^{-1} (which corresponds approximately to a frequency of 50 cps) and $\sigma_* =$

$= 100\ \text{kg/cm}^2 = 10^8\ \text{dyn/cm}^2$. As a result the parameter A in formula (3.10), where it is necessary to set $l = l_0$, is equal $A = [U/kT_0) - \beta]1/2\alpha\varkappa\tilde{\tau} = 0.25/\alpha$. From eq. (1.2), setting the last two terms to zero, we find the temperature increase with a uniform distribution σ_* through the sample to be

$$\Delta T' = \frac{1}{2}\frac{\omega I'' \sigma_*^2}{\alpha^2 \varkappa \varrho c}. \tag{4.4}$$

Taking the uniform temperature rise $\Delta T' = 5\,°\text{C}$, we find that corresponding to it $\alpha = 2\ \text{cm}^{-1}$. With such parameters we can use the approximate formula (4.1), from which

$$\frac{t_0}{t_0^v} = 1.7. \tag{4.5}$$

From (3.6) and (4.4) it becomes evident that

$$\frac{\Delta T_0}{\Delta T'} = 2\,l\alpha. \tag{4.6}$$

Since $\alpha \sim s^{-1}$ and the solution is valid for $l \gg s^{-1}$, the right hand side in (4.6) must be much greater than unity; for this reason the above calculation, in which $l_0 \sim 10^{-1}$ cm and $s = 0.5$ cm, represents only an estimate. For this case the problem should be actually solved under the assumption that the heating zone around a crack is much bigger that the crack size, and then it would be necessary to take into account the complete stress fields and not only the first terms of its asympotic expansion near the crack tips. It is clear that the effect for the case considered will be smaller.

The vibration heating effect should rise for bodies of large cross section if compared with the effect in thin plates. In this case the heat removal from the contours of propagating cracks will be reduced. The comparison of V. R. REGEL and A. M. LEKSOVSKY's results for films and fibres is significant in this case: while for films blowing of a sample results in disappearence of the effect, such disappearence remains impossible for fibres[1].

Let us note in conclusion that the principal assumption on which the theory proposed in the present paper is based is the hypothesis that vibrational loads have no direct force influence but only through visco-elastic heating. The other simplifying assumptions, as well as the selection of a concrete failure mechanism (purely fluctuational) are not

[1] Here only a quantitative comparison is possible since in [21] and [4] data referring to fatigue tests are presented, in which the condition that the vibrational load amplitude be small is not fulfilled.

essential ones and might be rejected if necessary (as to the failure mechanism, only the knowledge of the cohesion modulus $K(u, T)$ is required, and this might be determined experimentally). Within the framework of these concepts the theory is rigorous; it gives, as was demonstrated above, an expression for time of failure which permits both qualitative and quantative comparisons between theory and experiment. Thus it is possible to assume that the deviation from quantative coincidence, if it occurs, will allow conclusions to be drawn about the importance of other possible mechanisms through which the vibration effects influence the lifetime and strength of polymers.

Note added in proof: In connection with formula (3.2) see also the paper Entor V. M. and Salganik R. L., On the Prandtl model of brittle fracture, Inzh. Zhurn., Mekhanika Tverdogo Tela, No 6, (1968) [in Russian].

References

[1] Zhurkov, S. N., and B. N. Narzullaev: The Time-dependence of Solids Strength. Zhurn. Tekhn. Fiz. **23**, No. 10, 167 (1953). [in Russian]

[2] Zhurkov, S. N.: Kinetic Concept of the Strength of Solids. Intern. J. Fracture Mech. **1**, No. 4 (1965).

[3] Bailey, J.: Attempt to Correlate Some Tensile Strength Measurements on Glass. Glass Industry **20**, No. 1—4 (1939).

[4] Regel, V. R., and A. M. Leksovsky: A Study of Fatigue within the Framework of the Kinetic Concept of Fracture. Intern. J. Fracture Mech. **3**, No. 2, 99 (1967).

[5] Bartenev, G. M., B. I. Panshin, I. V. Razumovskaya and G. N. Finogenov: On Long Time Cyclic Strength of Organic Glass. Izv. AN SSSR, OTN, Mech. i mash. No. 6 (1960). [in Russian]

[6] Bland, D. R.: The Theory of Linear Viscoelasticity. London: Pergamon Press. 1960.

[7] Ferry, J.: Viscoelastic Properties of Polymers. New York: J. Wiley. 1965.

[8] Entov, V. M., and R. L. Salganik: On the Cracks in Viscoelastic Solids. Inzh. Zhurn., Mekhanika Tverdogo Tela No. 2 (1968). [in Russian]

[9] Sneddon, I. N.: The Distribution of Stress in the Neighborhood of a Crack in an Elastic Solid. Proc. Roy. Soc., A **187**, 229—260 (1946).

[10] Williams, M. L.: On the Stress Distribution at the Base of Stationary Crack. J. Appl. Mech. **24**, 109—114 (1957).

[11] Irwin, G. R.: Fracture. In: Handbuch der Physik (S. Flügge, ed.), Bd. VI. Berlin-Göttingen-Heidelberg: Springer, 551—590 (1958).

[12] Barenblatt, G. I.: The Mathematical Theory of Equilibrium Cracks in Brittle Fracture. Advances in Appl. Mechanics, 7. New York: Academic Press. 1962.

[13] Ryzhik, I. M., and I. S. Gradstein: Tables of Integrals, Sums, Series and Products. Gostekhisdat, 1951. [in Russian]

[14] Barenblatt, G. I.: The Effects of Small Vibrations at the Deformation of Polymers. PMM **30**, 1 (1966).

[15] Barenblatt, G. I., V. M. Entov and R. L. Salganik: On the Kinetics of the Crack Propagation. General Concepts. The Nearly Equilibrium Cracks. Inzh. Zhurn., Mekhanika Tverdogo Tela, No. 5 (1966). [in Russian]

[16] BARENBLATT, G. I., V. M. ENTOV, and R. L. SALGANIK: On the Kinetics of the Crack Propagation. Fluctuational Failure. Inzh. Zhurn., Mekhanika Tverdogo Tela No. 1 (1967). [in Russian]

[17] FRANK-KAMENETSKIJ, D. A.: Diffusion and Heat Transfer in the Chemical Kinetics. Izd. Nauka, 1967. [in Russian]

[18] BERRY, J. P.: Fracture Processes in Polymeric Materials, II. Tensile Strength of Polysterene. J. Polymer Sci. **50**, 313–321 (1961).

[19] BERRY, J. P.: Fracture Processes in Polymeric Materials, IV. Dependence of the Fracture Surface Energy on Temperature and Molecular Structure. J. Polymer Sci. Part A **1**, 993–1003 (1961).

[20] TAKAYANAGI, M.: Viscoelastic Properties of Crystalline Polymers. Mem. Fac. Engng Kyushu Univ., 23, No. 1 (1963).

[21] REGEL, V. R., and A. M. LEKSOVSKY: Time Dependence of Strength at Static and Cyclic Loads. Fiz. Tv. Tela **4**, No. 4 (1962). [in Russian]

Thermal Mechanism of Cold-Drawing of Polymers

By

G. I. Barenblatt, V. M. Entov and **A. E. Segalov**

Moscow (U.S.S.R.)

Introduction

The phenomenon of cold-drawing (formation and steady propagation of an abrupt contraction — "necking" — in stretching cylindrical samples) is one of the qualitative peculiarities of polymers and plays an important role in the production processes of strong fibres [1—4].

A theoretical scheme was proposed in [5] for a description of this phenomenon, in which the orientational deformation of the elements of supramolecular structure was considered as a conversion reaction with a rate strongly dependent on the stress. One of the basic conclusions of that paper is that the deformation zone should necessarily have an activating influence on the non-deformed parts of the sample adjacent to this zone. For this reason steady propagation of necks is possible as the deformation capacity is exhausted at a given place. Today, there is no single explanation which explains how the excitation is transferred to the non-deformed parts of the sample. However, since the necking phenomenon is of a very general nature and occurs both in crystalline and in amorphous polymers, the transfer mechanism should be sufficiently general. In [5], an isothermal process is considered and a peculiar quasidiffusion of oriented elements of the microstructure caused by microstresses is proposed as the mechanism for the transfer of excitation. In this paper, a different scheme is considered, in which necking is caused by the transfer of heat generated in the deformation zone along the sample. The thermal mechanism of cold-drawing was suggested first by Marshall and Thompson [6], and by Müller et al. [7, 8]. This mechanism, attractive because of its simplicity, is certainly not the only possible one: in the experiments of Yu. S. Lazurkin [2] and Vin-

CENT [9] with films, cold-drawing was so slowly carried out that the temperature could not have been significantly increased at any point of the sample; nevertheless necking was very clearly observed. Fig. 1 shows the temperature distribution along the length of the sample plotted against the distribution of cross-sectional area along the length, as obtained by D. YA. PAVLOV in stretching cylindrical caprolactam

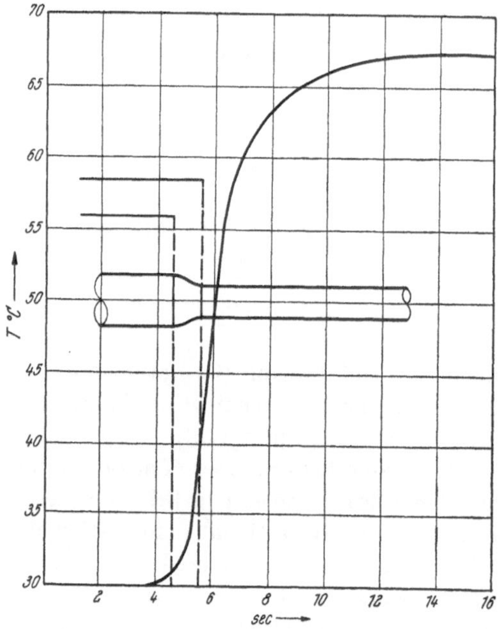

Fig. 1

samples at a rate of 400 mm/min and at an ambient temperature o 30 °C. It is evident from the figure that the temperature increase in the non-deformed part of the sample is insignificant and thus cannot be responsible for the whole effect.

It would not be difficult to indicate in which range of velocity of necks, the thermal conduction could affect necking. On the one hand, the neck should spread sufficiently fast so that the sample did not lose its heat into the surrounding medium. When the sample is cooled in stagnant air, the characteristic heat exchange time is $\tau \sim (10^2 - 10^3)$ d sec, where d is the cross-sectional size of sample expressed in centimeters. The time of heating τ_1 for a sample of thickness d is of the order $\tau_1 \sim d^2/\varkappa$ ($\varkappa$ is the thermal diffusivity of the substance, for polymers $\varkappa \sim 10^{-3}$ cm^2/sec). For the usual thin samples, the time of heat exchange is determined by τ. So we have

$$v \geq b/\tau$$

(b is the length of transition region). On the other hand, the temperature front, travelling at a rate v along the sample, increases the temperature at a distance $\sim \varkappa/v$ ahead of the front. Hence it follows that $\varkappa/v \sim b$. Since in experiments, the quantity d usually does not exceed b by more than an order, the last condition gives an upper bound for the velocity. Thus we can consider that the heat mechanism of necking, when a polymer sample is tested in air, may be of importance at a neck propagation velocity such that

$$10\varkappa/d \geq v \geq 10^{-2}\, b/d \text{ (cm/sec)},$$

which when $d \sim 10^{-1}$ cm becomes

$$10^{-1} \text{ cm/sec} \geq v \geq 10^{-3} \text{ cm/sec}.$$

There thus, exists a sufficiently wide range of rates, in which the heat mechanism of necking cannot be eliminated on the basis of general considerations. In this paper, a mathematical theory is proposed for the stationary necking, within the framework of the heat mechanism of the phenomenon. As in [5], the treatment of the problem is based on a mathematical analogy with the problems of flame propagation as treated by Y. B. Zelgovich [10]. In our case, the mathematical formulation of problem proved to be very close to the problems of combustion theory considered by S. S. Novikov and Yu. S. Ryazantsev in [11].

1. Formulation of the Problem and Fundamental Equations

Let us consider, as in [5], a one-dimensional cold-drawing, i.e. such that the changes in all quantities along the cross-section of the sample

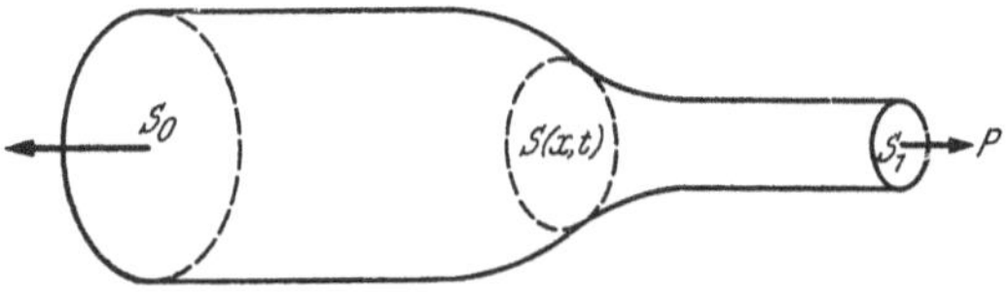

Fig. 2

are neglected; the axis of the sample is chosen as the x-axis. Supposing that the deformation is quasistatic, we have

$$\sigma S = \sigma_0 S_0 = P = \text{const}, \tag{1.1}$$

where $\sigma(x, t)$ is the stress at the cross-section at x at the instant t, S is the area of the section, σ_0 and S_0 are the same quantities for some fixed section. Furthermore the section S_0 is taken in the non-deformed part of the sample. The equation of continuity, supposing that the density does not change, is

$$\frac{\partial S}{\partial t} + \frac{\partial (S u)}{\partial x} = 0, \tag{1.2}$$

where u is the rate at which the cross-section shifts in the direction of drawing.

Let us choose the relative constraction of the sample as a measure of deformation, namely

$$\alpha = S/S^0, \tag{1.3}$$

where S^0 is the area of the cross-section of the sample in a state of tota isotropy of the substance; the initial state of the substance (in which the sample has the cross-section S_0) can possess some anisotropy. If the elements of micro-structure subjected to orientational transformation could be determined in the substance, then the quantity α could be expressed in terms of the concentration n of oriented elements of the microstructure, which is the fundamental dependent variable used in [5].

It is natural to suppose that the rate of strain at a given section (equal to $[-S^{-1}\, dS/dt]$ because of the assumed incompressibility of the material) depends on the stress, the strain and the temperature T, or

$$-\frac{1}{S}\frac{dS}{dt} = -\frac{1}{\alpha}\frac{d\alpha}{dt} = f(\alpha, \sigma, T). \tag{1.4}$$

We shall use only the following general assumptions on the nature of f.

At each temperature and deformation, there exists some critical stress $\sigma_*(\alpha, T)$, below which the strain rate of the material vanishes. The critical stress σ_* increases as α decreases (growth of deformation), and decreases as the temperature T increases. At $\sigma > \sigma_*$ the function $f(\alpha, \sigma, T)$ rapidly increases with the increase of σ and T. Thus,

$$f(\alpha, \sigma, T) \equiv 0 \quad \text{for} \quad \sigma \leq \sigma_*(\alpha, T); \quad f(\alpha, \sigma, T) > 0 \quad \text{for} \quad \sigma > \sigma_*(\alpha, T);$$

$$\frac{\partial \sigma_*}{\partial \alpha} < 0; \quad \frac{\partial \sigma_*}{\partial T} < 0. \tag{1.5}$$

The critical stress is the basic parameter characterizing the orientational strain-hardening of the material.

Furthermore, the equation of energy balance gives

$$\frac{\partial (U S)}{\partial t} + \frac{\partial (U S u)}{\partial x} = \frac{\partial}{\partial x}\left(kS\frac{\partial T}{\partial x}\right) + \sigma S \frac{\partial u}{\partial x} - 2h\sqrt{\pi S}(T - T_0). \tag{1.6}$$

Here U is the internal energy per unit volume; since elastic deformation is neglected, we can take U as independent of σ, or $U = U(\alpha, T)$. The quantity $(\sigma\, \partial u/\partial x)$ is rate of the specific work done by tensile stresses, k is the coefficient of thermal conductivity and h is coefficient of external heat exchange, and T_0 is the ambient temperature. Using (1.1) and (1.2) we can easily show that

$$\sigma \frac{\partial u}{\partial x} = \frac{\partial \sigma}{\partial t} + u \frac{\partial \sigma}{\partial x}. \tag{1.7}$$

Hence relation (1.6) is transformed into

$$S\left(\frac{\partial H}{\partial t} + u \frac{\partial H}{\partial x}\right) = \frac{\partial}{\partial x}\left(k S \frac{\partial T}{\partial x}\right) - \Delta_1 \sqrt{\alpha}(T - T_0), \tag{1.8}$$

where

$$H = U(\alpha, T) - \sigma; \quad \Delta_1 = 2h\sqrt{\pi S^0}.$$

As a result of the monotonic increase in internal energy with a rise in temperature, we have

$$\frac{\partial H}{\partial T} = \frac{\partial U}{\partial T} > 0. \tag{1.9}$$

We shall assume that the inequality

$$-\frac{\partial H}{\partial \alpha} < \sigma, \tag{1.10}$$

holds; hence it follows that

$$\frac{dH}{d\alpha} = \frac{\partial H}{\partial \alpha} + \frac{\partial H}{\partial \sigma}\frac{\partial \sigma}{\partial \alpha} = \frac{\partial H}{\partial \alpha} + \frac{\sigma}{\alpha} > 0.$$

The set of conditions (1.9) and (1.10) is sufficient for the heat generation and the rise of temperature which occur at any point where deformation takes place with an arbitrary initial deformation α_0. Thus, with the above assumptions, the process of cold-drawing is described by eqs. (1.1), (1.2), (1.4) and (1.8) with the corresponding boundary and initial conditions.

2. Steady Propagation of the Neck

Now we shall suppose that the sample is infinite, extending from $x = -\infty$ to $x = +\infty$, and that the neck moves from right to left with a constant velocity V. When $x = +\infty$, the substance is totally deformed; when $x = -\infty$ the deformation has not yet started and the substance is at rest. In a moving reference system $\xi = x + Vt$, the process will be a steady one. Eliminating S with the help of (1.1) and using the above-mentioned boundary conditions, we obtain a system of ordinary differential equations for T, namely:

$$V\frac{dH}{d\xi}=\frac{d}{d\xi}\left(a\frac{dT}{d\xi}\right)-\varDelta\sqrt{\alpha}(T-T_0), \tag{2.1}$$

$$u\frac{d\alpha}{d\xi}=-\alpha f(\alpha,\sigma,T), \tag{2.2}$$

$$\alpha\sigma=\alpha_0\sigma_0, \tag{2.3}$$

with the boundary conditions:

$$T=T_0,\quad \alpha=\alpha_0,\quad \sigma=\sigma_0\qquad (\xi\to-\infty),$$
$$\frac{dT}{d\xi}=\frac{d\alpha}{d\xi}=\frac{d\sigma}{d\xi}=0\qquad (\xi\to\infty), \tag{2.4}$$

where

$$a=k\alpha/\alpha_0;\qquad \varDelta=\varDelta_1/S_0=2h(\pi/S_0\alpha_0)^{1/2}.$$

The symbol u for the velocity is retained in the moving reference system also.

System (2.1) — (2.3) with boundary conditions (2.4) is overdetermined and has no non-trivial solution when the parameters are selected arbitrarily. The speed V of necking at a given stress σ_0 is found, as is usual in problems of similar type, as an eigen-value satisfying the condition of existence of solution of the overdetermined problem.

3. Adiabatic Necking

The heat mechanism appears in a most striking form when the heat loss from the surface is negligibly small and the necking process is close to an adiabatic one, in which case a complete analytical investigation is possible.

When this heat loss is neglected, eq. (2.1) takes the form

$$V\frac{dH}{d\xi}=\frac{d}{d\xi}\left(a\frac{dT}{d\xi}\right) \tag{3.1}$$

which clearly has a first integral

$$VH-a\frac{dT}{d\xi}=\text{const}. \tag{3.2}$$

Hence, we get

$$H(\alpha_1,\sigma_1,T_1)=H(\alpha_0,\sigma_0,T_0), \tag{3.3}$$

where the subscript 1 indicates the values of the parameters in a fully deformed sample ($\xi = \infty$). Moreover, from (2.2) and (2.4) we have

$$f(\alpha_1, \sigma_1, T_1) = 0. \tag{3.4}$$

From this equation and from eq. (1.5) we immediately get

$$\sigma_1 = \sigma_*(\alpha_1, T_1), \tag{3.5}$$

since, if we had $\sigma_1 < \sigma_*$, the contraction of the sample would have ceased at some finite point, which is impossible. The system of eqs. (2.3), (3.3) and (3.5) allows the determination of the final contraction α_1 and the temperature T_1 in terms of the known quantities α_0, σ_0, T_0.

In fact, from (2.3) and (3.3) the temperature T_1 is determined as a monotonically decreasing function of the contraction α_1. Let us denote this function by $T_*(\alpha)$ (Fig. 3). Evidently, $T_*(\alpha)$ is the temperature to which the element of the substance is heated, when subjected to an adiabatic extension with a constant force $\sigma_0 S_0$ from the initial contraction α_0 up to a contraction α, the initial temperature being T_0. Besides $T_*(\alpha)$, let us introduce an another temperature, $T^*(\alpha)$, at which the deformation starts at the section with contraction α and acting stress $\sigma = \alpha_0\sigma_0/\alpha$. The function $T^*(\alpha)$ is defined as the solution of the equation

$$\alpha\sigma_*(\alpha, T^*(\alpha)) = \alpha_0\sigma_0, \tag{3.6}$$

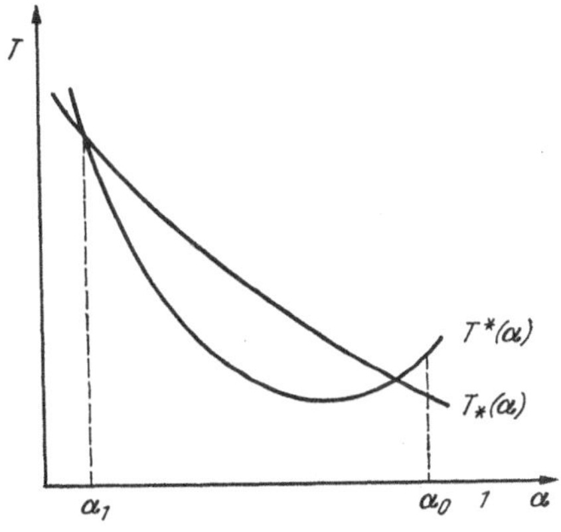

Fig. 3

uniquely due to the condition $\partial\sigma_*/\partial T < 0$. The intersection of the curves $T^*(\alpha)$ and $T_*(\alpha)$ determines the values α_1 and T_1 which correspond to the point where the deformation ceases. For the existence of the solution, it is necessary that the curves $T^*(\alpha)$ and $T_*(\alpha)$ should intersect in at least two points as shown in Fig. 3. At the point α_1, T_1 we have

$$\partial T^*/\partial\alpha < \partial T_*/\partial\alpha < 0,$$

which condition clearly follows from the condition of vanishing deformation at $\xi = \infty$. The initial point α_0, T_0 lies on the curve $T_*(\alpha)$ and at this point the substance does not get deformed so that $T_*(\alpha_0) \leq$

$\leq T^*(\alpha_0)$. Fig. 3 permits us to investigate what happens when the initial state of the sample, defined by the quantities α_0, σ_0 and T_0, changes. From the position of the curves for T_* and T^* shown in the figure, we can conclude that when the temperature T_0 increases, while α_0 and σ_0 do not undergo any changes, the $T_*(\alpha)$ curve shifts upwards and $T^*(\alpha)$ remains unchanged so that α_1 decreases. Thus, under the conditions of the thermal mechanism of necking, the degree of contraction should increase with an increase in the temperature of the experiment if the stress σ_0 remains constant.

Now, let the tensile stress σ_0 increase when the temperature T_0 remains constant. Let us consider two points lying on the curve $T = T_*(\alpha)$. Along the curve $H = \text{const}$ we have, according to formula (1.8),

$$H(\alpha, \sigma, T_*) = U(\alpha, T_*) - \alpha_0\sigma_0/\alpha = U(\alpha_0, T_0) - \sigma_0$$

or

$$U(\alpha, T_*) = U(\alpha_0, T_0) + \sigma_0(\alpha_0/\alpha - 1). \tag{3.7}$$

Since $\partial U/\partial T > 0$, it follows that when σ_0 increases, T_* increases for all $\alpha < \alpha_0$ and remains constant when $\alpha = \alpha_0$. For the curve $T^*(\alpha)$, according to (3.6), we have $\alpha_0\sigma_0 = \alpha\sigma_*(\alpha, T^*)$. Hence from (1.5) it follows that T^* decreases as σ_0 increases. As a result, we obtain the result that α_1 decreases with an increase in σ_0. Thus, under the conditions of thermal mechanism of necking, the greater the degree of contraction, the greater the acting stress.

The graphs of Fig. 3 undergo qualitative changes when the initial values of temperature and stresses are changed. If $T_0 = T_*(\alpha_0)$ exceeds $T^*(\alpha_0)$, as a result of the increase in temperature or increased stresses, a stationary spreading of the neck is not possible, since deformation can start without any additional heating. In this case, a non-steady flow of the sample takes place along its whole length.

If the temperature and the stress decrease, the points of intersection of the curves T_* and T^* approach each other and can finally disappear. In this case, the critical stress is not developed at any point and elastic deformation takes place, i.e. in our approximation the sample does not deform at all. In this case, it is natural to expect that the sample would be broken in a brittle manner when the stress is increased. On the other hand, by increasing the initial temperature T_0 and the tensile stress σ_0 (within the limits of the strength of the material) the polymer can be drawn with neck formation, provided the curves of T^* and T_* have the typical form shown in Fig. 3.

The same qualitative arguments could be applied to investigate the dependence of deformation on the initial strain α_0 (orientation) of the polymer. From (3.6), it follows that the curve $T^*(\alpha)$ shifts down when α_0 increases. From (3.7), it may be easily shown that $T_*(\alpha)$ increases for each value of α when α_0 increases. Thus, when α_0 increases (i.e. the initial orientation decreases) the final orientation is increased (α_1 decreases). Thus, the degree of contraction in the neck increases, if the tensile stress σ_0 or the temperature T_0 is increased or if the initial strain of the sample is decreased (increase in α_0).

4. Existence of the Solution

Until recently, it was assumed that a solution of the type considered does exist. It can be shown that if the pattern of the curves has the form shown in Fig. 3, determining α_1 and T_1, then the problem (3.2), (2.2)—(2.4) has a solution at a certain value of the parameter V. Eliminating σ from the equations, using (2.3) we obtain

$$a(\alpha, T)\frac{dT}{d\xi} = V[H(\alpha, T) - H_0], \tag{4.1}$$

$$V\frac{d\alpha}{d\xi} = -\frac{\alpha^2}{\alpha_0} f(\alpha, \sigma, T) = -q(\alpha, T), \tag{4.2}$$

$$\left.\begin{aligned} &T = T_0, \quad \alpha = \alpha_0 \quad \text{when} \quad \xi \to -\infty, \\ &T = T_1 \quad \text{when} \quad \xi \to \infty. \end{aligned}\right\} \tag{4.3}$$

On the phase plane (α, T) of the (4.1)—(4.2) system, the trajectory originating form the A, and ending at the singular (saddle) point B, corresponds to the solution to be found. Therefore the trajectory should consist of a vertical section AD (corresponding to the heating of the non-deformed part of the sample) and of the separatrix Γ of the saddle B. The condition that the separatrix should meet the point D gives the condition for the determination of V. This problem is close to the combustion problem considered by S. S. Novikov and Yu. S. Ryazanstev [11] and differs only in having a different dependence of q on α.

When V is sufficiently small, the separatrix lies arbitrarily close to the horizontal line $T = T_1$, and when V is sufficiently large it is arbitrarily close to the zero isocline BA. Hence it follows from continuity that there exists at least one required value of V. It is easy to prove that this value is unique.

By comparing the solutions corresponding to the various values of the drawing stress σ_0, we can establish the dependence of the neck velocity on σ_0. As shown above, at a given value of α, T_* increases, while T^* decreases with an increase in σ_0 and the pattern in the phase plane of the

(4.1)—(4.2) system changes correspondingly. Fig. 4 shows the phase diagrams at two values of the stress, $\sigma_0' < \sigma_0''$. From this figure it follows that the phase trajectories $AD'B'$ and $AD''B''$ have a point of intersection (α^+, T^+), at which

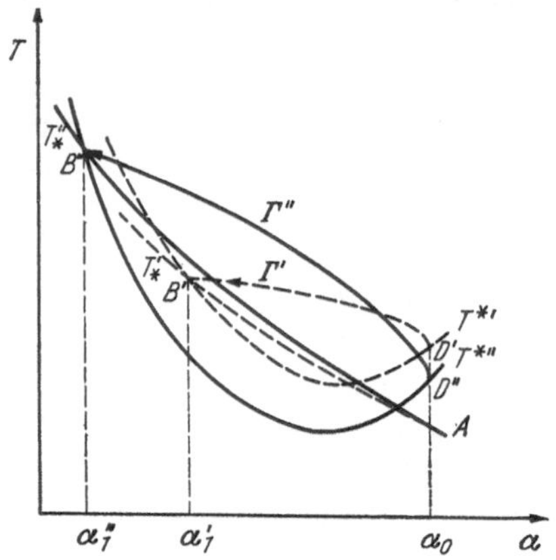

Fig. 4

$$\left(\frac{dT}{d\alpha}\right)'' \leq \left(\frac{dT}{d\alpha}\right)' < 0. \tag{4.4}$$

From the system (4.1)—(4.2) we have, at the point of intersection,

$$V''^2 \frac{H''(\alpha^+, T^+) - H_0''}{a(\alpha^+, T^+)\, q''(\alpha^+, T^+)} \leq$$

$$\geq V'^2 \frac{H'(\alpha^+, T^+) - H_0'}{a(\alpha^+, T^+)\, q'(\alpha^+, T^+)}, \tag{4.5}$$

On the other hand, according to the properties of the functions H and f stated earlier, we have

$$q''(\alpha^+, T^+) > q'(\alpha^+, T^+),$$

$$H''(\alpha^+, T^+) - H_0'' < H'(\alpha^+, T^+) - H_0'. \tag{4.6}$$

From (4.5) and (4.6), we have

$$V'' > V'.$$

Thus, we have established that the neck velocity increases with an increase in drawing stress. Similarly we can easily show that V increases with an increase in the initial temperature T_0.

5. Analysis of the Results

With fixed values of the initial orientation α_0 and temperature T_0, the drawing of the sample takes place with a constant velocity V at a given stress σ_0, and V corresponds to the rate v at which the sample ends are drawn apart, i.e.

$$v = V(\alpha_0/\alpha_1 - 1). \tag{5.1}$$

As shown above, with an increase in σ_0, V increases and the final contraction α_1 decreases, so that the rate increases. Because of the monotonic dependence of v on σ_0, the problem could have been formulated by considering the velocity of the grips v as fixed and σ_0 as unknown. Keeping in mind the comparison with the experimental data, let us briefly repeat the qualitative analysis of section 3, supposing however that the values of v are given.

From what has been said, it follows first of all that the stress σ_0 increases (at constant values of α_0 and T_0) when the rate of drawing v is increased. A weak increase of σ_0 was noticed by Yu. S. Lazurkin in his work referred to above; a number of authors, in particular F. H. Müller [3], conclude that σ_0 is independent of v.

Now, let the rate v and the initial contraction α_0 be fixed. Then, as the temperature T_0 increases, the drawing stress σ_0 should decrease. This obvious fact was observed many times experimentally. Because of the equality

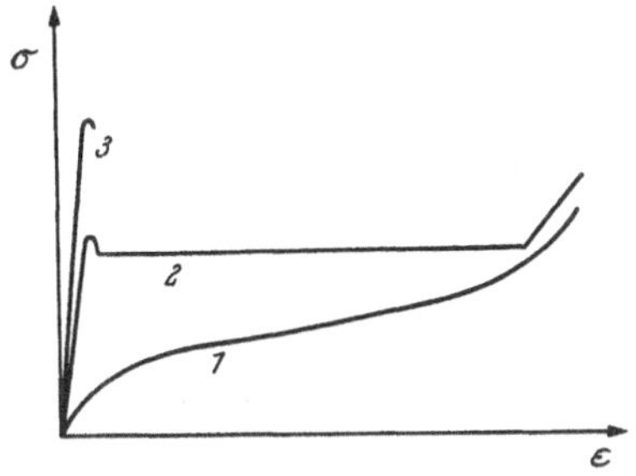

Fig. 5

$$\left(\frac{\partial \sigma_0}{\partial T_0}\right)_{\alpha_0, v} = - \frac{\left(\frac{\partial v}{\partial T_0}\right)_{\alpha_0, \sigma_0}}{\left(\frac{\partial v_0}{\partial \sigma_0}\right)_{\alpha_0, T_0}}$$

it follows from the results of section 3 and the curves of Fig. 3 change correspondingly. In particular, at sufficiently high temperatures the sample deforms uniformly along the entire length; and, when the temperature is decreased, deformation with neck takes place and at some sufficiently low temperatures the curves of Fig. 3 cease to touch each other and the sample is broken in a brittle manner after a small elastic deformation. Fig. 5, in which the curves are numbered according to decreasing temperature, shows qualitatively the changes in deformation with changes in T_0; similar experimental results are well known [1].

Increasing the initial strain of the sample (degree of its orientation), as shown in section 3, is equivalent to decreasing the temperature of drawing.

In conclusion, we can mention that, without limiting the general assumptions made in this paper, it is not possible to find the dependence of the contraction parameters on the conditions of drawing. According to experimental data, this dependence is qualitatively different for different polymers.

References

[1] KARGIN, V. A., and T. I. SOGOLOVA: Studies of Mechanical Properties of Crystalline Polymers. Zhurn. Fiz. Khim. **27**, No. 7—9 (1953). [in Russian]

[2] LAZURKIN, YU. S., and G. L. FOGELSON: On the Nature of Deformation of Highmolecular Substances in Glass-like State. Zhurn. Techn. Fiz. **21**, No. 3 (1951). [in Russian.]

LAZURKIN, YU. J. S.: Cold-drawing of Glass-like and Crystalline Polymers. J. Polymer Sci. **30**, No. 121 (1958).

[3] MÜLLER, F. H.: Kaltverstreckung von Kunststoffen. Materialprüfung. **5**, No. 9 (1963).

[4] JÄCKEL, K.: Ein Beitrag zur Kaltverstreckung von Kunststoffen. Kolloid-Z. **137**, No. 2—3 (1954).

[5] BARENBLATT, G. I.: On the Neck Formation in Drawing Polymers. PMM **28**, No. 6 (1964). [in Russian.]

BARENBLATT, G. I.: On the structure of fully developed plastic flow in polymers. Proc. XI Intern. Congress of Applied Mechanics, Munich, 1964.

[6] MARSHALL, J., and A. B. THOMPSON: The Cold Drawing of High Polymers. Proc. Roy. Soc., A, **221**, No. 1147 (1954).

[7] MÜLLER, F. H., and K. JÄCKEL: Energie-Bilanz bei Kaltverstreckung. Kolloid-Z. **129**, No. 2—3 (1952).

[8] BRAUER, P., and F. H. MÜLLER: Über die Temperaturüberhöhung in der Fließzone während der Kaltverstreckung. Kolloid-Z. **135**, No. 2 (1954).

[9] VINCENT, P. J.: The Necking and Cold-drawing of Rigid Plastics. Polymer **1**, No. 1 (1960).

[10] ZELDOVICH, YA. B.: Theory of Flame Propagation. Zhurn. Fiz. Khim. **22**, No. 1 (1948). [in Russian.]

[11] NOVIKOV, S. S., and YU. S. RYAZANTSEV: On the Existence and Uniqueness of Solution of the System of Equations of Thermal Theory of Combustion. J. Appl. Mech. and Techn. Phys., No. 4, 1965. [in Russian.]

On 'Thermo-Rheologically Simple' Solids

By

M. J. Crochet and **P. M. Naghdi**

Louvain (Belgium) and Berkeley, Calif. (U. S. A.)

Summary

Starting with the results for a non-isothermal finite linear theory of viscoelasticity, a systematic derivation of a linearized theory with infinitesimal deformation is presented for 'thermo-rheologically simple' solids. A comparison of the resulting constitutive equations and the internal dissipation with those given previously is included.

1. Introduction

In a recent paper by Crochet and Naghdi [5] dealing with a class of simple solids with fading memory, after some general considerations, detailed attention is given to the construction of a thermodynamical theory of finite linear viscoelasticity. Here, we take up the matter further; and, by specialization of the results in [5], consider a systematic development of a linearized theory with infinitesimal deformation for a class of materials (with temperature-dependent characteristics) called 'thermo-rheologically simple' by Schwarzl and Staverman [10]. The developments in [5] are carried out within the framework of the theory of simple materials with fading memory and begin with Coleman's results [1] on the thermodynamics of the subject.

The main feature which characterizes the mechanical behavior of 'thermo-rheologically simple' solids rests on the temperature-time equivalence postulate originally proposed by Leaderman and in a slightly different form by Ferry. According to this postulate, the mechanical response functions (e.g., the relaxation moduli in shear and dilatation for an isotropic material) are affected by a uniform temperature change only within a uniform change of time scale and are uniformly accelerated (or decelerated) if the temperature is increased (or decreased). The temperature-time equivalence postulate is usually expressed by means of a definition for a "reduced time" involving a "shift function", a nonlinear function of temperature, which represents the contraction (or the stretching) of the time scale due to temperature change. Considerable experimental evidence in support of the postulate, over fairly wide temperature ranges and for a variety of polymeric materials, has been reported in the literature and may be found in the works of Staverman and Schwarzl [11], Payne [9] and Tobolsky [13].

Morland and Lee [7], after generalizing the definition of the "reduced time" mentioned above, proposed a scheme whereby the temperature field (in the absence of thermo-mechanical coupling) is assumed to be governed by Fourier's heat conduction equation, while the constitutive relation for stress includes the effect of temperature variation on the mechanical properties of the viscoelastic solids. The development in [7], which was further supplemented by Muki and Sternberg [8], has since become the basis of analysis for thermal stresses in linear isotropic viscoelastic solids with temperature-dependent characteristics. Thermodynamical aspects of 'thermo-rheologically simple' solids with infinitesimal deformation have been considered and discussed previously by Staverman and Schwarzl [11] and by Hunter [6]; but the point of view of these authors is different from ours and their analyses are based on a "model" consisting of a combination of springs and dashpots. For additional background on 'thermo-rheologically simple' solids, reference may be made to an expository article by Sternberg [12].

In the present paper, after recalling some of the basic equations and certain additional preliminaries in Sec. 2, we summarize in Sec. 3 the main results for a non-isothermal finite linear theory of viscoelasticity with fading memory which were deduced previously in [5]. We also discuss in Sec. 3 an appropriate constitutive equation for the heat flux vector for materials which possess a center of symmetry in the initial reference configuration; this latter discussion of the constitutive functional for the heat flux was not considered in detail in [5]. We then discuss a linearized theory with infinitesimal deformation in Sec. 4 which is obtained by specialization of the results for finite linear viscoelasticity (Sec. 3) and after the introduction of suitable assumptions, including the smallness of displacements and strains. A comparison of some of the results of the linearized theory with those given previously for 'thermo-rheologically simple' solids by Hunter [6], and others, is included in Sec. 4.

In contrast to the usual linearization of the constitutive equations of a finite theory (e. g., nonlinear elasticity) in which the temperature difference (from the initial reference temperature) also is to be considered "small," our present task is less straightforward. Apart from the fact that we allow for the presence of an initial stress and do not restrict the discussion to isotropic materials, in the developments of Sec. 4, we assume the temperature history is such that it is "small" only in a certain sense (corresponding to the smallness of the Hilbert space norm of the difference history from the present value of temperature). Roughly speaking, this implies that the temperature variation is small only in the recent past but (due to the requirements of fading memory) does not place any restriction on the present value of the temperature or the temperature variation from the initial reference temperature.

2. Notation. General Background

The motion of a body may be described by a vector function $\boldsymbol{x}(\tau) = \chi(\boldsymbol{X}, \tau)$, $-\infty < \tau \leqq t$, where $\boldsymbol{x}(\tau)$ is the position of a material point at time τ which has the position $\boldsymbol{X}$ in a fixed reference configuration $\mathscr{R}$. The deformation gradient at a material point at time τ, relative to the configuration $\mathscr{R}$, is given by

$$\boldsymbol{F}(\tau) = \frac{\partial \chi(\boldsymbol{X}, \tau)}{\partial \boldsymbol{X}}, \qquad -\infty < \tau \leqq t. \tag{2.1}$$

We assume that $\chi(\boldsymbol{X}, \tau)$ is always smoothly invertible in $\boldsymbol{X}$, i.e., $\det \boldsymbol{F}(\tau) \neq 0$. In what follows, for convenience we do not explicitly exhibit the dependence of a function on the reference position $\boldsymbol{X}$ and use the notation

$$\boldsymbol{x} = \chi(t), \qquad \boldsymbol{F} = \boldsymbol{F}(t). \tag{2.2}$$

Also, $\boldsymbol{F}^{-1}$ and $\boldsymbol{F}^T$ will denote the inverse of $\boldsymbol{F}$ and the transpose of $\boldsymbol{F}$, respectively, while div and grad operators will be reserved for spatial derivative with respect to $\boldsymbol{x}$ keeping τ fixed.

For our present purpose, a convenient measure of deformation is

$$2\boldsymbol{E}(\tau) = \boldsymbol{F}^T(\tau)\,\boldsymbol{F}(\tau) - \boldsymbol{1}^+(\tau), \quad 2\boldsymbol{E} = \boldsymbol{F}^T\boldsymbol{F} - \boldsymbol{1}, \tag{2.3}$$

where $\boldsymbol{1}$ is the unit tensor and $\boldsymbol{1}^+(\tau)$ denotes the constant function with value $\boldsymbol{1}$. Let $\boldsymbol{L}$ and $\boldsymbol{D}$ stand, respectively, for the velocity gradient and the rate of deformation tensor. Then,

$$\begin{aligned} \boldsymbol{L} &= \dot{\boldsymbol{F}}\boldsymbol{F}^{-1} = \operatorname{grad} \boldsymbol{v}, \qquad \boldsymbol{v} = \dot{\boldsymbol{x}}, \\ 2\boldsymbol{D} &= \boldsymbol{L} + \boldsymbol{L}^T, \qquad \dot{\boldsymbol{E}} = \boldsymbol{F}^T\boldsymbol{D}\boldsymbol{F}, \end{aligned} \tag{2.4}$$

where a superposed dot denotes material time differentiation with respect to τ holding $\boldsymbol{X}$ fixed and $\boldsymbol{v}$ is the velocity at time t. Let $\boldsymbol{T}$ be the symmetric tensor of Cauchy and let ϱ denote the mass density at time t. It is convenient to define the (symmetric) Piola-Kirchhoff stress tensor $\boldsymbol{\Sigma}$ by the formula

$$\boldsymbol{\Sigma} = \frac{1}{\varrho} \boldsymbol{F}^{-1} \boldsymbol{T} (\boldsymbol{F}^{-1})^T, \tag{2.5}$$

which differs by a constant factor ϱ_0 (the mass density in the reference configuration) from that called the second Piola-Kirchhoff tensor by TRUESDELL and NOLL [14].

We also recall here the (local) equations of motion

$$\operatorname{div} \boldsymbol{T} + \varrho \boldsymbol{b} = \varrho \ddot{\boldsymbol{x}}, \tag{2.6}$$

the (local) balance of energy in the form

$$\varrho r - \varrho(\theta \dot{\eta} + \dot{\theta} \eta + \dot{\psi}) + tr\{\boldsymbol{T}\,\boldsymbol{L}\} - \operatorname{div} \boldsymbol{q} = 0 \tag{2.7}$$

and the Clausius-Duhem inequality

$$\dot{\eta} - \frac{r}{\theta} + \frac{1}{\varrho\theta} \operatorname{div} \boldsymbol{q} - \frac{1}{\varrho\theta^2} \boldsymbol{q} \cdot \operatorname{grad} \theta \geqq 0. \tag{2.8}$$

In (2.6) to (2.8), $\boldsymbol{b}$ is the body force per unit mass, r is the heat supply function per unit mass per unit time, θ (> 0) denotes the absolute temperature, η is the entropy per unit mass, ψ is the free energy per unit mass which is related to the internal energy ε per unit mass by $\psi = \varepsilon - \theta\eta$ and tr denotes the trace operator.

It is convenient to introduce the change of variable $\tau = t - s$ $(0 \leqq s < \infty)$ and let the histories up to time t, of the deformation gradient $\boldsymbol{F}(\tau)$ and of the temperature $\theta(\tau)$ at a fixed material point, be denoted by

$$\boldsymbol{F}^t(s) = \boldsymbol{F}(t - s), \quad \theta^t(s) = \theta(t - s), \quad 0 \leqq s < \infty. \tag{2.9}$$

The restrictions of these functions to the open interval $(0, \infty)$, called past histories, will be designated as $\boldsymbol{F}_r^t(\cdot)$ and $\theta_r(\cdot)$, while the members $\boldsymbol{F} = \boldsymbol{F}(t) = \boldsymbol{F}^t(0)$ and $\theta = \theta(t) = \theta^t(0)$ are the present values. A simple material is defined by a set of constitutive equations for ψ, η, $\boldsymbol{T}$ and $\boldsymbol{q}$ which depend on $\boldsymbol{F}^t(\cdot)$, $\theta^t(\cdot)$, as well as the present value of the temperature gradient[1]. By applying the principle of material frame-indifference, these equations can then be reduced to a set which is invariant under a change of frame. For our present purpose, we recall the

---

[1] As shown by COLEMAN [1], it follows from the energy equation (2.7) and the inequality (2.8) that ψ, η, $\boldsymbol{T}$ cannot depend on the present value of the temperature gradient even though $\boldsymbol{q}$ does.

constitutive equations of simple materials in the invariant forms[1]

$$\psi = p\big(\boldsymbol{E}_r^t(\cdot), \theta_r^t(\cdot); \boldsymbol{E}, \theta\big), \qquad \eta = \mathfrak{H}\big(\boldsymbol{E}_r^t(\cdot), \theta_r^t(\cdot); \boldsymbol{E}, \theta\big),$$
$$\boldsymbol{\Sigma} = \mathfrak{T}\big(\boldsymbol{E}_r^t(\cdot), \theta_r^t(\cdot); \boldsymbol{E}, \theta\big), \qquad \boldsymbol{F}^{-1}\boldsymbol{q} = \mathscr{K}\big(\boldsymbol{E}_r^t(\cdot), \theta_r^t(\cdot); \boldsymbol{E}, \theta, \boldsymbol{g}\big), \tag{2.10}$$

where $\boldsymbol{\Sigma}$ is the Piola-Kirchhoff stress tensor defined by (2.5), $\boldsymbol{E}^t(\cdot)$ denotes the strain history with the present value $\boldsymbol{E}$ given by $(2.3)_2$ and $\boldsymbol{g}$ in $(2.10)_4$ stands for the value at time t of the temperature gradient at X, i.e.,

$$\boldsymbol{g} = \frac{\partial \theta(\boldsymbol{X}, t)}{\partial \boldsymbol{X}} = \boldsymbol{F}^T \operatorname{grad} \theta. \tag{2.11}$$

For background information, in the remainder of this Section we first recall some aspects of thermodynamics of simple materials with fading memory as developed by COLEMAN [1, 2] and then briefly summarize certain results for a class of non-isothermal simple solids discussed recently by CROCHET and NAGHDI [5]. Let $h(\cdot)$ be a fixed influence function which is positive, monotone decreasing, continuous over $[0, \infty)$ and decays to zero fast enough to be square integrable. Let $\mathfrak{S}$ be the space of all second order symmetric tensors $\boldsymbol{S}$ and define a norm on $\mathfrak{S}$ by

$$|\boldsymbol{S}| = \{tr\, \boldsymbol{S}^2\}^{1/2}, \qquad \boldsymbol{S} \in \mathfrak{S}. \tag{2.12}$$

We introduce a Hilbert space $\mathfrak{S}_h$ of tensor-valued functions $\boldsymbol{S}(\cdot)$ defined over $(0, \infty)$ with a finite h-norm

$$\|\boldsymbol{S}(\cdot)\|_h = \left\{\int_0^\infty |\boldsymbol{S}(s)|^2 h(s)^2 ds\right\}^{1/2}. \tag{2.13}$$

The inner product in $\mathfrak{S}_h$ is

$$\langle \boldsymbol{S}_1(\cdot), \boldsymbol{S}_2(\cdot) \rangle_h = \int_0^\infty \operatorname{tr}\{\boldsymbol{S}_1(s)\, \boldsymbol{S}_2(s)\}\, h(s)^2 ds, \quad \boldsymbol{S}_1(\cdot), \boldsymbol{S}_2(\cdot) \in \mathfrak{S}_h. \tag{2.14}$$

We also need a Hilbert space $\mathfrak{R}_h$ of real-valued functions $R(\cdot)$ defined over $(0, \infty)$ with a finite h-norm

$$\| R(\cdot)\|_h = \left\{\int_0^\infty R^2(s)\, h(s)^2\, ds\right\}^{1/2}, \tag{2.15}$$

and the associated scalar product.

[1] We use here the principle of material frame-indifference in order to maintain continuity with COLEMAN's work [1]. Alternatively, we can require that the constitutive equations must be invariant under superposed rigid body motions.

We assume that each of the functionals in (2.10), for fixed $\boldsymbol{E}$ and θ, is sufficiently smooth in $\boldsymbol{E}_r^t(\cdot)$ and $\theta_r^t(\cdot)$ so that it is Fréchet differentiable for all past histories of strain and temperature which belong to $\mathfrak{S}_h$ and $\mathfrak{R}_h$, respectively. We also assume that the free energy functional in $(2.10)_1$ is twice continuously differentiable with respect to $\boldsymbol{E}$ and θ; that $\boldsymbol{E}^t(\cdot)$ and $\theta^t(\cdot)$ are bounded; and that $\frac{d\boldsymbol{E}_r^t(\cdot)}{d(\cdot)}$ and $\frac{d\theta_r^t(\cdot)}{d(\cdot)}$ exist and belong to $\mathfrak{S}_h$ and $\mathfrak{R}_h$, respectively. Under less restrictive assumptions, COLEMAN [1] has shown that for simple materials with fading memory, the functionals $\mathfrak{H}$ and $\mathfrak{T}$ are fully determined from the knowledge of the functional p, i.e.,

$$\mathfrak{H} = -\frac{\partial}{\partial\theta}\, p\big(\boldsymbol{E}_r^t(\cdot), \theta_r^t(\cdot); \boldsymbol{E}, \theta\big), \qquad \mathfrak{T} = \frac{\partial}{\partial \boldsymbol{E}}\, p\big(\boldsymbol{E}_r^t(\cdot), \theta_r^t(\cdot); \boldsymbol{E}, \theta\big). \quad (2.16)$$

In addition we need the expression for the internal dissipation which, together with the heat flux, must obey certain inequalities deduced from (2.7), (2.8) and (2.16). We postpone recalling these results, however, until Sec. 3.

The constitutive equations in (2.10) have been defined in terms of the strain and temperature histories $\boldsymbol{E}^t(\cdot)$ and $\theta^t(\cdot)$. For later reference, we assume that the constitutive relation $(2.10)_3$ is smoothly invertible in $\boldsymbol{\Sigma}$ and $\boldsymbol{E}$, in the sense that a functional ε' exists such that

$$\boldsymbol{E}(t) = \varepsilon'\big(\boldsymbol{\Sigma}_r^t(\cdot), \theta_r^t(\cdot); \boldsymbol{\Sigma}, \theta\big), \quad (2.17)$$

and further assume that ε' is smooth with respect to $\boldsymbol{\Sigma}_r^t(\cdot)$ and $\theta_r^t(\cdot)$ belonging to $\mathfrak{S}_h$ and $\mathfrak{R}_h$, respectively[1]. A necessary condition for the existence of an inverse of $(2.10)_3$ is that $\det\left|\frac{\partial\mathfrak{T}}{\partial\boldsymbol{E}}\right|$ does not vanish, or else that the material has a non-vanishing elastic response to any instantaneous strain at time t, whatever the past history of strain and temperature.

Suppose the continuum is initially in the state of rest at a constant temperature θ_0 and a constant entropy η_0, and suppose that $\boldsymbol{\Sigma}_0$ is the state of stress at the temperature θ_0 and zero strain in the reference configuration which we take to be the initial configuration. Consider now the constant histories

$$\boldsymbol{E}^t(s) = \boldsymbol{O}^+(s) = \boldsymbol{O}, \ \ \theta^t(s) = \theta_0 1^+(s), \ \ \boldsymbol{\Sigma}^t(s) = \boldsymbol{\Sigma}_0 1^+(s), \ \ (\boldsymbol{O} \leqq s < \infty), \quad (2.18)$$

[1] As will become apparent later, an explicit knowledge of the functional ε' will not be required in discussing the main features of the constitutive equations for a class of simple solids considered in [5].

where $O^+(s)$ stands for the zero function, O for the zero tensor and $1^+(s)$ is the function that maps the positive real axis into the real number 1, i.e., $1^+(s) = 1,\ 0 \leqq s < \infty$. The histories in (2.18) designate the zero strain history, a constant temperature history and a constant stress history, respectively. Corresponding to the constant history $\theta^t(\cdot) = \theta_0 1^+(\cdot)$, the functional $\mathfrak{p}$ in $(2.10)_1$ and $\mathfrak{T}$ in $(2.16)_2$ will be denoted by $\mathfrak{p}^*$ and $\mathfrak{T}^*$, i.e.,

$$\mathfrak{p}\left(E_r^t(\cdot),\, \theta_0 1^+(\cdot);\, E,\, \theta_0\right) = \mathfrak{p}^*\left(E_r^t(\cdot);\, E\right),$$

$$\frac{\partial}{\partial E}\, \mathfrak{p}\left(E_r(\cdot),\, \theta_0 1^+(\cdot);\, E,\, \theta_0\right) = \frac{\partial}{\partial E}\, \mathfrak{p}^*\left(E_r^t(\cdot);E\right) = \mathfrak{T}^*\left(E_r^t(\cdot);E\right), \tag{2.19}$$

where $\theta_0 1^+(\cdot)$ is specified by $(2.18)_2$. Similarly, the functionals on the right-hand sides of $(2.16)_1$ and (2.17), when evaluated at the constant history $\theta^t(\cdot) = \theta_0 1^+(\cdot)$ will be designated by $\mathfrak{H}^*$ and ε^*, respectively. In subsequent developments, the starred functionals will be called *isothermal functionals.*

Before proceeding further, we dispose of one additional preliminary matter and introduce a time scale which depends on the history of temperature. For a fixed t, associate with each value of $s\,(0 \leqq s < \infty)$ a number $\xi^{(t)}(s)$ and define a *modified time scale* by the operator

$$\xi^{(t)}(s) = \mathop{\mathfrak{Z}}_{\sigma=0}^{\infty}\left(\theta_r^t(\sigma);s\right), \quad 0 \leqq s < \infty, \tag{2.20}$$

where $\mathfrak{Z}$ is a functional which depends on the history of temperature $\theta_r^t(\sigma)$, and also impose the conditions[1]

$$\xi^{(t)}(0) = 0, \qquad \frac{\partial \xi^{(t)}(s)}{\partial s} > 0. \tag{2.21}$$

For convenience, in the rest of this paper, we use the notation ξ_s in place of $\xi^{(t)}(s)$ and require that the operator ξ_s reduce to the identity operator when the temperature history is specified by $(2.18)_2$. Thus, we specify the operator characterizing the modified time scale by

$$\xi_s = \mathop{\mathfrak{Z}}_{\sigma=0}^{\infty}\left(\theta_r^t(\sigma);s\right), \quad \mathop{\mathfrak{Z}}_{\sigma=0}^{\infty}\left(\theta_0 1^+(\sigma);s\right) = s, \tag{2.22}$$

and we also assume that the functional $\mathfrak{Z}$ is smooth with respect to the norm

$$\| G(\cdot) \|_h = \left\{\int_0^\infty G(\sigma)^2\, h(\sigma)^2\, d\sigma\right\}^{1/2}, \quad G(\cdot) \in \mathfrak{R}_h, \tag{2.23}$$

[1] A more detailed discussion of the operator $\xi^{(t)}(s)$ is given in [5].

over a domain of past histories of temperature which is a neighborhood of the constant history $(2.18)_2$.

For a prescribed s, the right-hand side of $(2.22)_1$ is a functional over a (restricted) past history of temperature $\theta_r^t(\sigma)$. Our notation in (2.22) is simple and quite clear; a more precise notation for the functional $\mathfrak{Z}$ would be cumbersome and too elaborate for our purpose. In specifying the operator (2.22), we have been guided by the observations reported in the literature from which it can be inferred that the modified time scale should exhibit nonlinear dependence on the past history of temperature. Later we will have occasions to use isothermal functionals [such as those on the right-hand sides of (2.19)] with the modified time scale $(2.22)_1$. Such functionals will be referred to as the *modified isothermal functionals.*

We assume that the functional $\mathfrak{Z}$ in $(2.22)_1$ has an inverse $\mathfrak{U}$ such that

$$u = \mathop{\mathfrak{U}}_{\tau=0}^{\infty}\left(\theta_r^t(\tau); \xi_u\right). \tag{2.24}$$

An example of the functional $\mathfrak{U}$ which will be utilized subsequently is given by[1]

$$\mathop{\mathfrak{U}}_{\tau=0}^{\infty}\left(\theta_r^t(\tau); \xi_u\right) = \int_0^{\xi_u} \Phi\left(\theta^t(\tau)\right) d\tau, \tag{2.25}$$

where

$$\Phi(\theta^t(\tau)) = \begin{cases} \varphi(\theta^t(\tau)) & \text{if} \quad \tau < \bar{\xi}, \\ 1 & \text{if} \quad \tau > \bar{\xi}, \end{cases} \tag{2.26}$$

with

$$\varphi(\theta) > 0, \qquad \varphi(\theta_0) = 1. \tag{2.27}$$

The response of the strain functional ε' in (2.17) under the constant stress history $(2.18)_3$ can be expressed in terms of another functional $\bar{\varepsilon}$ which depends only on the temperature history, since Σ_0 is fixed. Let the value of $\bar{\varepsilon}$ at time t be denoted by $\alpha(t)$ so that

$$\alpha(t) = \bar{\varepsilon}\left(\theta_r^t(\cdot); \theta\right) = \varepsilon'\left(\Sigma_0 1^+(\cdot), \theta_r^t(\cdot); \Sigma_0, \theta\right). \tag{2.28}$$

[1] It may be recalled that in the literature on 'thermo-rheologically simple' materials the "reduced time" (corresponding to the inverse of our modified time scale) is measured from zero to t and usually the function φ in (2.26) is specified by the form $\varphi(\theta) = \exp[\Gamma(\theta)]$; see, e. g., HUNTER [6]. Here, a restriction of the type (2.26) on the function Φ is necessary in order to ensure that the free energy satisfies the constitutive assumption of fading memory. In contrast, in the literature on 'thermo-rheologically simple' solids and in connection with the "reduced time" it is assumed that the temperature is held constant prior to a specified instant of time.

It is convenient to decompose the strain functional ε' in (2.17) into two parts such that

$$\varepsilon'\big(\Sigma_r^t(\cdot), \theta_r^t; \Sigma, \theta\big) = \bar{\varepsilon}\big(\theta_r^t(\cdot); \theta\big) + \varepsilon\big(\Sigma_r^t(\cdot), \theta_r^t(\cdot); \Sigma, \theta\big),$$
$$\bar{\varepsilon}\left(\theta_0 1^+(\cdot); \theta_0\right) = 0, \quad \varepsilon\left(\Sigma_0 1^+(\cdot), \theta_r^t; \Sigma_0, \theta\right) = 0. \tag{2.29}$$

Then, by (2.17), (2.28) and $(2.29)_1$, we have

$$\boldsymbol{E}(t) - \alpha(t) = \varepsilon(\Sigma_r^t(\cdot), \theta_r^t(\cdot); \Sigma, \theta), \tag{2.30}$$

where ε satisfies the condition $(2.29)_3$. We remark here that the decomposition of the strain functional in (2.29) is unique in that $\bar{\varepsilon}$ depends only on the temperature history, while the dependence of ε' on stress history is represented entirely by ε on the right-hand side of (2.30). If the temperature history is always specified by $(2.18)_2$ throughout the motion, then $\alpha \equiv 0$ and ε in (2.30) reduces to the isothermal strain functional ε^*.

We define a restricted class of simple solids with fading memory by the assumption that the functional relationship between $[\boldsymbol{E}(\cdot) - \alpha(\cdot)]$ and $\Sigma(\cdot)$ is such that the constitutive functional ε in (2.30) is equivalent to the corresponding modified isothermal functional, i.e., the isothermal functional ε^* in terms of the same history of stress but with a modified time scale defined by (2.22). From this assumption, together with the properties of ξ_s and the invertibility of $(2.16)_2$ and (2.17) — for the class of simple solids under discussion — we have shown elsewhere [5] that the constitutive equations for ψ, Σ and η have the forms

$$\psi = \underset{s=0}{\overset{\infty}{p^*}} \left([\boldsymbol{E}_r^t(\xi_s) - \alpha_r^t(\xi_s)]; \boldsymbol{E} - \alpha\right) + \underset{s=0}{\overset{\infty}{\mathscr{B}}}\big(\theta_r^t(s); \theta\big), \tag{2.31}$$

$$\Sigma = \frac{\partial}{\partial \boldsymbol{E}} \underset{s=0}{\overset{\infty}{p^*}} \left([\boldsymbol{E}_r^t(\xi_s) - \alpha_r^t(\xi_s)]; \boldsymbol{E} - \alpha\right) = \underset{s=0}{\overset{\infty}{\mathfrak{T}^*}}\big([\boldsymbol{E}_r^t(\xi_s) - \alpha_r^t(\xi_s)]; \boldsymbol{E} - \alpha\big), \tag{2.32}$$

$$\eta = \operatorname{tr}\left\{\Sigma \frac{\partial}{\partial \theta} \bar{\varepsilon}\right\} - \frac{\partial}{\partial \theta} \underset{s=0}{\overset{\infty}{\mathscr{B}}}\big(\theta_r^t(s); \theta\big). \tag{2.33}$$

In the above equations, the modified functionals p^* and $\mathfrak{T}^*$ are obtained from the isothermal functionals on the right-hand sides of (2.19) with the history $\boldsymbol{E}^t(s)$ replaced by $[\boldsymbol{E}^t(\xi_s) - \alpha^t(\xi_s)]$. Also, in (2.32) and (2.31), $\frac{\partial}{\partial \boldsymbol{E}} p^*$ is obtained from the isothermal functional $\underset{s=0}{\overset{\infty}{p^*}}\big(\boldsymbol{E}_r^t(s); \boldsymbol{E}\big)$

and the functional $\mathscr{B}$ is defined by

$$\mathop{\mathscr{B}}_{s=0}^{\infty}\left(\theta_r^t(s);\theta\right) = \mathop{\overline{\mathscr{B}}}_{s=0}^{\infty}\left(\theta_r^t(s);\theta\right) - \mathop{p^*}_{s=0}^{\infty}\left(-\alpha_r^t(\xi_s);-\alpha\right),$$

$$\mathop{\mathscr{B}}_{s=0}^{\infty}\left(\theta_0 1^+(s);\theta_0\right) = 0, \tag{2.34}$$

where the functional $\overline{\mathscr{B}}$ is the response of the functional p in $(2.10)_1$ when the strain history is specified by $(2.18)_1$.

We observe that the constitutive equations (2.31)—(2.33) require the knowledge of four functionals; one of these is the modified isothermal free energy functional in (2.31) and the remaining three, namely $\mathfrak{Z}$, $\bar{\varepsilon}$ and $\mathscr{B}$, depend only on the temperature history. Moreover, once $\mathfrak{Z}$ and $\bar{\varepsilon}$ are specified, the modified isothermal free energy functionals can be obtained directly from the knowledge of the isothermal free energy functional. Also, the stress functional on the right-hand side of (2.32) and the first term on the right-hand side of (2.33) are determined from the knowledge of the functional p^* in (2.31).

3. Finite Linear Viscoelasticity

We outline here the main results from a theory of finite linear viscoelasticity obtained in [5] with the use of suitable integral representations for the functionals in (2.31)—(2.33). We also discuss an appropriate form for the heat flux functional in $(2.10)_4$ in terms of integrals. For additional details we refer the reader to the paper by Crochet and Naghdi [5]. However, it is desirable to indicate briefly the manner in which the functionals on the right-hand side of (2.31) are represented in terms of integrals. For this purpose, consider a continuous scalar-valued functional $\mathfrak{F}\left(\Gamma_r^t(\cdot);\Gamma\right)$ defined on the Hilbert space to which $\Gamma_r^t(\cdot)$ belongs. If $\mathfrak{F}$ is twice Fréchet differentiable at the constant history $\Gamma_r^t(\cdot) = \Gamma 1^+(\cdot)$, where Γ is the present value, then $\mathfrak{F}$ may be represented in terms of a polynomial functional in the form[1]

$$\mathfrak{F}\left(\Gamma_r^t(\cdot);\Gamma\right) = \mathfrak{F}\left(\Gamma 1^+(\cdot);\Gamma\right) + \mathscr{L}\left(\Gamma_r^t(\cdot)-\Gamma;\Gamma\right) + \mathscr{Q}\left(\Gamma_r^t(\cdot)-\Gamma;\Gamma\right) + $$
$$+\, o\left(\|\Gamma_r^t(\cdot)-\Gamma;\Gamma\|_h^2\right), \tag{3.1}$$

[1] In (3.1) and the rest of the paper, for convenience we write $[\Gamma_r^t(\cdot)-\Gamma]$ in place of the difference history $[\Gamma_r^t(\cdot)-\Gamma\, 1^+(\cdot)]$.

where the order symbol o is defined by

$$\lim_{\|\Gamma_r^t(\cdot)-\Gamma\|_h \to 0} \frac{o\left(\|\Gamma_r^t(\cdot)-\Gamma\|_h^n\right)}{\|\Gamma_r^t(\cdot)-\Gamma\|_h^n} = 0. \tag{3.2}$$

By Riesz theorem, the linear functional $\mathscr{L}$ has an integral representation. We further assume that the operator corresponding to $\mathscr{D}$ in (3.1) is completely continuous so that it can also be represented by integrals.

The free energy in (2.31) is specified in terms of two functionals. One of these, namely $\mathscr{B}$, depends only on temperature history $\theta^t(s)$ and the other is obtained directly from the knowledge of the isothermal functional p^* in (2.19) with the history $\boldsymbol{E}^t(s)$ replaced by $[\boldsymbol{E}^t(\xi_s) - \alpha^t(\xi_s)]$. Keeping this in mind, it suffices to obtain a suitable representation for p^* in (2.19) and then deduce the corresponding representation for the first functional on the right-hand side of (2.31). A representation for the functional $\mathscr{B}$ which depends only on the temperature history can be obtained separately, but in a similar manner. Provided the Hilbert space norm $\|\Gamma_r^t(\cdot) - \Gamma\|_h$ is small, terms of $o\left(\|\Gamma_r^t(\cdot) - \Gamma\|_h^2\right)$ are negligible in comparison with the third term in (3.1) and the functional $\mathfrak{F}$ may be represented by the first three terms on the right-hand side of (3.1). In what follows, we assume that the two functionals on the right-hand side of (2.31) can be represented (in the sense just stated) by integrals in terms of the difference histories

$$[\boldsymbol{E}^t(\cdot) - \alpha^t(\cdot) - (\boldsymbol{E} - \alpha)], \qquad \left(\theta^t(\cdot) - \theta\right). \tag{3.3}$$

The constitutive equations for the non-isothermal finite linear viscoelasticity under discussion were obtained in [5] on the basis of the following assumptions: (i) The free energy is bilinear in the difference history (3.3); (ii) the functional $\bar{\mathfrak{e}}$ in (2.28) is linear in the difference history $(3.3)_2$; (iii) the stress[1] is linear in the difference history $(3.3)_1$; (iv) the entropy functional is such that $\mathfrak{H} - \operatorname{tr}\left\{\Sigma \frac{\partial}{\partial\theta}\bar{\mathfrak{e}}\right\}$ is linear in the difference history $(3.3)_2$; and (v) the functional $\mathscr{K}$ in $(2.10)_4$ is linear in the difference histories $[\boldsymbol{E}^t(s) - \boldsymbol{E}\,1^+(s)]$ and $(3.3)_2$. With these assumptions, we have shown in [5] that the functional $\bar{\mathfrak{e}}$ reduces to a function of temperature alone, i.e.,

$$\bar{\mathfrak{e}} = \hat{\alpha}(\theta), \tag{3.4}$$

and this gives rise to an effective simplification of some of the results, particularly in the constitutive relation for entropy and the expression for the internal dissipation.

---

[1] The assumption that the stress be linear in $(3.3)_1$ conforms to the previous development of the (purely mechanical) theory of finite linear viscoelasticity by Coleman and Noll [3].

With the above background, we now summarize the constitutive equations of the finite linear theory. The free energy is given by (2.31) with the modified isothermal free energy functional in the form[1]

$$\mathop{p^*}_{s=0}^{\infty}([\boldsymbol{E}_r^t(\xi_s) - \boldsymbol{\alpha}_r^t(\xi_s)];\ \boldsymbol{E} - \boldsymbol{\alpha}) = \Phi(\boldsymbol{E} - \boldsymbol{\alpha}) + \frac{1}{2}\int_0^\infty\int_0^\infty \operatorname{tr}\Big\{\big[[\boldsymbol{E}^t(\xi_s) - \boldsymbol{\alpha}^t(\xi_s)] - \boldsymbol{E} + \boldsymbol{\alpha}\big]\,\boldsymbol{M}(s, u; \boldsymbol{E} - \boldsymbol{\alpha})\,\big[[\boldsymbol{E}^t(\xi_u) - \boldsymbol{\alpha}^t(\xi_u)] - \boldsymbol{E} + \boldsymbol{\alpha}\big]\Big\}\,ds\,du, \tag{3.5}$$

and with

$$\mathop{\mathscr{B}}_{s=0}^{\infty}\big(\theta_r^t(s);\ \theta\big) = B(\theta) + \frac{1}{2}\int_0^\infty\int_0^\infty \frac{\partial^2 m(s,u;\theta)}{\partial s\,\partial u}\,[\theta^t(s) - \theta]\,[\theta^t(u) - \theta]\,ds\,du. \tag{3.6}$$

The material function $\boldsymbol{M}(s, u;\ \boldsymbol{E} - \boldsymbol{\alpha})$ in (3.5), a fourth order tensor with Cartesian components M_{ijkl}, must satisfy

$$\int_0^\infty |\boldsymbol{M}(s, u; \boldsymbol{E} - \boldsymbol{\alpha})|^2\, h(u)^{-2}\, du < \infty \quad \text{for all} \quad (\boldsymbol{E} - \boldsymbol{\alpha}) \text{ and for all } 0 \leqq s < \infty, \tag{3.7}$$

where

$$|\boldsymbol{M}(s, u; \boldsymbol{E} - \boldsymbol{\alpha})|^2 = M_{ijkl}(s, u; \boldsymbol{E} - \boldsymbol{\alpha})\, M_{ijkl}(s, u; \boldsymbol{E} - \boldsymbol{\alpha}) \tag{3.8}$$

and without loss of generality we set

$$M_{ijkl}(s, u; \boldsymbol{E} - \boldsymbol{\alpha}) = M_{jikl}(s, u; \boldsymbol{E} - \boldsymbol{\alpha}) = M_{ijlk}(s, u; \boldsymbol{E} - \boldsymbol{\alpha}) = M_{klij}(u, s; \boldsymbol{E} - \boldsymbol{\alpha}). \tag{3.9}$$

In (3.8)—(3.9) and elsewhere in this paper, Latin indices have the range 1, 2, 3 and the usual summation convention is implied over repeated indices unless stated otherwise.

For later convenience, we define a fourth order tensor function $G(s, u; \boldsymbol{E} - \boldsymbol{\alpha})$ with the same symmetry properties as $\boldsymbol{M}(s, u; \boldsymbol{E} - \boldsymbol{\alpha})$ by

$$G(s, u; \boldsymbol{E} - \boldsymbol{\alpha}) = \int_s^\infty\int_u^\infty \boldsymbol{M}(\sigma, \varrho; \boldsymbol{E} - \boldsymbol{\alpha})\, d\sigma\, d\varrho, \tag{3.10}$$

---

[1] We have shown in [5] that the free energy cannot contain a linear functional in $[\boldsymbol{E}^t(\cdot) - \boldsymbol{\alpha}^t(\cdot) - (\boldsymbol{E} - \boldsymbol{\alpha})]$ or $[\theta^t(\cdot) - \theta]$.

and note that

$$\frac{\partial^2 G(s, u; \boldsymbol{E} - \alpha)}{\partial s \partial u} = \boldsymbol{M}(s, u; \boldsymbol{E} - \alpha),$$

$$G(s, u; \boldsymbol{E} - \alpha) = 0 \quad \text{as } s \text{ or } \quad u \to \infty, \tag{3.11}$$

where $(3.11)_2$ follows from (3.7). The material coefficient in (3.6), i.e., the scalar function $m(s, u; \theta) = m(u, s; \theta)$ must satisfy a condition similar to (3.7) with $\boldsymbol{M}(s, u; \boldsymbol{E} - \alpha)$ replaced with $\frac{\partial^2 m(s, u; \theta)}{\partial s \partial u}$. From this latter condition follows that

$$m(s, u; \theta) = 0 \quad \text{as } s \text{ or } \quad u \to \infty. \tag{3.12}$$

The constitutive equations for stress and entropy are given by

$$\boldsymbol{\Sigma} = \frac{\partial \Phi(\boldsymbol{E} - \alpha)}{\partial \boldsymbol{E}} + \int_0^\infty \frac{\partial}{\partial u} G(0, u; \boldsymbol{E} - \alpha) \{[\boldsymbol{E}^t(\xi_u) - \alpha^t(\xi_u) - \boldsymbol{E} + \alpha\} du, \tag{3.13}$$

and

$$\eta = \operatorname{tr}\left\{\boldsymbol{\Sigma} \frac{\partial \hat{\alpha}(\theta)}{\partial \theta}\right\} - \frac{\partial}{\partial \theta} \mathop{\mathscr{B}}_{s=0}^{\infty}\left(\theta_r^t(s); \theta\right),$$

$$\frac{\partial}{\partial \theta} \mathop{\mathscr{B}}_{s=0}^{\infty}\left(\theta_r^t(s); \theta\right) = \left\{\frac{\partial B}{\partial \theta} - \int_0^\infty m(s, 0; \theta) \frac{d\theta^t(s)}{ds} ds\right\}. \tag{3.14}$$

For completeness, we should also discuss the form of the constitutive equation for $\mathscr{K}$; but we defer this until later in this Section.

The first two terms on the right-hand side of (3.5) approximate the modified isothermal free energy functional in (2.31) in the limit as the norm $\|\boldsymbol{E}_r^t(\cdot) - \alpha_r^t(\cdot) - (\boldsymbol{E} - \alpha)\|_h \to 0$ with an error which approaches zero faster than the square of the norm. Similarly, the right-hand side of (3.6) approximates the functional $\mathscr{B}$ in (2.31) in the limit, as the norm $\|\theta_r^t(\cdot) - \theta\|_h \to 0$ with an error which approaches zero faster than $\|\theta_r^t(\cdot) - \theta\|_h^2$. Because of our assumption that the constitutive equation for stress is linear in the difference history $(3.3)_1$, the expression (3.13) approximates the right-hand side of (2.32) in the limit as the norm $\|\boldsymbol{E}_r^t(\cdot) - \alpha_r^t(\cdot) - (\boldsymbol{E} - \alpha)\|_h \to 0$ with an error which approaches zero faster than the norm (rather than the square of the norm). A parallel remark may be made about $\mathfrak{H} - \operatorname{tr}\left\{\boldsymbol{\Sigma} \frac{\partial}{\partial \theta} \bar{\varepsilon}\right\}$ which occurs in (3.14). It should be emphasized that in obtaining the representations (3.5) and (3.6) it is assumed that the Hilbert space norms be small but no restriction is placed on the present value or the variation of $\boldsymbol{E}$ and θ from their initial reference values. Moreover, as noted in [5], in obtaining the above results we have not assumed that the functional $\mathfrak{Z}$ in (2.22) can be expanded in terms of a polynomial functional similar to $\mathfrak{B}$. The modified time scale functional $\mathfrak{Z}$ or its inverse $\mathfrak{U}$ [as exemplified by the special form (2.25)] is such that the error inherent in a polynomial expansion of $\mathfrak{Z}$ should decrease slowly with $\|G(\cdot)\|_h$ in (2.23) as compared to the error in $\mathscr{B}$ with the representation (3.6).

We now recall the restrictions placed on various quantities by the Clausius-Duhem inequality (2.8). One of these is the requirement that the internal dissipation, denoted by σ, be non-negative, i.e.,

$$\sigma \geqq 0. \tag{3.15}$$

In addition the functional $\mathscr{K}$ must obey the inequality

$$\sigma - \frac{1}{\varrho\,\theta^2}\,\mathscr{K}\cdot\boldsymbol{g} \geqq 0, \tag{3.16}$$

and the residual energy equation is

$$\varrho\dot{\eta} = \frac{\varrho r - \operatorname{div}\boldsymbol{q}}{\theta} + \varrho\,\sigma. \tag{3.17}$$

tf we adopt the form (2.25) for the inverse of the modified time scale, Ihen it is shown in [5] that the internal dissipation is given by

$$\begin{aligned}\theta\,\sigma = -\int_0^\infty\!\!\int_0^\infty \operatorname{tr}\Bigg\{\Bigg[\frac{\partial \boldsymbol{E}^t(\xi_s)}{\partial \xi_s} - \\ - \frac{\partial\hat{\alpha}(\theta^t(\xi_s))}{\partial\theta^t(\xi_s)}\,\frac{\partial\theta^t(\xi_s)}{\partial\xi_s}\Bigg]\frac{\partial}{\partial u}\,G(s,u;\boldsymbol{E}-\alpha)\Bigg[\frac{\partial\boldsymbol{E}^t(\xi_u)}{\partial\xi_u} - \\ - \frac{\partial\hat{\alpha}(\theta^t(\xi_u))}{\partial\theta^t(\xi_u)}\,\frac{\partial\theta^t(\xi_u)}{\partial\xi_u}\Bigg]\Bigg\}\,\Phi(\theta)\,\frac{\partial\xi_u}{\partial u}\,\frac{\partial\xi_s}{\partial s}\,ds\,du + \delta\,\underset{s=0}{\overset{\infty}{\mathscr{B}}}\Big(\theta_r^t(s);\theta\,\Big|\,\frac{d\theta_r^t(s)}{ds}\Big),\end{aligned} \tag{3.18}$$

where the notation $\delta\mathscr{B}$ which stands for the Fréchet derivative of $\mathscr{B}$ is introduced for later convenience. By (3.6) and after an integration by parts

$$\delta\,\underset{s=0}{\overset{\infty}{\mathscr{B}}}\Big(\theta_r^t(s);\theta\,\Big|\,\frac{d\theta_r^t(s)}{ds}\Big) = -\int_0^\infty\!\!\int_0^\infty \frac{\partial m(s,u;\theta)}{\partial u}\,\frac{d\theta^t(s)}{ds}\,\frac{d\theta^t(u)}{du}\,ds\,du, \tag{3.19}$$

where (3.12) has been used.

The above results have been deduced without any restriction on material symmetry. We now limit the discussion to materials having a center of symmetry in the reference configuration, mainly for the purpose of recording a constitutive equation for the heat flux vector. Let $\mathcal{G}$ be the symmetry group of the material in the reference configuration. The orthogonal (time-independent) tensor $\boldsymbol{Q}$, which characterizes a change of frame in the reference configuration, belongs to $\mathcal{G}$ if the constitutive functionals $\mathfrak{p}$ and $\mathcal{K}$ in $(2.10)_{1,4}$ obey the identities

$$\underset{s=0}{\overset{\infty}{\mathfrak{p}}}\left(\boldsymbol{E}_r^t(s), \theta_r^t(s); \boldsymbol{E}, \theta\right) = \underset{s=0}{\overset{\infty}{\mathfrak{p}}}\left(\boldsymbol{Q}\boldsymbol{E}_r^t(s)\,\boldsymbol{Q}^T, \theta_r^t(s);\ \boldsymbol{Q}\boldsymbol{E}\boldsymbol{Q}^T, \theta\right),$$

$$\boldsymbol{Q}\,\underset{s=0}{\overset{\infty}{\mathcal{K}}}\left(\boldsymbol{E}_r^t(s), \theta_r^t(s); \boldsymbol{E}, \theta, \boldsymbol{g}\right) = \underset{s=0}{\overset{\infty}{\mathcal{K}}}\left(\boldsymbol{Q}\,\boldsymbol{E}_r^t(s)\boldsymbol{Q}^T, \theta_r^t(s); \boldsymbol{Q}\boldsymbol{E}\boldsymbol{Q}^T, \theta, \boldsymbol{Q}\boldsymbol{g}\right), \tag{3.20}$$

for all $\boldsymbol{E}^t(\cdot)$, $\theta^t(\cdot)$ and $\boldsymbol{g}$ and for all orthogonal tensors belonging to $\mathcal{G}$.

The unit tensor $\mathbf{1}$ is always in $\mathcal{G}$. If the material has a center of symmetry, the identities (3.20) have to be satisfied also for $\boldsymbol{Q} = -\mathbf{1}$. We note that if $-\mathbf{1}$ is also in $\mathcal{G}$, the temperature gradient and the heat flux vector transform according to

$$\boldsymbol{g} \to -\boldsymbol{g}, \quad \boldsymbol{F}^{-1}\boldsymbol{q} \to -\boldsymbol{F}^{-1}\boldsymbol{q}, \tag{3.21}$$

while the scalars ψ, θ, η and the tensors $\boldsymbol{E}$ and $\boldsymbol{\Sigma}$ are unaffected under the transformation. Hence, for $\boldsymbol{Q} = -\mathbf{1}$, from $(3.20)_2$ and $(2.10)_4$ we have

$$\mathcal{K}\left(\boldsymbol{E}_r^t(\cdot), \theta_r^t(\cdot); \boldsymbol{E}, \theta, \boldsymbol{g}\right) = -\mathcal{K}\left(\boldsymbol{E}_r^t(\cdot), \theta_r^t(\cdot); \boldsymbol{E}, \theta, -\boldsymbol{g}\right), \tag{3.22}$$

for all $\boldsymbol{E}^t(\cdot)$ and $\theta^t(\cdot)$, i.e., for materials with a center of symmetry, the heat flux must be an odd function of the temperature gradient. If $\mathcal{K}$ is a continuous function of $\boldsymbol{g}$, an immediate consequence of (3.22) is that

$$\mathcal{K}\left(\boldsymbol{E}_r^t(\cdot), \theta_r^t(\cdot); \boldsymbol{E}, \theta, \boldsymbol{O}\right) = \boldsymbol{O}, \tag{3.23}$$

for all $\boldsymbol{E}^t(\cdot)$ and $\theta^t(\cdot)$. The result (3.23) was noted by COLEMAN [1].

Assuming now that $\mathcal{K}$ is a polynomial in $\boldsymbol{g}$, in the expansion of $\mathcal{K}$ there will be no term of degree zero in $\boldsymbol{g}$, in view of (3.23), and we may write $\mathcal{K}$ (with Cartesian components $\mathcal{K}_i$) in the form[1]

$$\mathcal{K}_i\left(\boldsymbol{E}_r^t(\cdot), \theta_r^t(\cdot); \boldsymbol{E}, \theta, \boldsymbol{g}\right) = \mathfrak{M}_{ij}\left(\boldsymbol{E}_r^t(\cdot), \theta_r^t(\cdot); \boldsymbol{E}, \theta, \boldsymbol{g}\right)g_j, \tag{3.24}$$

where $g_j = \dfrac{\partial\theta}{\partial X_j}$ are the Cartesian components of $\boldsymbol{g}$ and $\mathfrak{M}_{ij}$ is a tensor functional which is jointly continuous over the history of strain and

[1] If $\mathfrak{M}_{ij}$ does not depend on the temperature gradient, then $\mathcal{K}$ is linear in $\boldsymbol{g}$ as in Fourier's theory of heat conduction.

temperature and is an even function of the temperature gradient. If $\mathfrak{M}_{ij}$ is continuous and Fréchet differentiable at the constant histories $\boldsymbol{E}_r^t(\cdot) = \boldsymbol{E}\,1^+(\cdot)$ and $\theta_r^t(\cdot) = \theta\,1^+(\cdot)$, then with the help of (3.1) and in view of the assumption for $\mathscr{K}$ stated earlier we may represent $\mathfrak{M}_{ij}$ in terms of linear continuous functionals by

$$\mathop{\mathfrak{M}_{ij}}_{s=0}^{\infty}\left(\boldsymbol{E}_r^t(s), \theta_r^t(s); \boldsymbol{E}, \theta, \boldsymbol{g}\right) = \boldsymbol{M}_{ij}^{(0)}(\boldsymbol{E}, \theta, \boldsymbol{g}) + $$
$$+ \int_0^\infty \boldsymbol{M}_{ijkl}^{(1)}(s; \boldsymbol{E}, \theta, \boldsymbol{g})\,[E_{kl}^t(s) - E_{kl}]\,ds +$$
$$+ \int_0^\infty \boldsymbol{M}_{ij}^{(2)}(s; \boldsymbol{E}, \theta, \boldsymbol{g})\,[\theta^t(s) - \theta]\,ds. \qquad (3.25)$$

In (3.25), $\boldsymbol{M}_{ij}^{(0)}$, $\boldsymbol{M}_{ijkl}^{(1)}$ and $\boldsymbol{M}_{ij}^{(2)}$ are tensor functions of their arguments and the last two satisfy

$$\int_0^\infty |\boldsymbol{M}^{(1)}(s; \boldsymbol{E}, \theta, \boldsymbol{g})|^2\, h(s)^{-2}\,ds < \infty,$$
$$\int_0^\infty |\boldsymbol{M}^{(2)}(s; \boldsymbol{E}, \theta, \boldsymbol{g})|^2\, h(s)^{-2}\,ds < \infty, \qquad (3.26)$$

for all $\boldsymbol{E}, \theta, \boldsymbol{g}$ and for all $0 \leqq s < \infty$, where we have also used the notation of (3.8). The expression (3.25) approximates the functional $\mathfrak{M}$ in (3.24) in the limit as the Hilbert space norms of the difference histories of strain and temperature tend to zero and the error approaches zero faster than the norms $\|\boldsymbol{E}_r^t(s) - \boldsymbol{E}\|_h$, $\|\theta_r^t(s) - \theta\|_h$.

Here, we make an additional observation regarding the representation (3.25) for $\mathfrak{M}_{ij}$ which will be used in our later discussion of the constitutive equation for the heat flux vector. Consider the first integral on the right-hand side of (3.25). For each pair of indices (i, j), this integral is an inner product over the Hilbert space $\mathfrak{S}_h$ and by (2.14) and Schwarz's inequality we have

$$< \boldsymbol{M}_r^{(1)}(\cdot) h(s)^{-2}, \boldsymbol{E}_r^t(\cdot) - \boldsymbol{E} >_h = \int_0^\infty \boldsymbol{M}_{ijkl}^{(1)}(s; \boldsymbol{E}, \theta, \boldsymbol{g})\, h(s)^{-2}\,[E_{kl}^t(s) -$$
$$- E_{kl}]\, h(s)^2\, ds \leqq \left\{ \int_0^\infty \boldsymbol{M}_{ijkl}^{(1)}(\cdot)\, \boldsymbol{M}_{ijkl}^{(1)}(\cdot)\, h(s)^{-4}\, h(s)^2\, ds \right\}^{1/2} \times$$
$$\times \left\{ \int_0^\infty [E_{pq}^t(s) - E_{pq}][E_{pq}^t(s) - E_{pq}]\, h(s)^2\, ds \right\}^{1/2},$$
$$\text{(no summation over } i, j). \qquad (3.27)$$

Hence,

$$\left|\int_0^\infty M_{ijkl}^{(1)}(s;\boldsymbol{E},\theta,\boldsymbol{g})\,h(s)^{-2}\,[E_{kl}^t(s)-E_{kl}]\,h(s)^2\,ds\right| \leqq$$
$$\leqq \|\boldsymbol{M}_r^{(1)}(\cdot)\,h(s)^{-2}\|_h\,\|\boldsymbol{E}_r^t(\cdot)-\boldsymbol{E}\|_h. \tag{3.28}$$

Similarly, by considering the second integral on the right-hand side of (3.25), we find that

$$\left|\int_0^\infty M_{ij}^{(2)}(s;\boldsymbol{E},\theta,\boldsymbol{g})\,[\theta^t(s)-\theta]\,ds\right| \leqq \|\boldsymbol{M}_r^{(2)}(\cdot)\,h(s)^{-2}\|_h\|\,\theta_r^t(\cdot)-\theta\|_h. \tag{3.29}$$

As will become apparent later, in view of the smallness of the norms $\|\boldsymbol{E}_r^t(s)-\boldsymbol{E}\|_h$ and $\|\theta_r^t(s)-\theta\|_h$, the results (3.28) and (3.29) will be helpful in obtaining further approximations for the heat flux functional in Sec. 4.

Before closing this Section, we record the expressions for the free energy and stress functionals appropriate to the *isothermal* finite linear theory. When the temperature history is specified by $(2.18)_2$ throughout the motion, then $\alpha \equiv 0$ and $\xi_s \equiv s$ by $(2.29)_2$ and $(2.22)_2$. Hence, for an isothermal process, (3.5) and the right-hand side of (3.13) reduce to

$$\underset{s=0}{\overset{\infty}{p^*}}\left(\boldsymbol{E}_r^t(s);\boldsymbol{E}\right)=\Phi(\boldsymbol{E})+$$
$$+\frac{1}{2}\int_0^\infty\int_0^\infty \operatorname{tr}\{[\boldsymbol{E}^t(s)-\boldsymbol{E}]\,M(s,u;\boldsymbol{E})\,[\boldsymbol{E}^t(u)-\boldsymbol{E}]\}\,ds\,du, \tag{3.30}$$

and

$$\underset{s=0}{\overset{\infty}{\mathfrak{T}^*}}\left(\boldsymbol{E}_r^t(s);\boldsymbol{E}\right)=\frac{\partial\Phi(\boldsymbol{E})}{\partial\boldsymbol{E}}+\int_0^\infty\frac{\partial}{\partial u}G(0,u;\boldsymbol{E})\,\{\boldsymbol{E}^t(u)-\boldsymbol{E}\}\,du, \tag{3.31}$$

respectively. The above expressions will be useful in our discussion of a linearized theory in Sec. 4.

4. A Linearized Theory with Infinitesimal Deformation

In this Section, we specialize the results of Sec. 3 and consider a non-isothermal linearized theory in which the history of deformation is "small." Let

$$\chi(\tau)=\boldsymbol{X}+\delta\boldsymbol{u}(\tau),\quad \boldsymbol{v}(\tau)=\delta\dot{\boldsymbol{u}}(\tau),\quad -\infty<\tau\leqq t, \tag{4.1}$$

where δ is a non-dimensional parameter and $\boldsymbol{v}(\tau)$ is the velocity at time τ. Now put

$$\boldsymbol{H}(\tau) = \boldsymbol{F}(\tau) - \mathbf{1}^+(\tau). \tag{4.2}$$

Then, it follows from (2.1) and (2.3) that

$$\boldsymbol{H}(\tau) = \delta \frac{\partial \boldsymbol{u}(\tau)}{\partial \boldsymbol{X}}, \tag{4.3}$$

and

$$\boldsymbol{E}(\tau) = \frac{1}{2}[\boldsymbol{H}(\tau) + \boldsymbol{H}^T(\tau)] + 0(\delta^2). \tag{4.4}$$

We say the motion characterized by $(4.1)_1$ describes *infinitesimal* deformation if

$$\delta \ll 1. \tag{4.5}$$

We shall be concerned in the following with functions of τ (or s, $0 \leqq s < \infty$), determined by the histories $\boldsymbol{H}(\tau)$ and $\theta(\tau)$ [or $\boldsymbol{H}^t(s)$ and $\theta^t(s)$], which are such that for each τ (or s) their magnitude is less than $C\delta^n$, i.e.,

$$|0(\delta^n)| < C\delta^n, \tag{4.6}$$

C being a number independent of τ, $\boldsymbol{H}(\tau)$, $\theta(\tau)$ and δ. Moreover, as remarked by COLEMAN and NOLL [3], any function of $0(\delta^n)$ in the sense just described is also a function of $0(\delta^n)$ with respect to the Hilbert space norms (2.13) and (2.15). This observation will enable us to use the customary order symbol $0(\delta^n)$ in terms of the convergence in the Hilbert space of histories $\mathfrak{S}_h$ and $\mathfrak{R}_h$.

We have already seen that the strain tensor given by (2.17) is fully determined by the histories of stress and temperature $\boldsymbol{\Sigma}^t(\cdot)$ and $\theta^t(\cdot)$. In this Section, we assume that the stress history is confined to a neighborhood V of the constant stress history $(2.18)_3$ in $\mathfrak{S}_h \times \mathfrak{S}$ (the Cartesian product of $\mathfrak{S}_h$ and $\mathfrak{S}$) and the temperature history is confined to a neighborhood I of the constant temperature history $(2.18)_2$ in $\mathfrak{R}_h \times \mathfrak{R}$. In addition, for given values of $\boldsymbol{\Sigma}$ and θ, we assume that $\boldsymbol{\varepsilon}'$ in (2.17) is defined over the Cartesian product of V and I, i.e., $V \times I$. Thus, if the deformation is infinitesimal in the sense of $(4.1)_1$ and (4.5), the strain must be small of $0(\delta)$ for all pairs $\left(\boldsymbol{\Sigma}^t(\cdot), \theta^t(\cdot)\right)$ belonging to $V \times I$; and, in particular $\boldsymbol{E}^t(\cdot)$ must be of $0(\delta)$ for all pairs $\left(\boldsymbol{\Sigma}_0 \mathbf{1}^+(\cdot), \theta^t(\cdot)\right)$, where $\theta^t(\cdot) \in I$. Recalling (2.28) and (3.4), it follows that the neighborhood I of $\theta_0 \mathbf{1}^+(\cdot)$ must be such that

$$\forall\, \theta^t(\cdot) \in I, \quad \sup_{\tau} |\hat{a}(\theta(\tau))| = 0(\delta), \tag{4.7}$$

where the norm $|\,|$ is defined by (2.12).

Let $\hat{\alpha}(\theta)$ in (3.4) be a continuous function of its argument and consider its Taylor expansion in the form

$$\hat{\alpha}(\theta) = \gamma_1 \left(\frac{\theta - \theta_0}{\theta_0}\right) + \frac{1}{2} \gamma_2 \left(\frac{\theta - \theta_0}{\theta_0}\right)^2 +, \ldots \tag{4.8}$$

where the coefficients $\gamma_1, \gamma_2, \ldots$ stand for $\left.\frac{\partial \hat{\alpha}}{\partial \theta}\right|_{\theta_0} \theta_0, \left.\frac{\partial^2 \hat{\alpha}}{\partial \theta^2}\right|_{\theta_0} \theta_0^2, \ldots$ respectively, and we have used the fact that $\hat{\alpha}(\theta_0) = 0$. If the coefficients γ_n $(n = 1, 2, \ldots)$ are of $0(1)$ in the expansion (4.8), then (4.7) will be satisfied if $\frac{\theta - \theta_0}{\theta_0}$ is of $0(\delta)$ and $\hat{\alpha}(\theta)$ can then be replaced by the leading term in (4.8). In this case, all equations of Sec. 3 can be linearized in terms of an infinitesimal strain tensor and an infinitesimal temperature variation; and the resulting constitutive equations will correspond to a special case of those given previously by CHRISTENSEN and NAGHDI [4]. In the rest of this Section, however, we assume that the temperature history belongs to a neighborhood of the constant history $\theta_0 1^+(\cdot)$ such that (4.7) is satisfied. More specifically, if $\hat{\alpha}(\theta)$ is a polynomial in θ, then its constitutive coefficients must be such that (4.7) is satisfied. For example, the use of the special form[1]

$$\hat{\alpha}(\theta) = \beta\theta \tag{4.9}$$

can be justified if we assume that there exists a constant K of $0(1)$ such that $|\beta| < K\delta'$, $\delta' \leqq \delta$.

Let $\overline{\boldsymbol{E}}(\tau)$ and $\overline{\boldsymbol{W}}(\tau)$ denote, respectively, the infinitesimal strain and the infinitesimal rotation defined by

$$\overline{\boldsymbol{E}}(\tau) = \frac{1}{2}[\boldsymbol{H}(\tau) + \boldsymbol{H}^T(\tau)], \qquad \overline{\boldsymbol{W}}(\tau) = \frac{1}{2}[\boldsymbol{H}(\tau) - \boldsymbol{H}^T(\tau)], \tag{4.10}$$

where $\boldsymbol{H}(\tau)$ given by (4.3) is of $0(\delta)$. Returning to the notation of (2.9), it follows from (4.2), (4.4) and (4.10) that

$$\boldsymbol{F} = \boldsymbol{1} + \overline{\boldsymbol{E}} + \overline{\boldsymbol{W}} = \boldsymbol{1} + 0(\delta), \quad \boldsymbol{F}^T = \boldsymbol{1} + \overline{\boldsymbol{E}} - \overline{\boldsymbol{W}} = \boldsymbol{1} + 0(\delta), \tag{4.11}$$

$$\boldsymbol{E}^t(\cdot) = \overline{\boldsymbol{E}}^t(\cdot) + 0(\delta^2) = 0(\delta), \quad \boldsymbol{E} = \overline{\boldsymbol{E}} + 0(\delta^2) = 0(\delta), \tag{4.12}$$

and

$$\varrho = \varrho_0[1 - \operatorname{tr} \overline{\boldsymbol{E}}] + 0(\delta^2) = \varrho_0[1 + 0(\delta)], \tag{4.13}$$

---

[1] The linear form (4.9) is sometimes assumed in the theory of 'thermeo-rheologically simple' solids. It is known that for a class of polymeric materials under "small" deformation, the magnitude of a constitutive coefficient such as β (in suitable non-dimensional form) is small compared to one; in this connection, see for example TOBOLSKY [13, p. 62].

where ϱ_0 is the initial mass density. Also, by (2.5), (4.11) and (4.13), we have

$$T = \varrho_0(1 - \operatorname{tr}\bar{E})(1 + \bar{E} + \bar{W})\,\Sigma(1 + \bar{E} - \bar{W}) =$$
$$= \varrho_0\Sigma - \varrho_0(\operatorname{tr}\bar{E})\Sigma + \varrho_0(\bar{E} + \bar{W})\,\Sigma + \varrho_0\Sigma(\bar{E} - \bar{W}) + 0(\delta^2). \quad (4.14)$$

With the above background, we now state what we mean by a linearized theory with infinitesimal deformation discussed in this Section. It is clear from (4.1), (4.3), (4.5) and (4.10) that the displacement, velocity, $\bar{E}$ and $\bar{W}$, as well as their rates, are at all times "small" of $0(\delta)$. By (4.7), the temperature history [confined to a neighborhood of the constant history $(2.18)_2$] is such that at all times $\alpha = 0(\delta)$ and we also assume that $\frac{\partial\hat{\alpha}(\theta(\tau))}{\partial\theta(\tau)}$, as well as $\frac{\partial^2\hat{\alpha}(\theta(\tau))}{\partial\theta(\tau)^2}$, is of $0(\delta)$; the reasonableness of this latter assumption is particularly evident when $\hat{\alpha}(\theta)$ is a polynomial as in (4.9), where the constitutive coefficient (in suitable non-dimensional form) is of $0(\delta)$. Let Σ_0 be the value of the Piola-Kirchhoff stress tensor in the initial reference configuration and put

$$\Sigma = \Sigma_0 + \Sigma'. \quad (4.15)$$

We assume that the stress Σ' when expressed in suitable non-dimensional form (e.g., with the use of the initial value of one of the isothermal relaxation moduli associated with the "rest" history) is of $0(\delta)$. It follows from this assumption and (4.14) that

$$T = \varrho_0\Sigma_1 + \varrho_0\Sigma' + 0(\delta^2),$$
$$\Sigma_1 = \Sigma_0 - (\operatorname{tr}\bar{E})\Sigma_0 + \bar{E}\Sigma_0 + \bar{W}\Sigma_0 + \Sigma_0\bar{E} - \Sigma_0\bar{W},$$
$$\Sigma - \Sigma_0 = \Sigma' + 0(\delta^2) = 0(\delta), \qquad \Sigma_0 = 0(1). \quad (4.16)$$

Also, consistent with the above, in the expansion of the modified isothermal free energy functional (or the corresponding isothermal free energy functional) we retain up to and including terms of $0(\delta^2)$. These assumptions pertain only to the "mechanical" variables and will enable us to systematically linearize those parts of the constitutive equations (of the theory of Sec. 3) which involve the modified isothermal free energy functional. As far as the functional $\mathscr{B}$ or its derivative (which occur in some of the constitutive equations) are concerned we make no further assumption beyond that already made in Sec. 3 for the smallness of the norm $\| \theta_r^t(\cdot) - \theta \|_h$ and in deducing the representation (3.6).

We defer the related assumption for the heat flux vector until further developments.

In view of (4.1), (4.5), (4.7) and (4.16), as well as other assumptions stated above, in the field equations such as (2.6), (2.7) and (3.17) all partial derivatives with respect to $\boldsymbol{x}$ are now replaced with partial derivatives with respect to $\boldsymbol{X}$ in the initial reference configuration and all material time derivatives are now partial derivatives with respect to t. In particular, since the development in Sec. 3 represents the material description of the theory, the material derivative of entropy which occurs in (3.17) is simply the partial derivative of η with respect to t holding $\boldsymbol{X}$ fixed. We also adopt the special form (2.25) for the inverse of the modified time scale.

By (4.12)—(4.16), the isothermal functionals (3.30) and (3.31) become

$$\underset{s=0}{\overset{\infty}{p^*}}\left(\boldsymbol{E}_r^t(s);\boldsymbol{E}\right)=\operatorname{tr}\{\boldsymbol{\Sigma}_0\overline{\boldsymbol{E}}\}+\frac{1}{2}\operatorname{tr}[\overline{\boldsymbol{E}}H\overline{\boldsymbol{E}}\}+$$

$$+\frac{1}{2}\int_0^\infty\int_0^\infty\operatorname{tr}\{[\overline{\boldsymbol{E}}^t(s)-\overline{\boldsymbol{E}}]\,M(s,u)\,[\overline{\boldsymbol{E}}^t(u)-\overline{\boldsymbol{E}}]\}\,ds\,du+0(\delta^3),\qquad(4.17)$$

$$\underset{s=0}{\overset{\infty}{\mathfrak{T}^*}}\left(\boldsymbol{E}_r^t(s);\boldsymbol{E}\right)=\boldsymbol{\Sigma}_0+H\overline{\boldsymbol{E}}+\int_0^\infty\frac{\partial}{\partial u}G(0,u)\,\{\overline{\boldsymbol{E}}^t(u)-\overline{\boldsymbol{E}}\}\,du+0(\delta^2),\qquad(4.18)$$

where in expanding (3.30) about $\boldsymbol{E}=0$ we have set

$$\left.\frac{\partial\Phi(\boldsymbol{E})}{\partial\boldsymbol{E}}\right|_{\boldsymbol{E}=0}=\Sigma_0,\quad\left.\frac{\partial^2\Phi(\boldsymbol{E})}{\partial\boldsymbol{E}^2}\right|_{\boldsymbol{E}=0}=H.\qquad(4.19)$$

In (4.17)—(4.19), H is a fourth order tensor with constant coefficients and[1] M (as well as G) is now a material function which depends only on s and u. Consistent with our assumptions for the linearized theory under consideration, in what follows, we omit terms of $0(\delta^3)$ from (4.17) and terms of $0(\delta^2)$ from (4.18).

For the isothermal theory, it follows from $(4.16)_1$, (4.18) and the notations of $(2.10)_3$ and $(2.19)_2$ that the constitutive relation for stress, to the order of approximation considered, is given by

$$\frac{1}{\varrho_0}\boldsymbol{T}=\boldsymbol{\Sigma}_1+\boldsymbol{\Sigma}'=\boldsymbol{\Sigma}_1+H\overline{\boldsymbol{E}}+\int_0^\infty\frac{\partial}{\partial u}G(0,u)\,\{\overline{\boldsymbol{E}}^t(u)-\overline{\boldsymbol{E}}\}\,du,\quad(4.20)$$

[1] Strictly speaking M and G are now different functions of s and u, but we use the same letters to designate $M(s,u;O)$ and $G(s,u;O)$.

where the expression for $\boldsymbol{\Sigma}' = \frac{1}{\varrho_0}\boldsymbol{T} - \boldsymbol{\Sigma}_1 = 0(\delta)$ is easily determined from the comparison of $(4.20)_{1,2}$. Now, put

$$\frac{1}{\varrho_0}\boldsymbol{T}^{(\infty)} = \boldsymbol{\Sigma}_1 + \boldsymbol{\Sigma}'^{(\infty)} = \boldsymbol{\Sigma}_1 + H\overline{\boldsymbol{E}}, \tag{4.21}$$

which defines the "equilibrium" value of the stress corresponding to the "rest" history $\overline{\boldsymbol{E}}^t(s) = \overline{\boldsymbol{E}}\,1^+(s)$. Then, (4.20) can be rewritten as

$$\frac{1}{\varrho_0}\boldsymbol{T} = \frac{1}{\varrho_0}\boldsymbol{T}^{(\infty)} - \int_0^\infty G\;(0,\;u)\,\frac{\partial}{\partial u}\{\overline{\boldsymbol{E}}^t(u)\}\,du, \tag{4.22}$$

where the last term is obtained with the help of (3.11) after an integration by parts. Apart from the difference in notation, the result (4.22) is equivalent to that given by Coleman and Noll [3] and Coleman [2]. The latter author has shown that the initial value of the stress relaxation modulus, namely $G(0, 0)$ in our notation, is both positive and semi-definite. If the initial stress $\boldsymbol{\Sigma}_0$ is zero, then $\boldsymbol{\Sigma}_1$ vanishes and (4.22) reduces to the usual constitutive equation for the stress in the linear isothermal viscoelasticity.

We are now in a position to record the free energy and the stress for the non-isothermal theory with infinitesimal deformation. Keeping in mind our assumptions concerning $\hat{\alpha}(\theta)$ and $\frac{\partial \hat{\alpha}(\theta)}{\partial \theta}$, and recalling the result [stated in Sec. 2, after (2.33)] that the modified isothermal free energy and stress functionals can be obtained directly from the knowledge of the corresponding isothermal functionals, we replace $\overline{\boldsymbol{E}}^t(s)$ by $[\overline{\boldsymbol{E}}^t(\xi_s) - \boldsymbol{\alpha}^t(\xi_s)]$ in (4.17) and with the help of (2.31) deduce

$$\begin{aligned}\psi = \operatorname{tr}\{\Sigma_0(\overline{\boldsymbol{E}} - \boldsymbol{\alpha}\} + \frac{1}{2}\operatorname{tr}\{(\overline{\boldsymbol{E}} - \boldsymbol{\alpha})\,H(\overline{\boldsymbol{E}} - \boldsymbol{\alpha})\} + \frac{1}{2}\int_0^\infty\!\!\int_0^\infty \operatorname{tr}\{\overline{\boldsymbol{E}}^t(\xi_s) -\\ - \hat{\boldsymbol{\alpha}}\big(\theta^t(\xi_s)\big) - \overline{\boldsymbol{E}} + \boldsymbol{\alpha}\}\,M(s,u)\,\{\overline{\boldsymbol{E}}^t(\xi_u) - \hat{\boldsymbol{\alpha}}\big(\theta^t(\xi_u)\big) -\\ - \overline{\boldsymbol{E}} + \boldsymbol{\alpha}\}\,ds\,du + \mathop{\mathscr{B}}_{s=0}^{\infty}\big(\theta_r(s);\theta\big),\end{aligned} \tag{4.23}$$

where the functional $\mathscr{B}$ in (4.23) is given by (3.6). In a similar manner, we obtain an expression for the Piola-Kirchhoff stress tensor which with the help of (4.15) we write as

$$\begin{aligned}\boldsymbol{\Sigma}' = \boldsymbol{\Sigma} - \boldsymbol{\Sigma}_0 = H(\overline{\boldsymbol{E}} - \boldsymbol{\alpha}) + \int_0^\infty \frac{\partial G\,(0,u)}{\partial u}\,\{[\overline{\boldsymbol{E}}^t(\xi_u) -\\ - \hat{\boldsymbol{\alpha}}\big(\theta^t(\xi_u)\big)] - \overline{\boldsymbol{E}} + \boldsymbol{\alpha}\}\,du.\end{aligned} \tag{4.24}$$

Also, by $(4.16)_1$ and (4.24), we have

$$\frac{l}{\varrho_0} \boldsymbol{T} = \boldsymbol{\Sigma}_1 + H(\overline{\boldsymbol{E}} - \alpha) + \int_0^\infty \frac{\partial G\,(0, u)}{\partial u} \{[\overline{\boldsymbol{E}}^t\,(\xi_u) - \hat{\alpha}\big(\theta^t\,(\xi_u)\big)] - \overline{\boldsymbol{E}} + \alpha\}\, du =$$

$$= \boldsymbol{\Sigma}_1 + H(\overline{\boldsymbol{E}} - \alpha) - \int_0^\infty G\,(0, u)\, \frac{\partial}{\partial \xi_u} \times \{\overline{\boldsymbol{E}}^t(\xi_u) - \hat{\alpha}\big(\theta^t\,(\xi_u)\big)\}\, d\xi_u, \tag{4.25}$$

where $\boldsymbol{\Sigma}_1$ is given by $(4.16)_2$ and the first two terms on the right-hand side of $(4.25)_1$ correspond to (4.21) when the histories are specified by $\overline{\boldsymbol{E}}^t(\cdot) = \overline{\boldsymbol{E}}\,1^+(\cdot)$ and $\theta^t(\cdot) = \theta\,1^+(\cdot)$. Similarly the expression for entropy can be obtained from the corresponding result in Sec. 3 after introducing the required approximation which occurs only in the first term on the right-hand side of $(3.14)_1$. Thus,

$$\eta = \operatorname{tr}\left\{(\boldsymbol{\Sigma}_0 + \boldsymbol{\Sigma}')\, \frac{\partial\,\hat{\alpha}\,(\theta)}{\partial \theta}\right\} - \frac{\partial}{\partial \theta} \mathop{\mathscr{B}}_{s=0}^{\infty} \big(\theta_r^t(s);\, \theta\big), \tag{4.26}$$

where $\boldsymbol{\Sigma}'$ and $\frac{\partial\,\hat{\alpha}\,(\theta)}{\partial \theta}$ are both of $0(\delta)$ and the second term on the right-hand side of (4.26) is still given by $(3.14)_2$.

We now turn our attention to the heat flux vector which is given by (3.24) and (3.25) but with $\boldsymbol{E}(\cdot)$ replaced by $\overline{\boldsymbol{E}}(\cdot)$. By expanding the functions $M_{ij}^{(0)}$, $M_{ijkl}^{(1)}$, $M_{ij}^{(2)}$ about $\boldsymbol{E} = \boldsymbol{O}$ and neglecting terms of $0(\delta^2)$ when these can be compared with those of $0(\delta)$, (3.25) becomes

$$\mathop{\mathfrak{M}_{ij}}_{s=0}^{\infty}\big(\boldsymbol{E}_r^t(s), \theta_r^t(s);\; \boldsymbol{E}, \theta, \boldsymbol{g}\big) = M_{ij}^{(0)}(\boldsymbol{O}, \theta, \boldsymbol{g}) + \operatorname{tr}\left\{\frac{\partial M_{ij}^{(0)}}{\partial \boldsymbol{E}}\bigg|_{\boldsymbol{E}=0} \overline{\boldsymbol{E}}\right\} +$$

$$+ \int_0^\infty M_{ijkl}^{(1)}(s;\; \boldsymbol{O}, \theta, \boldsymbol{g})\, [\overline{\boldsymbol{E}}_{kl}^t(s) - \overline{\boldsymbol{E}}_{kl}]\, ds$$

$$+ \int_0^\infty M_{ij}^{(2)}(s;\; \boldsymbol{O}, \theta, \boldsymbol{g})\, [\theta^t(s) - \theta]\, ds, \tag{4.27}$$

in view of (3.28)—(3.29) and since $\|\overline{\boldsymbol{E}}_r^t(s) - \overline{\boldsymbol{E}}\|_h$ and $\|\theta_r^t(s) - \theta\|_h$ are of $0(\delta)$. Recalling that the integrals in (4.27) are scalar products and using the Schwarz inequality in (3.27), it follows that if each of the quantities

$$\| M^{(1)}_{ijkl}(s;\, \boldsymbol{O}, \theta, \boldsymbol{g})\, h(s)^{-2} \|_h, \quad \| M^{(2)}_{ij}(s;\, \boldsymbol{O}, \theta, \boldsymbol{g})\, h(s)^{-2} \|_h, \tag{4.28}$$
$$\left| \frac{\partial M^{(o)}_{ij}}{\partial \boldsymbol{E}} \right|_{\boldsymbol{E}=0}$$

can be considered to have the same order of magnitude as $|M^{(o)}_{ij}(\boldsymbol{O}, \theta, \boldsymbol{g})|$, then (3.24) can be put in the form[1]

$$\underset{s=0}{\overset{\infty}{\mathscr{K}_i}}\left(\boldsymbol{E}^t_r(s), \theta^t_r(s);\, \boldsymbol{E}, \theta, \boldsymbol{g}\right) = K_{ij}(\theta, \boldsymbol{g})\,[1 + 0(\delta)]\, g_j, \tag{4.29}$$

where we have introduced the notation $K_{ij}(\theta, \boldsymbol{g}) = M^{(o)}_{ij}(\boldsymbol{O}, \theta, \boldsymbol{g})$.

The function $\hat{\alpha}(\theta(\tau))$ and its derivative $\frac{\partial \hat{\alpha}(\theta(\tau))}{\partial \theta(\tau)}$ for all $\tau \leqq t$ are of $0(\delta)$, provided the temperature history belongs to a neighborhood I of the constant history $\theta_0 1^+(\cdot)$. The modified temperature history $\theta^t(\xi_s)$ belongs also to I and therefore $\frac{\partial \hat{\alpha}(\theta^t(\xi_s))}{\partial \theta^t(\xi_s)}$ is of $0(\delta)$.
Keeping this in mind, the internal dissipation (3.18) up to and including terms of $0(\delta^2)$ becomes

$$\theta\sigma = -\int_0^\infty\!\!\int_0^\infty \operatorname{tr}\left\{\frac{\partial \overline{\boldsymbol{E}}(\xi_s)}{\partial \xi_s} - \frac{\partial \hat{\alpha}(\theta^t(\xi_s))}{\partial \theta^t(\xi_s)} \frac{\partial \theta^t(\xi_s)}{\partial \xi_s}\right\} \frac{\partial G(s,u)}{\partial u} \left\{\frac{\partial \overline{\boldsymbol{E}}^t(\xi_u)}{\partial \xi_u} - \right.$$
$$\left. - \frac{\partial \hat{\alpha}(\theta^t(\xi_u))}{\partial \theta^t(\xi_u)} \frac{\partial \theta^t(\xi_u)}{\partial \xi_u}\right\} \Phi(\theta) \frac{\partial \xi_u}{\partial u} \frac{\partial \xi_s}{\partial s}\, ds\, du + \delta \underset{s=0}{\overset{\infty}{\mathscr{B}}}\left(\theta^t_r(s);\, \theta, \left|\frac{d\theta^t_r(s)}{ds}\right.\right), \tag{4.30}$$

and $\delta\mathscr{B}$ is given by (3.19)[2]. In obtaining (4.30) from (3.18) we have replaced $\boldsymbol{E}^t(\cdot)$ by $\overline{\boldsymbol{E}}^t(\cdot)$ and $G(s, u;\, \boldsymbol{E} - \alpha)$ by[3] $G(s, u)$, but the coefficient $m(s, u;\, \theta)$ in (3.19) still depends on θ.

An equation for the determination of temperature can be obtained from the residual energy equation upon substitution of (4.26), (4.29) and (4.30) into (3.17). Recalling that the material derivative of entropy [as given by (4.26)] is simply the partial derivative of η with respect to t

---

[1] We have retained terms of $0(\delta)$ in (4.29) in order to clearly exhibit the nature of approximation in later developments; otherwise, terms of $0(\delta)$ may be neglected in comparison with one on the right-hand side of (4.29).

[2] The last term on the right-hand side of (4.30), namely δB, can be seen to be of $0(\delta^2)$ after an integration by parts of (3.19) and use of (3.12).

[3] Strictly speaking we replace $G(s, u;\, \boldsymbol{E} - \alpha)$ by $G(s, u;\, \boldsymbol{O})$ or another function of s and u. But we use the same letter and designate the latter by $G(s, u)$.

holding $\boldsymbol{X}$ fixed, we have

$$\dot{\eta} = \operatorname{tr}\left\{\dot{\boldsymbol{\Sigma}}' \frac{\partial \hat{\alpha}(\theta)}{\partial \theta}\right\} + \operatorname{tr}\left\{(\boldsymbol{\Sigma}_0 + \boldsymbol{\Sigma}') \frac{\partial^2 \hat{\alpha}(\theta)}{\partial \theta^2}\right\} \frac{\partial \theta}{\partial t} -$$

$$- \frac{\partial^2 B(\theta)}{\partial \theta^2} \frac{\partial \theta}{\partial t} + \frac{\partial}{\partial t} \int_0^\infty m(s, 0; \theta) \frac{\partial \theta^t(s)}{ds}\, ds. \tag{4.31}$$

Introducing (4.31) into (3.17), the residual energy equation becomes

$$-\varrho_0 \theta \frac{\partial^2 B(\theta)}{\partial \theta^2} \frac{\partial \theta}{\partial t} + \varrho_0 \theta \frac{\partial}{\partial t} \int_0^\infty m(s,\, 0;\, \theta) \frac{d\theta^t(s)}{ds}\, ds + \operatorname{tr}\left\{\Sigma_0 \frac{\partial^2 \hat{\alpha}(\theta)}{\partial \theta^2}\right\} \frac{\partial \theta}{\partial t} +$$

$$+ \operatorname{div} \boldsymbol{q} - \varrho_0 r + S = 0, \tag{4.32}$$

where

$$S = \varrho_0 \theta \left[\operatorname{tr}\left\{\dot{\boldsymbol{\Sigma}}' \frac{\partial \hat{\alpha}}{\partial \theta}\right\} + \operatorname{tr}\left\{\boldsymbol{\Sigma}' \frac{\partial^2 \hat{\alpha}(\theta)}{\partial \theta^2}\right\} \frac{\partial \theta}{\partial t}\right] - \varrho_0 \theta \sigma, \tag{4.33}$$

the componente of $\boldsymbol{q}$ are given by (4.29), K_{ij} is a function of θ and $\boldsymbol{g}$ and we have made no assumption regarding the magnitude of θ, g_j and $\frac{\partial g_j}{\partial X_i}$. If $\hat{\alpha}(\theta)$ is assumed to be of the form (4.9) with β being a constant tensor, then the third term in (4.32) and the second term on the right-hand side of (4.33) vanish identically. We also observe that the left-hand side of (4.32), apart from S, involves only the temperature field and its derivatives and that the integral

$$\int_0^\infty m(s, 0; \theta) \frac{d\theta^t(s)}{ds}\, ds, \tag{4.34}$$

which occurs in (4.31) and (4.32), is of[1] $0(\delta)$.

By virtue of our assumptions stated after (4.14), as well as the smallness of the norm $\|\theta_r^t(\cdot) - \theta\|_h$, the terms given by (4.33) are of $0(\delta^2)$. If we also assume that the partial derivative with respect to t of (4.34) and the partial derivative with respect to X of terms of $0(\delta)$ in $\boldsymbol{q}$ [or in (4.29)] are also of $0(\delta)$, then for fairly large temperature variation (from the initial reference temperature) terms of $0(\delta)$ and those given by (4.33) are negligible in comparison with other terms in (4.32). The corresponding residual energy equation, although nonlinear in θ, will then involve

[1] Although no restriction is placed on $\frac{d\theta^t(s)}{ds}$, the fact that (4.34) is of $0(\delta)$ follows from an integration by parts and use of (3.12).

only the temperature field and is free from the presence of any "mechanical" variable. Thus, if we neglect terms of $0(\delta)$ in comparison with other terms in (4.32), we obtain[1]

$$\frac{r}{\theta} - \frac{1}{\varrho_0 \theta} \frac{\partial}{\partial X_i} (K_{ij} g_j) + \frac{\partial^2 B}{\partial \theta^2} \frac{\partial \theta}{\partial t} = 0 . \qquad (4.35)$$

The basic equations of the linearized theory will then consist of the linearized equations of motion, the constitutive equations (4.23), (4.25), (4.26), (4.29) and the equation (4.35) for the determination of temperature. An examination of this system of equations reveals that the thermomechanical problem is uncoupled in the sense that (4.35) involves only the temperature field. Although the displacement, strain and $\hat{\alpha}$ are infinitesimal in the sense of (4.1), (4.5) and (4.7), we have made no special assumption regarding the smallness of temperature θ relative to the initial reference temperature θ_0. The system of partial differential equations characterizing the problem is nonlinear in temperature. However, for purposes of stress analysis, once the solution of (4.35) is known [or if the temperature field which satisfies the energy equation is assumed at the outset], then the system of equations consisting of the equations of motion, the constitutive relation (4.25), and the kinematical results (4.10) and (4.3) is linear in the components of displacement. The complete theory, however, remains nonlinear in temperature.

In the rest of this section, for simplicity, we put

$$\Sigma_0 = 0 , \qquad (4.36)$$

and consider a comparison of the above constitutive relations and the internal dissipation with previously known results. In order to provide a basis for such a comparison with those usually given for 'thermo-rheologically simple' solids, we introduce the following change of variables:

$$t - \xi_s = t', \quad t - \xi_u = t'' \qquad (4.37)$$

$$q = \int_0^t \Phi(\theta(\tau))\, d\tau , \quad q' = \int_0^{t'} \Phi(\theta(\tau))\, d\tau , \quad q'' = \int_0^{t''} \Phi(\theta(\tau))\, d\tau , \qquad (4.38)$$

where Φ is defined in (2.26). It follows from (2.25), (4.37) and (4.38) that

$$s = q - q' , \quad u = q - q'' . \qquad (4.39)$$

[1] Equation (4.35) can be further simplified if special assumptions are introduced for the function K_{ij} in (4.29) and the functional $\mathscr{B}$ in (3.6). In fact, a special form of (4.35) can be made to correspond to the Fourier's heat conduction equation often employed in problems of stress analysis for 'thermo-rheologically simple' solids.

With the help of (4.37)—(4.39) and (4.36), (4.23) and $(4.25)_2$ become

$$\psi = \frac{1}{2} \operatorname{tr} \{(\bar{\boldsymbol{E}} - \alpha) H (\bar{\boldsymbol{E}} - \alpha)\} + \frac{1}{2} \int_{-\infty}^{t} \int_{-\infty}^{t} \operatorname{tr} \left\{ \frac{d\bar{\boldsymbol{E}}(t')}{dt'} - \frac{\partial \hat{\alpha}}{\partial \theta(t')} \frac{d\theta(t')}{dt'} \right\}$$

$$G(q - q', q - q'') \left\{ \frac{d\bar{\boldsymbol{E}}(t'')}{dt''} - \frac{\partial \hat{\alpha}}{\partial \theta(t'')} \frac{d\theta(t'')}{dt''} \right\} dt' \, dt'' +$$

$$+ B(\theta) + \frac{1}{2} \int_{-\infty}^{t} \int_{-\infty}^{t} m(t - \tau', \; t - \tau''; \theta) \frac{d\theta(\tau')}{d\tau'} \frac{d\theta(\tau'')}{d\tau''} d\tau' \, d\tau'', \tag{4.40}$$

$$\frac{1}{\varrho_0} \boldsymbol{T} = H[\bar{\boldsymbol{E}} - \hat{\alpha}] + \int_{-\infty}^{t} G(0, q - q'') \frac{\partial}{\partial t''} \left[\bar{\boldsymbol{E}}(t'') - \hat{\alpha}\big(\theta(t'')\big) \right] dt'', \tag{4.41}$$

and a similar expression for entropy follows from (4.26). Also, from (4.30) and (3.19), we obtain

$$\theta \sigma' = \theta \sigma - \int_{-\infty}^{t} \int_{-\infty}^{t} \frac{\partial m(t - t', \; t - t''; \theta)}{\partial t''} \frac{d\theta(t')}{dt'} \frac{d\theta(t'')}{dt''} dt' \, dt'' =$$

$$= - \Phi(\theta) \int_{-\infty}^{t} \int_{-\infty}^{t} \operatorname{tr} \left\{ \left[\frac{\partial \bar{\boldsymbol{E}}(t')}{\partial t'} - \frac{\partial \hat{\alpha}(\theta(t'))}{\partial \theta(t')} \frac{\partial \theta(t')}{\partial t'} \right] \frac{\partial G(q - q', q - q'')}{\partial (q - q'')} \times \right.$$

$$\left. \times \left[\frac{\partial \bar{\boldsymbol{E}}(t'')}{\partial t''} - \frac{\partial \hat{\alpha}(\theta(t''))}{\partial \theta(t'')} \frac{\partial \theta(t'')}{\partial t''} \right] \right\} dt' \, dt''. \tag{4.42}$$

The constitutive relation (4.41) has the same form as that used by Morland and Lee [7] and Muki and Sternberg [8]; however, (4.41) is in agreement with Hunter's [6] expression for stress if $\hat{\alpha}$ is assumed to have the form (4.9). Also, the expression $(4.42)_2$ is of the same form as that called dissipation by Hunter [6] if $\hat{\alpha}$ is taken to be linear in θ. Hunter's dissipation function does not contain a term corresponding to the second term in $(4.42)_1$.

Acknowledgement

The work of one of us (P. M. N.) was supported by the U. S. Office of Naval Research under Contracts Nonr-222(69) and N00014-67-A-0114-0021 with the University of California, Berkeley.

References

[1] Coleman, B. D.: Thermodynamics of Materials with Memory. Arch. Ratl Mech. Anal. **17**, 1–46 (1964).

[2] Coleman, B. D.: On Thermodynamics, Strain Impulses and Viscoelasticity. Arch. Ratl Mech. Anal. **17**, 230–254 (1964).

[3] Coleman, B. D., and W. Noll: Foundations of Linear Viscoelasticity. Rev. Mod. Phys. **33**, 239–249 (1961).

[4] Christensen, R. M., and P. M. Naghdi: Linear Non-isothermal Viscoelastic Solids. Acta Mech. **3**, 1–12 (1967).

[5] Crochet, M. J., and P. M. Naghdi: A Class of Simple Solids with Fading Memory. Int. J. Engng. Sci. (1969). To appear.

[6] Hunter, S. C.: Tentative Equations for the Propagation of Stress, Strain and Temperature Fields in Viscoelastic Solids. J. Mech. Phys. Solids **9**, 39–51 (1961).

[7] Morland, L. W., and E. H. Lee: Stress Analysis for Linear Viscoelastic Materials with Temperature Variation. Trans. Soc. Rheology **4**, 233–263 (1960).

[8] Muki, R., and E. Sternberg: On Transient Thermal Stresses in Viscoelastic Materials with Temperature-dependent Properties. J. Appl. Mech. **28**, 193–207 (1961).

[9] Payne, A. R.: Temperature-Frequency Relationships of Dielectric and Mechanical Properties of Polymers. The Rheology of Elastomers (edited by P. Mason and N. Wookey), pp. 86–110. London: Pergamon Press. 1958.

[10] Schwarzl, F., and A. J. Staverman: Time-temperature Dependence of Linear Viscoelastic Behavior. J. Appl. Phys. **23**, 838–843 (1952).

[11] Staverman, A. J., and F. Schwarzl: Linear Deformation Behavior of High Polymers. The Physics of High Polymers, Vol. 4, Chapter 1. Berlin-Göttingen-Heidelberg: Springer. 1956.

[12] Sternberg, E.: On the Analysis of Thermal Stresses in Viscoelastic Solids. Proc. 3rd Symp. on Naval Structural Mechanics, pp. 348–382 (1964).

[13] Tobolsky, A. V.: Properties and Structure of Polymers. New York: John Wiley & Sons. 1962.

[14] Truesdell, C., and W. Noll: The Non-Linear Field Theories of Mechanics. Encyclopedia of Physics, Vol. III/3 (edited by S. Flügge). Berlin-Heidelberg-New York: Springer. 1965.

Creep Instability of Aluminum under Varying Temperature

By

Alfred M. Freudenthal

New York, N.Y. (U.S.A.)

Summary

Creep under unsteady temperature is affected by changes in the mobility of elements of the micro-structural configuration produced by temperature variation, such as local density changes, changes in mobile dislocations due to their thermal break-away and grain-boundary migration. Since these changes are not instantaneous, it is to be expected that, following a temperature variation, a condition of steady-state creep will not immediately establish itself.

Results of elevated temperature creep experiments on pure aluminum indicate that a period of sharp creep-acceleration follows a rapid temperature increase until steady state creep establishes itself after several hours. The implications of this phenomenon on the results of creep and stability analysis of simple structural elements are discussed.

1. Introduction

In the stress and deformation analysis of linear viscoelastic media with temperature-gradients the simplifying assumption is made that these media are "thermo-rheologically simple" so that the relaxation modulus in shear $G(T, t) = G(t)$ at temperature T, is related to this modulus $G(T_0, t) = G_0(t)$ at the base temperature T_0 by the equation

$$G(T, t) = G_0[\varphi(T). t] = G_0(\xi), \tag{1}$$

where $\xi = t.\varphi(T)$ denotes the "reduced time" and $\varphi(T)$ represents the (steadily increasing) "shift function", characteristic of the temperature sensitivity of the medium and subject to the obvious conditions

$$\varphi(T_0) = 1, \quad \varphi(T) > 0 \quad \text{and} \quad d\varphi(T)/dt > 0,$$

within the range of the assumed validity of "thermo-rheologically simple" behavior [1]. Similarily the creep compliance $J(t)$ at temperature T is related to the compliance $J_0(t)$ at T_0 by the equation

$$J(t) = J_0[\varphi(T) \cdot t] = J_0(\xi), \tag{2}$$

because of the well-known relation between the Laplace transforms $\bar{G}(p)$ and $\bar{J}(p)$ of, respectively, $G(t)$ and $J(t)$

$$p^2 \bar{G}(p) \bar{J}(p) = 1, \tag{3}$$

which follows from the Laplace transformation of the deviatoric stress-strain relations expressed in the form of Stieltjes convolutions.

The assumption of thermo-rheologically simple behavior implies that the relaxation moduli or the creep compliances at various temperatures T, plotted on a logarithmic time scale $G[\ln t]$ or $J[\ln t]$ are obtained from the respective functions $G_0[\ln t]$ and $J_0[\ln t]$ at the reference temperature T_0 by translational shift, without change of shape, in the direction of the time axis; they depend, therefore, on time and temperature only through the "reduced time"

$$\ln \xi = \ln t + f(T - T_0) = \ln t + \ln \varphi(T) = \ln [t \cdot \varphi(T)], \tag{4}$$

where the shift function $\varphi(T) = \exp [f(T - T_0)]$ and the function

$$\ln \varphi(T) = f(T - T_0) = f(T) - f(T_0), \tag{5}$$

is obtained from the observed shift of the visco-elastic response functions at various temperatures with respect to their location at the reference temperature T_0. The effect of temperature change thus consists simply in multiplying all relaxation or retardation times by the common factor $\varphi(T)$. The function $f(T)$ which, in general, is a decreasing function of T or an increasing function of $(1/T)$ fixes the relative positions of the modulus or compliance curves on the logarithmic time scale. For simple Newtonian flow with the exponential temperature dependence of the coefficient of viscosity

$$\ln \left(\frac{\eta}{\eta_0}\right) = \frac{Q}{R}\left(\frac{1}{T} - \frac{1}{T_0}\right) = \ln \varphi(T), \tag{6}$$

the function

$$f(T) = \frac{Q}{RT}, \tag{7}$$

is a linear function of $(1/T)$. For various rubbers equations of the form

$$f(T) = c/(T - \text{const}),$$

were found [2], while in many cases the dependence of $f(T)$ on T is fairly well represented by the form $(1/T)^{\alpha}$.

2. Time Dependence of Viscosity Change

Determination of the function $\ln \varphi(T)$ by observation of the shift of the relaxation modulus with temperature presupposes conditions of thermal equilibrium which do not, in general, establish themselves immediately after a change of temperature. This implies that the experimentally determined functions $f(T)$ are associated with times elapsed after the temperature change that are long enough to permit reestablishment of thermal equilibrium. Since changes with temperature of the relaxation mechanisms in viscoelastic materials are necessarily associated with changes in the mobility of elements of the micro-structure by configurational changes, by changes in the concentration and distribution of defects, and by local density changes due to changes in composition, and since such changes are not instantaneous, it is to be expected that the equilibrium function $\varphi(T)$ does not adequately describe the temperature shift of the relaxation modulus during rapid temperature changes, but only reflects the equilibrium conditions that are asymptotically attained as the micro-structure reaches a quasi-stationary configuration under the applied combination of stress and temperature. Before such conditions are attained, time-dependent shift functions are necessarily of the form $\varphi(t, T)$, provided the concept of shifting of the modulus without change of form can be retained.

Experimental evidence for the existence of such functions is extremely scarce. Only for thin glass fibers has the time dependence of the change of the coefficient of viscosity following a sudden change of temperature been demonstrated by experiment [3]. It was found that near the annealing point the "equilibrium viscosity" $\eta_\infty(T) = \eta(T, \infty)$ is approached from below (temperature drop) or from above (temperature surge) at a rate given by the expression

$$\frac{d\eta}{dt} = k\,\frac{\eta_\infty - \eta}{\eta}. \tag{8}$$

Introducing the abbreviation $y = \eta/\eta_\infty = \lambda/\lambda_\infty$, where λ denotes the coefficient of viscous traction, the time dependence of y is therefore of the form

$$e^{y}(y - 1) = e^{y_0}(y_0 - 1)e^{-kt/\lambda_\infty}, \tag{9}$$

where $y_0 = \lambda_0/\lambda_\infty = \varphi^{-1}$ is the ratio between the equilibrium values of the viscosity at temperature T_0 and T. In this relation the instantaneous drop of λ accompanying an instantaneous temperature increase implied by eq. (6) is replaced by a gradual, initially linear, decrease. Since eq. (9) cannot be solved explicitely for y, its use for the solution of specific boundary value problems is limited to numerical integration procedures. The decrease of viscosity with time is much less rapid than the exponential decrease from either of the limiting approximations of eq. (8)

$$k\,\frac{\eta_\infty - \eta}{\eta_0} < \frac{d\eta}{dt} < k\,\frac{\eta_\infty - \eta}{\eta_\infty},$$

which produce the expressions

$$e^{-kt/\lambda_\infty} < \frac{y-1}{y_0-1} < e^{-kt/\lambda_0}, \tag{10}$$

with initial (negative) slopes (at $t = 0$)

$$\frac{k}{\lambda_0}(y_0 - 1) < \left(\frac{dy}{dt}\right)_0 < \frac{k}{\lambda_\infty}(y_0 - 1).$$

The decrease in the vicinity of $t = 0$ is, however, fairly close to the upper limit $k(y_0 - 1)/\lambda_0$ of eq. (10). Thus, for instance, in this range the rate of stress relaxation in a simple linear visco-elastic (relaxing) medium the temperature of which in increased at $t = 0$ from T_0 to T would be governed by the equation

$$\frac{d\sigma}{dt} + \frac{E}{\lambda}\,\sigma = \frac{d\sigma}{dt} + \frac{E\sigma}{\lambda_\infty}\,[1 + (\lambda_0/\lambda_\infty - 1)e^{-kt/\lambda_0}]^{-1} = 0, \tag{11}$$

with the solution

$$(\sigma/\sigma_0) = \exp\left[-\frac{E(a+1)}{k}\ln\frac{e^{kt/\lambda_0} + a}{1+a}\right], \tag{12}$$

where $a = (\lambda_0/\lambda_\infty) - 1 = y - 1$; it is therefore considerably slower than if it were immediately governed by the equilibrium relaxation time $\tau_\infty = \lambda_\infty/E$. The slope of eq. (12) at $t = 0$ is $\tau_0^{-1} = E/\lambda_0 \ll \tau_\infty^{-1}$ when $\lambda_0 \gg \lambda_\infty$; however when $\lambda_\infty \gg \lambda_0$ it is $\gg \tau_\infty^{-1}$. The relaxation process immediately after the temperature change is thus governed by the initial temperature and approaches the equilibrium relaxation function $\exp\,[-t/\tau_\infty]$ only as $t \gg (\lambda_0/k)\ln a$. A somewhat better approximation of eq. (9) than the simple exponential form, eq. (10), is provided by the relation

$$\ln y = \ln\left(\frac{\lambda}{\lambda_\infty}\right) = \ln\left(\frac{\lambda_0}{\lambda_\infty}\right)e^{-\beta t} = \ln y_0 \cdot e^{-\beta t} = -\ln\varphi \cdot e^{-\beta t}, \tag{13}$$

with $\beta \ll k/\lambda_\infty$.

3. Creep of Aluminum under Rapid Temperature Change

The effect of a rapid temperature increase on the creep rate of a pure polycrystalline metal is the result of the combined temperature effects on high-angle grain-boundaries and on sub-boundary and crystal regions. The generally held idea that the high-angle grain-boundary resembles a liquid layer a few atoms thick [4] would justify the belief that the time dependence of the effect of a temperature change on this layer is similar to that observed in glass; however, the relaxation times involved are likely to be several orders of magnitude shorter in view of the fact that the particle or vacancy motion in a metal is much faster than that of the molecular units of the glass structure. Moreover, the grain-boundary contributions to the total creep strain seem to be small [5] in comparison to the contribution from the various low activation energy processes of multiplication and motion of mobile dislocations, especially in high stacking fault metals, such as aluminum. While a substantial part of these dislocations reside in subgrain boundaries, the rest is in the subgrain volume. The gradually decreasing creep rate during primary creep is directly related to the decreasing density of mobile dislocations. The number of mobile dislocations decreases either by their migrating to the subgrains boundaries, or by their being repinned by point defects (such as vacancies or impurities), as extensive entanglements produced in the course of the creep deformation provide more effective barriers to dislocation motion [6]. The steady, secondary stage of creep is attained as balance is achieved between the rate of formation and the rate of annihilation of such barriers by recovery processes in the structure, so that the density of mobile dislocations in the subgrain volume remains constant.

A sudden temperature increase from T_0 to $T = T_0 + \Delta T$ at this stage produces a practically instantaneous thermal breakaway [7] of pinned dislocations; the density of mobile dislocations is thus suddenly increased beyond the steady state level associated with secondary creep at the applied stress-intensity, resulting in a sudden sharp increase in the creep rate. The rate of repinning of the mobile dislocations determines the length of this transient stage of creep which, proceeding at decreasing creep-rate, resembles the stage of primary creep, and is followed by a new steady secondary stage at a rate associated with the increased temperature. When the temperature increase is at a finite rate rather than sudden it is to be expected, on the basis of the concept of thermal breakaway, that production of mobile dislocations by such breakaway will continue as long as the temperature is increasing; it will stop only when the new temperature level has been attained. For a fast temperature increase the effect of the process of repinning in the course of creep deformation is likely to vanish during the short period of temperature increase. The creep rate will, therefore, increase at least as long as the temperature increases, reaching a maximum at the time when or very shortly after the new temperature level has been reached. When the temperature increase is slow, the stages of thermal breakaway and repinning may overlap considerably, with the result that the increase of the creep rate will no longer follow the temperature increase throughout, but slow down already before the new temperature is attained; thus the maximum of the creep rate may, occur before the completion of the temperature increase.

The ratio between the stationary creep rates at the two temperatures can be obtained from the relation

$$\dot{\varepsilon}_C = A f(\sigma) \exp\left[-Q/RT\right] = \lambda^{-1} f(\sigma), \tag{14}$$

the validity of which, at low and moderate stress-intensities, has been established for a number of metals [8]. For constant stress therefore

$$\ln\left[\frac{\dot{\varepsilon}_C(T_0)}{\dot{\varepsilon}_C(T)}\right] = \ln\left(\frac{\lambda}{\lambda_0}\right) = Q/R(1/T - 1/T_0) = \ln\varphi(T), \tag{15}$$

which is identical with eq. (6). However, since the stationary state does not establish itself at the time of the temperature increase, the validity of eq. (15) is limited to times beyond the transient stage of thermal breakaway and subsequent repinning of dislocations. During this stage the ratio $\ln(\lambda/\lambda_0)$ is a function of both temperature increase and time.

In a recent investigation of creep of an annealed aluminum alloy under transient temperatures [9] the form of this function was studied under various linear rates of temperature increase and decrease for a suitably selected single temperature difference of 40 °C at a sustained constant tensile stress $\sigma = 10{,}000$ psi. The temperature levels of 423 °K and 463 °K at which the creep tests were conducted, were chosen so as to produce a difference of roughly one order of magnitude between the steady-state creep rates at these two temperatures. Isothermal creep tests at stresses between 9,000 and 12,000 psi demonstrated the validity, in this temperature and stress range, of eq. (14) with $A = 1.75 \times 10^{-3}$, $f(\sigma) = \sigma^6$ and $Q = 32{,}000$ cal./mole (creep rate in in./in hr.).

The results of creep tests under four different rates of temperature increase are summarized in Fig. 1 in the form of the ratio $(\lambda/\lambda_\infty) = \dot{\varepsilon}_{C\infty}/\dot{\varepsilon}_C(t) = f(t)$. The results at the two highest rates of temperature increase of 0.5 °C per sec. and 0.1 °C per sec. cannot be distinguished from each other: the creep rates have increased by about 3 orders of magnitude at the time when the maximum temperature difference has been attained or shortly after. At the temperature rates of 0,01 °C per sec.

and 0,0011° per sec. the maximum rate of transient creep is attained at about the same time or somewhat before the maximum temperature has been reached; because of the overlapping breakaway and repinning processes the maximum creep rate at the slowest temperature rate is, however, considerably smaller than at the higher rates. Nevertheless the experimental relation for the very slow rate 0.0011 °C per sec. shows

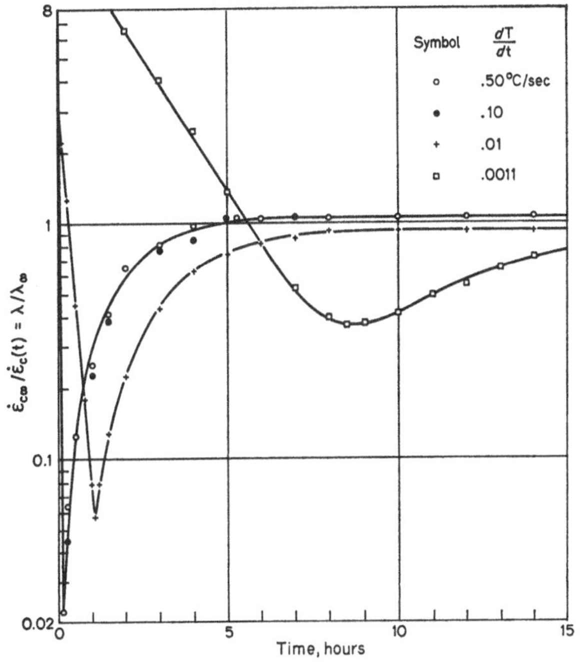

Fig. 1. Inverse of stationary creep rates at constant rates of temperature increase dT/dt of annealed aluminum

a peak of the creep rate substantially above its equilibrium rate, and the time dependence of the logarithm of the creep rate differs significantly from that under equilibrium temperature change.

The duration of the transient stage of creep produced by a temperature increase is not at all short, as can be seen from Fig. 1. Even in the case of rapid temperature changes (≥ 0.1 °C per sec.) the time to attain the second stage creep at the increased temperature is of the order of several hours. An analytic representation of the time dependence of the ratio (λ/λ_0) in this stage obtained from the creep tests under constant rate of temperature increase must contain at least two terms: one term representing the effect of the grain boundaries, which is similar to that discussed in Section 2, another representing the sharp increase and subsequent decrease of the creep rate due to thermal breakaway and repinning dislocations. An equation of the form

$$\log (\lambda/\lambda_\infty) = \log (\lambda_0/\lambda_\infty) \, [e^{-\beta t x} - A x^\alpha e^{-\alpha x}], \tag{16}$$

where $x = t/\bar{t}$ is a dimensionless time parameter $\bar{t}$ is the time interval between the start of the temperature increase $(t = 0)$ and the point of maximum creep rate, and α, β and A are empirical constants.

Equation (16) is valid only for a temperature increase $\Delta T = (T - T_0) > 0$. Since a temperature decrease has little effect on pinned dislocations, a transient stage of creep can only be associated with the grain-boundary response of the approximate form of eq. (13). The transient temperature experiments on the aluminum alloy have shown that the time scale of this response is extremely short: the steady state creep rate establishes itself as soon as the reduced steady temperature level is reached. This is to be expected in the temperature range in which the tests were performed and in which the relaxation times of the grain-boundaries are of the order of seconds or shorter. Thus, in eq. (16) $A > 0$ for temperature increase, while $A = 0$ for temperature decrease. Since the time decay of the first term is extremely fast, its effect can be neglected: eq. (15) with $\lambda = \lambda_\infty$ replaces eq. (16) for temperature decrease. Writing eq. (15) in the form

$$\log \left(\frac{\lambda_0}{\lambda_\infty}\right) = \frac{Q}{R} \left(\frac{1}{T_0} - \frac{1}{T}\right) (\ln 10)^{-1} = - \log \varphi(T), \tag{17}$$

eq. (16) becomes

$$\log (\lambda/\lambda_\infty) = - \log \varphi \cdot f(x), \tag{18}$$

with

$$\begin{aligned} f(x) = \left(e^{-\beta \bar{t} x} - A x e^{-\alpha x}\right) &= e^{-\beta \bar{t} x} - A e^{-\alpha(x - \ln x)} \quad \text{for } \Delta T > 0 \\ &= H(x) \qquad\qquad\qquad\qquad \text{for } \Delta T < 0, \end{aligned} \tag{19}$$

where $H(x)$ denotes the unit step function. Hence the combined time-temperature dependence of λ following a temperature change at $x = 0$

$$\lambda = \lambda_\infty \varphi^{-f(x)} = \lambda_\infty e^{-f(x) \ln \varphi} \tag{20}$$

The experimental data can be reasonably well fitted by the function $f(x)$ with the values of the parameters $A = 2.5$, $\alpha = 0.25$ or 0.35, $\beta \bar{t} = 50$ for $\varphi = 10^{-1}$. These functions have been plotted in Fig. 2.

In view of the fact that the experiments are of a rather exploratory nature in that only a single temperature step was studied so that the results cannot be generalized to an arbitrary interval ΔT, it does not seem too important to use more elaborate forms of the function $f(x)$ in order to obtain a more close fit. Since the selected form and values of the parameters reproduce the trend of the experimental results fairly

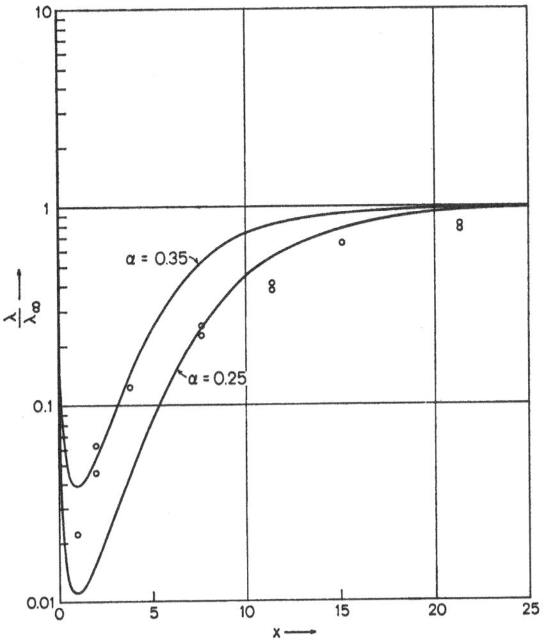

Fig. 2. Fitting of date of Fig. 1 by eq. (20)

well, eq. (20) can be used to study the effect of the stage of transient creep following a temperature increase on deformation and stress in simple metallic structural elements at temperature levels at which creep is a significant feature.

4. Effect of Thermal Transients on Creep in Bars and Thin Shells

a) Creep. Assuming the simplest one-dimensional non-linear visco-elastic constitutive equation of the form

$$\dot{\varepsilon} = \dot{\sigma}/E + \frac{\sigma}{\lambda}\left(\frac{\sigma}{\sigma^*}\right)^n \qquad \text{for} \quad \sigma > \sigma^* \tag{21}$$

where σ^* is a reference stress-intensity delimiting the non-linear range and $n > 0$ for $\sigma > \sigma^*$ and $n = 0$ for $\sigma < \sigma^*$, isothermal creep at

$T = T_0$ under constant stress of a bar is governed by the relation

$$\left(\frac{d\varepsilon}{dt}\right) = \frac{d\varepsilon}{dx} \cdot \frac{dx}{dt} = \frac{d\varepsilon}{dx} \bar{t}^{-1} = \frac{\sigma}{\lambda_0} \left(\frac{\sigma}{\sigma^*}\right)^n \qquad \sigma > \sigma^*. \tag{22}$$

If at $x = 0$ the temperature is rapidly increased by ΔT at a linear rate, the creep rate at any time $x > 0$

$$\left(\frac{d\varepsilon}{dt}\right)_T = \frac{d\varepsilon}{dx} \bar{t}^{-1} = \frac{\sigma}{\lambda} \left(\frac{\sigma}{\sigma^*}\right)^n = \frac{\sigma}{\lambda_\infty} e^{f(x)\ln\varphi} \left(\frac{\sigma}{\sigma^*}\right)^n. \tag{23}$$

The temperature increase therefore magnifies the creep rate at $T = T_0$ in the ratio

$$\nu(x) = \frac{\lambda_0}{\lambda_\infty} e^{f(x)\ln\varphi} = \frac{1}{\varphi} e^{f(x)\ln\varphi} = \varphi^{[f(x)-1]}.$$

Since $f(x) = 1$ and $x = 0$ and $f(x) \to 0$ as $x \to \infty$ the ratio $1.0 \leq \nu(x) \leq \varphi^{-1}$. The total creep between $x = 0$ and x according to eq. (23)

$$\varepsilon(x) = \int\limits_0^x \frac{d\varepsilon}{dx} = \bar{t}\,\dot{\varepsilon}_\infty \int\limits_0^x e^{f(\xi)\ln\varphi}\, d\xi = \dot{\varepsilon}_\infty \cdot \bar{t} \cdot F(x), \tag{24}$$

where $\dot{\varepsilon}_\infty$ is the steady state creep rate at $T = T_0 + \Delta T$ and $F(x) = \int\limits_0^x e^{f(\xi)\ln\varphi}\, d\xi$ has been evaluated for the parameters specified in Section 3 and is presented in Fig. 3. Since the creep function in which the transient effect is disregarded $\bigl(f(x) = 0\bigr)$ is simply

$$\varepsilon(x) = \dot{\varepsilon}_\infty \cdot \bar{t} \cdot x, \tag{25}$$

the ratio $[F(x)/x]$ is a measure of the magnification at x of the steady state creep by the thermal transient.

The effect of the transient becomes particularly significant if the increased temperature is not continuously sustained for a period x but intermittently applied for k shorter periods x_i such that $\sum\limits_1^k x_i = x$. In this case the magnification of the steady-state creep by the thermal transients is expressed by the sum $\left[\sum\limits_1^k F(x_i)/x_i\right]$, and the ratio between creep under intermittently applied and sustained temperature increase by $\left\{\left[x \sum\limits_1^k F(x_i)/x_i\right]/F(x)\right\}$. Thus, for instance, for $x = 20$ and $x_i =$

$= 0.1x = 2$ it follows from Fig. 3 that the magnification factor of the steady state creep for a single transient is about $[F(20)/20] \sim 14$, while for 10 intermittent temperature increases of duration $x = 2$ the factor is $10[F(2)/2] \sim 700$. Obviously, if much longer total and intermittent periods are considered the overall magnification is smaller.

Results of creep experiments on lead and steel under single and intermittent temperature increases [10] confirms the trends obtained from Fig. 3.

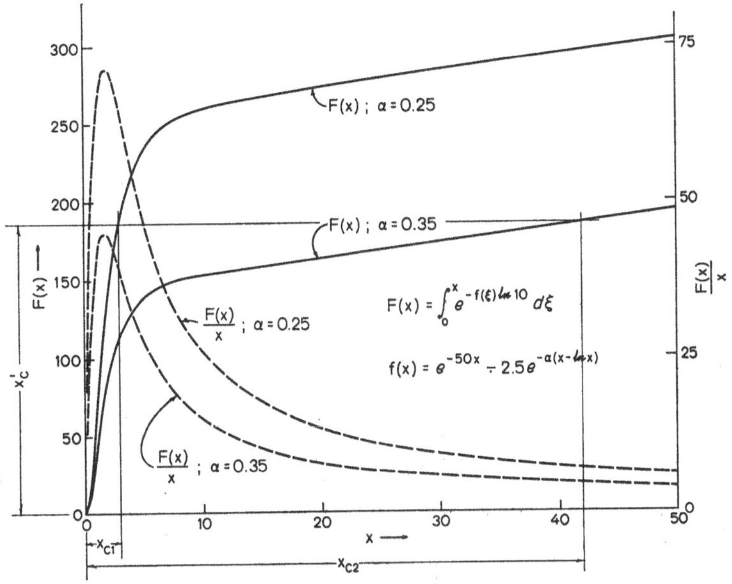

Fig. 3. Functions $F(x)$ and $F(x)/x$ according to eq. (24)

b) Stress Relaxation. The solution of eq. (21) for $\varepsilon = \varepsilon(0)$ and $d\varepsilon/dt = d\varepsilon/dx = 0$

$$-\frac{1}{E}\frac{d\sigma}{dx} = \bar{t}\,\frac{\sigma}{\lambda_\infty}\left(\frac{\sigma}{\sigma^*}\right)^n e^{f(x)\ln\varphi},$$

or (26)

$$\frac{d\sigma}{\sigma^{n+1}} = \frac{\bar{t}}{\tau_\infty}(\sigma^*)^{-n}\, e^{f(x)\ln\varphi}\, dx$$

where $\tau_\infty = \lambda_\infty E$, is of the form

$$\frac{1}{n}\sigma^{-n} = \frac{\bar{t}}{t_\infty}(\sigma^*)^{-n}\int_0^x e^{f(\xi)\ln\varphi}\, d\xi + C. \tag{27}$$

From the initial condition $\sigma = \sigma_0 = \varepsilon(0)$. E at $\lambda = 0$ the integration const $C = (n\sigma^n)^{-1}$. Hence

$$\sigma(x) = \sigma_0 \left[1 + n \frac{\bar{t}}{\tau_\infty} \left(\frac{\sigma_0}{\sigma^*}\right)^n F(x)\right]^{-1/n}. \tag{28}$$

Comparison of this expression with that obtained when the thermal transient is disregarded

$$\sigma(x) = \sigma_0 \left[1 + n \frac{\bar{t}}{\tau_\infty} \left(\frac{\sigma_0}{\sigma^*}\right)^n x\right]^{-1/n}, \tag{28a}$$

shows that the effect of a thermal transient on stress relaxation depends strongly on the ratio $(\bar{t}/\tau_\infty)$ and the exponent n. When the stress relaxation is very slow because $\bar{t}/\tau_\infty \ll 1$, the large difference between $F(x)$ and x for small values of x will have a rather small effect on $\sigma(x)$, while for very large values of x the ratio $[F(x)/x]$ tends towards one.

c) *Creep Expansion of Spherical Shell.* If a volume constant thin-walled spherical shell of constitutive eq. (21), radius R_0 and wall-thickness δ_0, under steady internal pressure p at temperature T_0 at which the creep rate is infinitesimal, is subject to a rapid linear temperature increase ΔT, the circumferential strain rate

$$\frac{d\varepsilon_\theta}{dx} = \bar{t}\, \frac{\sigma_\theta}{2\lambda} \left(\frac{\sigma_\theta}{\sigma^*}\right)^n = \frac{\sigma_\theta}{2\lambda_\infty} \left(\frac{\sigma_\theta}{\sigma^*}\right)^n e^{f(x)\ln\varphi} \tag{29}$$

where $\sigma_\theta = pR/2\delta$. Introducing the definition of strain $\varepsilon_\theta = \ln(R/R_0)$ and therefore

$$\frac{R}{R_0} = e^{\varepsilon_\theta}; \ \frac{R}{\delta} = \frac{R_0}{\delta_0} e^{3\varepsilon_\theta}; \ \sigma_\theta = \sigma_{\theta 0} e^{3\varepsilon_\theta}, \tag{30}$$

it follows that

$$\frac{d\varepsilon_\theta}{dx} = \bar{t}\dot{\varepsilon}_{\theta\infty}\, e^{3\varepsilon_\theta(n+1)}\, e^{f(x)\ln\varphi}, \tag{31}$$

where $\dot{\varepsilon}_{\theta\infty} = \sigma_\infty/2\lambda_\infty\, [(\sigma_{\theta 0}/\sigma^*)_n]$. Hence

$$e^{-3\varepsilon_\theta(n+1)}\, d\varepsilon_\theta = \bar{t}\,\dot{\varepsilon}_{\theta\infty} e^{f(x)\ln\varphi}\, dx, \tag{31a}$$

or

$$-\frac{1}{3(n+1)} e^{-3\varepsilon_\theta(n+1)} = \bar{t}\dot{\varepsilon}_{\theta\infty} \int_0^x e^{f(\xi)\ln\varphi}\, d\varphi + C. \tag{32}$$

From the initial condition $\varepsilon_\theta = 0$ at $x = 0$ follows

$$C = -[3(n+1)]^{-1},$$

and therefore

$$(R/R_0)^{-3(n+1)} = e^{-3\varepsilon_\theta(n+1)} = 1 - 3(n+1)\,\bar{t}\dot{\varepsilon}_{\theta\infty}\,F(x),$$

or

$$R/R_0 = [1 - 3(n+1)\,\bar{t}\dot{\varepsilon}_{\theta\infty}\,F(x)]^{-1/3(n+1)}. \tag{33}$$

The time x_c to creep instability is determined by the condition $R \to \infty$ or

$$F(x_c) = [3(n+1)\,\bar{t}\dot{\varepsilon}_{\theta\infty}]^{-1}, \tag{34}$$

while this time without consideration of the thermal transient is [11]

$$x_c' = [3(n+1)\,\bar{t}\dot{\varepsilon}_{\theta\infty}]^{-1}. \tag{34a}$$

Hence

$$R/R_0 = [1 - F(x)/x_c']^{-1/3(n+1)}, \tag{33a}$$

within the range $F(x) < x_c'$. The abscissa x_c associated with the value $F(x_c) = x_c'$ is the time to creep instability and can be obtained directly from Fig. 3. It is easily seen that the shorter x_c' the sharper its reduction due to the transient temperature effect.

If the temperature increase ΔT is applied intermittently for k shorter periods of duration $x_i = x/k$, the integral on the right side of eq. (32) takes the form $\bar{t}\dot{\varepsilon}_{\theta\infty} k F(x_i)$ and therefore eq. (33) becomes

$$R/R_0 = [1 - 3(n+1)\,\bar{t}\dot{\varepsilon}_{\theta\infty}\,k F(x_i)], \tag{35}$$

and the condition of creep instability

$$k F(x_i) = x_c' \quad \text{or} \quad F(x_c) = x_c'/k. \tag{36}$$

Thus even if x_c' is long, the effect of repeated transients will quite sharply reduce the time to creep instability. Thus, for instance, for $x_c' = 300$ the effect of the single transient starting at $x = 0$ will reduce this time according to Fig. 3 to $x_c = 46$, while the intermittent application of 10 temperature steps each of duration $x_i = 30$ is expected to produce a reduction to $x_c < 1.0$.

5. Conclusions

The results presented in Section 4 show that the effects of the transient creep acceleration produced by a temperature increase are quite significant even in the considered simple cases of uniform stressing. Those effects may be still more pronounced when a rapid uniform temperature increase is imposed on a non-uniform stress field, or when the imposed temperature increase produces a thermal gradient. Since the form of the function $f(x)$ depends not only on the extent of the thermal breakaway of pinned dislocations but also on their rate of repinning during creep, which is stress-dependent, these functions may be different for different points in the presence of stress gradients or temperature gradients or both.

If in the case of tension or pure bending a temperature gradient is rapidly induced, the well-known redistribution of stresses towards the cool fiber due to a stationary gradient [12] will be further intensified by the non-uniform transient creep acceleration. In the case of creep buckling of a strut the time dependent lateral deflection will be sharply accelerated during every temperature-increase; a number of intermittent temperature increases are therefore likely to reduce the time to failure to a fraction of the time estimated on the basis of steady-state creep.

However, in order to analyze these effects much more extensive observations of the transient creep acceleration due to temperature increase must be available. If this phenomenon is as pronounced as it has been found to be in the aluminum alloy investigated as well as in some steels, it appears that evaluation of the creep performance of metals for purposes of creep design will have to be based on the observation of creep rates under short or moderately long intermittent temperature increases rather than on the current observations of steady-state creep rates.

References

[1] STAVERMAN, A. J., and F. SCHWARZL: Linear Deformation Behavior of High Polymers. In: Physik der Hochpolymeren, **4**, 56–63. Berlin-Göttingen-Heidelberg: Springer. 1956.

[2] CONANT, F. S., G. L. HALL, and W. J. LYONS: Equivalent Effects of Time and Temperature in the Shear Creep and Recovery of Elastomers. J. Appl. Phys. **21**, 499 (1950).

[3] LILLIE, H. R.: Viscosity-Time-Temperature Relation in Glass At Annealing Temperatures. J. Amer. Ceramic Soc. **16**, 619 (1933).

[4] MCLEAN, D.: Grain Boundaries in Metals, 277. Oxford: Clarendon Press. 1957.

[5] Ibid. 198.

[6] DORN, J. E., and J. D. MOTE: Physical Aspects of Creep. In: High Temperature Structures and Materials, 109–118. London: Pergamon Press. 1964.

[7] TEUTONICO, L. J., A. V. GRANATO, and K. LUECKE: Theory of the Thermal Breakaway of a Pinned Dislocation Line with Application to Damping Phenomena. J. Appl. Phys. **35**, 220–231 (1964).

[8] Loc. cit. [6], 116.

[9] MURRO, R. P., and A. M. FREUDENTHAL: Creep Behavior of an Aluminum Alloy under Transient-Temperatures. Nuclear Eng. and Design **5**, 405–425 (1967).

[10] CARREKER, R. P., J. G. LESCHEN, and J. D. LUBAHN: Transient Plastic Deformation. Trans. AIME **180**, 139 (1949).
BROPHY, G. R., and D. E. FURMAN: Cyclic Temperature Acceleration of Strain in Heat Resisting Materials. Trans. Amer. Soc. Mech. Eng. **30**, 115 (1942).
NISHIHARA, T., S. TAIRA, K. TANAKA, and M. OHNAMI: Creep of Low Carbon Steel Under Varying Temperatures. Proc. First Jap. Cong. Test. Mat. 48 (1958).
[11] FREUDENTHAL, A. M.: Introduction to the Mechanics of Solids, 314. New York: J. Wiley and Sons. 1966.
[12] FREUDENTHAL, A. M.: Problems of Structural Design for Elevated Temperatures. Trans. New York Acad. Sci. **19**, 328 (1957).

Cyclic Thermal Loading in Viscoelastic Pressure Vessels

By

J. Hult

Gothenburg (Sweden)

Summary

Spherical and cylindrical vessels are considered, subjected to the simultaneous action of constant internal pressure, constant external temperature, and constant internal temperature with a superimposed cyclic temperature fluctuation. The vessel material is nonlinearly thermoviscoelastic with temperature dependent creep modulus. The influence of thermal frequency is studied for various cases of interest. A multimembrane analog of the real shell is shown to be useful in examining these effects.

1. Nomenclature

Symbols retaining their meaning through the paper are defined only once.

Symbol	Definition	Appears first on page
Section 3		
A	amplitude attenuation factor	106
a	area	105
E	Young's modulus	106
H	function	107
h	wall thickness	106
K	factor	108
L	length of bar	105
n	Norton's exponent (creep exp.)	106
P	loading force	105
T	temperature	106

Symbol	Definition	Appears first on page
$T_m = (T_{1m} + T_{2m})/2$		107
$T_d = T_{1m} - T_{2m}$		107
α	coefficient of thermal expansion	106
β	temperature coefficient of creep modulus	106
$\gamma = T_{1a}/T_d$		107
$\varDelta$	displacement	105
$\delta = T_d/T_m$		107
ε	strain	105
$\varepsilon_0 = P/3aE$		107
$\dot{\varepsilon}_0 = (P/3a\sigma_{n0})^n/\tau$		107
$\varTheta$	small number	106
θ	small number	105
$\varkappa$	thermal diffusivity	110
$\lambda = n\beta/T_m$		107
$\mu = \alpha T_m$		107
σ	stress	105
σ_n	creep modulus	106
τ	standard time unit	106
$\varPhi$	phase angle	106
ψ	phase angle	108
ω	angular frequency	106
$\omega_\infty = \varkappa/h^2$		110
$\varOmega = (h/2)\sqrt{\omega/2\varkappa}$		106

Section 4

Symbol	Definition	Appears first on page
a, b	inner and outer radii	112
F_0, F_1, F_2	functions	112
$g = n/2\nu\varepsilon_0$		115
K_s	factor	115
$M = (\alpha T_a/\varDelta)(\sigma_n/s)^n\tau$		114
p	pressure	112
r	radius	112
$s = pa/2h$	nominal hoop stress	113
$t' = t(s/\sigma_n)^n/\tau$		114
u	radial displacement	112
$\varDelta = h/a$		113
$\varepsilon_0 = s/2E$		114
$\nu = \omega(\sigma_n/s)^n\tau$		114
φ_1, φ_2	phase angles	114

Section 5

Symbol	Definition	Appears first on page
a_i	mean radius of i:th membrane	116
h_i	thickness . . .	116
u_i	radial displacement . . .	116
T_i	temperature . . .	116
$T_{i+1/2}$	temperature between i:th and $(i+1)$:th shell	116

Symbol	Definition	Appears first on page
w	axial displacement	116
L	length of shell	116
Section 6		
I_r	integral	117
δ_σ, δ_T	distances	118

2. Introduction

Pressure vessels in modern power plants often carry steam at such high pressures and temperatures, that significant creep deformations may develop. Smaller or larger, usually cyclic, temperature fluctuations are always present in such systems, causing a pulsating temperature field in the pressure vessel walls. This gives rise to cyclic thermal strains and also to cyclic changes in the temperature dependent constitutive parameters. As a result a stress redistribution occurs, which must be taken into account by the designer.

Thermal stresses in viscoelastic bodies have been studied by STERNBERG et al. in several papers [1—4]. These all deal with linearly thermoviscoelastic materials, considering both temperature independent and dependent mechanical response, and results of universal validity have been derived.

In this study thermally induced states of stress will be considered for spherical and cylindrical pressure vessels in the nonlinearly viscoelastic creep range. Due to this nonlinearity the governing equations become rather unwieldy, and a simplifying treatment is called for. A standard procedure is to replace the shell by a number of concentric or coaxial membrane shells. By such a discretization the continuous stress distribution through the shell wall is replaced by a finite number of membrane forces. Hence the nonlinear partial differential equation governing the original shell problem will be replaced by a finite number of ordinary differential equations. These may be solved for various cases of interest, and thereby some general insight into the mechanics of pressure vessels under cyclic thermal loading may be gained.

The main point in this study is the influence of cyclic temperature variations. The temperatures at the inner and outer shell walls are prescribed functions of time, and the temperature within the shell is governed by the law of heat conduction. For the multi-membrane shell, therefore, the layers intersecting the membranes are assigned the same heat conduction properties as the original shell material. The temperatures of the various membranes may then be determined by standard techniques, cf. CARSLAW and JAEGER [5].

All the main features of the multi-membrane shell subject to steady internal pressure and fluctuating internal temperature are present also in the one-dimensional counterpart shown in Fig. 1. It consists of an

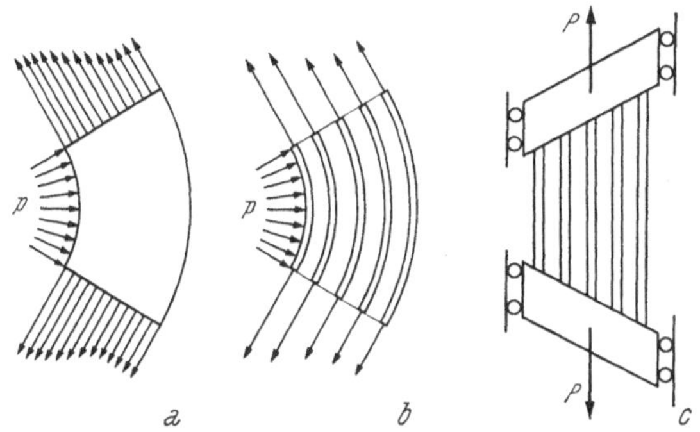

Fig. 1. *a* Real shell wall, *b* multi-membrane shell wall, *c* multi-bar model

assembly of parallel bars of varying length subject to a steady tensile force and a fluctuating temperature in the shortest bar. The temperatures in the other bars are governed by the heat conduction taking place in the heads. In the next section this system will be analyzed in some detail, and various conclusions will be drawn, which relate also to the original pressure vessel.

The following sections will then deal with thinwalled spherical shells, where a more direct analysis is possible, with multi-membrane shell model analyses, and with spherical cavities as a limiting case of interest.

3. Elementary Model

The simplest multi-membrane shell or multi-bar model which displays the particular thermo-inelastic properties of the real shell is of the type shown in Fig. 2 consisting of three members. The head, constrained to move without rotation, is loaded by a constant force P.

Equilibrium then requires

$$\sigma' + \sigma'' + \sigma''' = P/a, \tag{1}$$

and the strains in the three bars are

$$\varepsilon' = \Delta/L, \qquad \varepsilon'' = (1 - \theta)\Delta/L, \qquad \varepsilon''' = (1 - 2\theta)\Delta/L. \tag{2}$$

The bar material is assumed to be nonlinearly thermoviscoelastic with the constitutive equation

$$\dot{\varepsilon} = \dot{\sigma}/E + (\sigma/\sigma_n)^n/\tau + \alpha \dot{T}. \tag{3}$$

The material parameters E (Young's modulus), n (Norton's exponent) and α (coefficient of thermal expansion) are taken as constants for the rather moderate temperature fluctuations considered here. The creep modulus σ_n, however, is more strongly temperature dependent. Here will be used the empirical law

$$\sigma_n(T) = \sigma_{n0} \exp(-\beta T), \tag{4}$$

where σ_{n0} denotes the magnitude of σ_n at a reference temperature, and T denotes the temperature related to that reference.

The temperature T' of the shortest bar, corresponding to the inner wall of the pressure vessel, is assumed to vary harmonically around its mean level, i.e.

$$T' = T_{1m} + T_{1a} \sin \omega t, \tag{5}$$

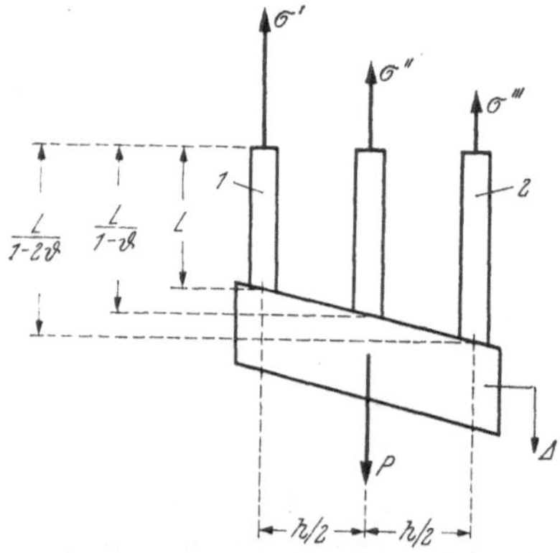

Fig. 2. Three bar model, cut at symmetry plane

while the temperature T''' of the longest bar, corresponding to the outer wall, is assumed to be constant

$$T''' = T_{2m}. \tag{6}$$

When all temperature transients have vanished, the temperature of the center bar will have the form

$$T'' = (T_{1m} + T_{2m})/2 + A\,T_{1a} \sin(\omega t + \Phi), \tag{7}$$

where the amplitude attenuation factor A and the phase lag angle Φ depend on the frequency ω, the heat conduction properties, and the distance between the bars.

The problem to be considered is now to study the stresses σ', σ'', and σ''' for various cases of interest. The equilibrium equation (1) is automatically satisfied by writing the stresses as

$$\begin{cases} \sigma' = (1 + \Theta_1) P/3a \\ \sigma'' = (1 - \Theta_1 - \Theta_2) P/3a \\ \sigma''' = (1 + \Theta_2) P/3a. \end{cases} \tag{8}$$

Introducing the dimensionless quantities ε_0, $\dot{\varepsilon}_0$, T_m, T_d, δ, γ, λ, and μ defined in Section 1, eqs. (2) through (8) yield, after linearization for small Θ_1, Θ_2, and δ

$$\begin{cases} \dot{\varepsilon}' = \varepsilon_0 \dot{\Theta}_1 + \dot{\varepsilon}_0 (1 + n\Theta_1)\,[1 + \lambda(1 + \delta/2 + \gamma\delta \sin \omega t)] + \omega\mu\gamma\delta \cos \omega t \\ (1 - \theta)\dot{\varepsilon}' = -\varepsilon_0(\dot{\Theta}_1 + \dot{\Theta}_2) + \\ \qquad + \dot{\varepsilon}_0(1 - n\Theta_1 - n\Theta_2)\,\{1 + \lambda[1 + A\gamma\delta \sin(\omega t + \Phi)]\} + \\ \qquad + \omega\mu A\gamma\delta \cos(\omega t + \Phi) \\ (1 - 2\theta)\dot{\varepsilon}' = \varepsilon_0 \dot{\Theta}_2 + \dot{\varepsilon}_0 (1 + n\Theta_2)\,[1 + \lambda(1 - \delta/2)]. \end{cases} \tag{9}$$

This linearization implies that the analysis will be restricted to moderate temperature fluctuations and moderate wall thicknesses, i.e. small θ-values. Taking the solutions for Θ_1 and Θ_2 as

$$\begin{cases} \Theta_1 = \Theta_{1m} + \Theta_{1a} \sin \omega t + \Theta_{1b} \cos \omega t \\ \Theta_2 = \Theta_{2m} + \Theta_{2a} \sin \omega t + \Theta_{2b} \cos \omega t \end{cases}$$

the constants $\Theta_{1m}, \ldots, \Theta_{2b}$ are obtained by insertion into (9). If, finally, the stresses are written as

$$\begin{cases} \sigma' = \sigma'_m + \sigma'_a \sin(\omega t + \psi') \\ \sigma'' = \sigma''_m + \sigma''_a \sin(\omega t + \psi'') \\ \sigma''' = \sigma'''_m + \sigma'''_a \sin(\omega t + \psi''') \end{cases}$$

the following mean stresses and stress amplitudes are obtained

$$\begin{cases} \sigma'_m = [1 + \theta(1 + \lambda)/n]\, P/3a \\ \sigma''_m = P/3a \\ \sigma'''_m = [1 - \theta(1 + \lambda)/n]\, P/3a \end{cases}$$

and

$$\begin{cases} \sigma'_a = (1 + \theta)\sigma_\infty \cdot H'(A, \Phi; \theta) \\ \sigma''_a = \sigma_\infty \cdot H''(A, \Phi) \\ \sigma'''_a = (1 - \theta)\sigma_\infty \cdot H'''(A, \Phi) \end{cases}$$

where

$$\sigma_\infty = \frac{1}{3} E \alpha T_{1a} \sqrt{\frac{\omega^2 + \dot{\varepsilon}_0^2 n^2 \beta^2/\alpha^2}{\omega^2 + \dot{\varepsilon}_0^2 n^2/\varepsilon_0^2}} \tag{10}$$

and

$$\begin{cases} H' = \sqrt{4 - 4A\cos\Phi + A^2 - 6\theta(2 - A\cos\Phi)} \\ H'' = \sqrt{1 - 4A\cos\Phi + 4A^2} \\ H''' = \sqrt{1 + 2A\cos\Phi + A^2}. \end{cases} \quad (11)$$

The phase angles have the magnitudes

$$\begin{cases} \psi' = \tan^{-1}[(A\sin\Phi)/(-2 + A\cos\Phi)] \\ \psi'' = \tan^{-1}[(2A\sin\Phi)/(-1 + 2A\cos\Phi)] \\ \psi''' = \tan^{-1}[(A\sin\Phi)/(1 + A\cos\Phi)]. \end{cases} \quad (12)$$

If no creep deformation were present, i.e. if $\dot{\varepsilon}_0 = 0$, the quantity σ_∞ in eq. (10) would reduce to

$$\sigma_\infty^{(e)} = \frac{1}{3} E\alpha T_{1a}$$

and the stress amplitudes in a linearly elastic bar system would therefore be

$$\begin{cases} \sigma_a^{(e)\prime} = (1 + \theta)\sigma_\infty^{(e)} \cdot H' \\ \sigma_a^{(e)\prime\prime} = \sigma_\infty^{(e)} \cdot H''. \\ \sigma_a^{(e)\prime\prime\prime} = (1 - \theta)\sigma_\infty^{(e)} \cdot H'''. \end{cases}$$

Accordingly, for all bars

$$\sigma_a = \sigma_a^{(e)} \cdot K, \quad (13)$$

where

$$K = \sqrt{\frac{\omega^2 + \dot{\varepsilon}_0^2 n^2 \beta^2/\alpha^2}{\omega^2 + \dot{\varepsilon}_0^2 n^2/\varepsilon_0^2}}. \quad (14)$$

Hence the stress amplitudes in a nonlinearly thermoviscoelastic system are obtained from those in a linearly thermoelastic system by multiplication with a factor K, given by eq. (14). At large frequencies, when

$$\omega \gg \omega_0 = \max[\dot{\varepsilon}_0 n\beta/\alpha,\ \dot{\varepsilon}_0 n/\varepsilon_0] \quad (15)$$

he factor K is close to unity, i.e. the stress amplitudes are then nearly qual to those in a linearly thermoelastic she ll, cf. Fig. 3. Characteristic data, pertaining to a standard carbon steel, would be of order

$\dot{\varepsilon}_0 = 10^{-5}(h^{-1})$; $n = 5$; $\beta = 10^{-2}(^\circ C^{-1})$; $\alpha = 10^{-5}(^\circ C^{-1})$; $\varepsilon_0 = 10^{-4}$

i.e. from eq. (15)

$$\omega_0 = \max\,[0.05, 0.5]\,h^{-1} = 0.5\,h^{-1}.$$

Hence the viscoelastic properties of the material are not being felt, if $\omega > 5\ (h^{-1})$, say, i.e. if the period of temperature fluctuations is shorter than one hour. From eqs. (13) and (14) follows that $\sigma_a < \sigma_a^{(e)}$ for all ω, if $\varepsilon_0 < \alpha/\beta \simeq 10^{-3}$, i.e. if the applied load is small. Hence large pressure in a pressure vessel will increase the thermal stress amplitudes, if creep deformation occurs.

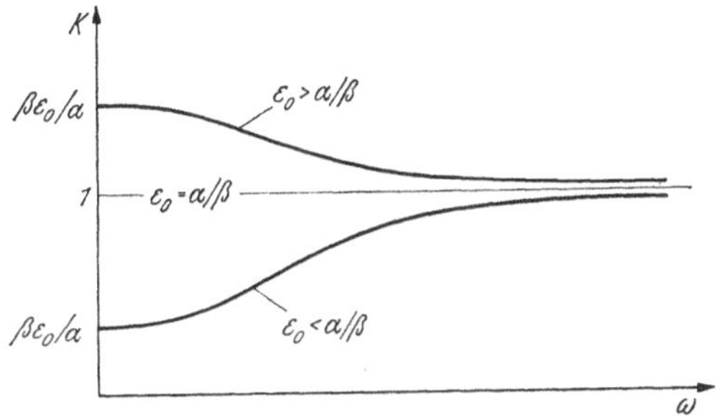

Fig. 3. Factor K, defined by eq. (14)

If the coefficient of thermal expansion were zero $(\alpha = 0)$, a thermal stress amplitude would still be found since eq. (10) then yields

$$\sigma_\infty = \frac{1}{3}\,E\,T_{1a}\,\frac{\dot{\varepsilon}_0 n\beta}{\sqrt{\omega^2 + \dot{\varepsilon}_0^2 n^2/\varepsilon_0^2}}.$$

This stress fluctuation is due to β, i.e. to the temperature dependence of the creep modulus σ_n.

If the frequency is very small $(\omega \to 0)$, eq. (10) yields

$$\sigma_\infty = \beta\,T_{1a}\,P/9a$$

i.e. the factor $E\alpha$ in eq. (10) then disappears. Hence at low frequencies the linearly thermoelastic stresses, proportional to $E\alpha$, relax completely due to the presence of creep. The remaining fluctuating stress is then again due entirely to the temperature dependence of the creep modulus σ_n.

To estimate the dependence of temperature distribution on the frequency ω, the head in Fig. 2 is considered as a slab of thickness h, with the surfaces maintained at the temperatures T' and T''', given by eqs. (5) and (6) respectively. The magnitudes of A and Φ, defined by eq. (7), are then, cf. CARSLAW and JAEGER [5], p. 105

$$\begin{cases} A = 1/\sqrt{2(\cosh 2\Omega + \cos 2\Omega)} \\ \Phi = -\tan^{-1}(\tanh\Omega + \tan\Omega) \end{cases}$$

where

$$\Omega = (h/2)\sqrt{\omega/2\varkappa}.$$

Numerical values of A and Φ as well as H', H'', and H''' defined by eqs. (11) and ψ', ψ'', and ψ''' defined by eqs. (12) are given in Table 1. A graphical representation of $A(\Omega)$ is given in Fig. 4.

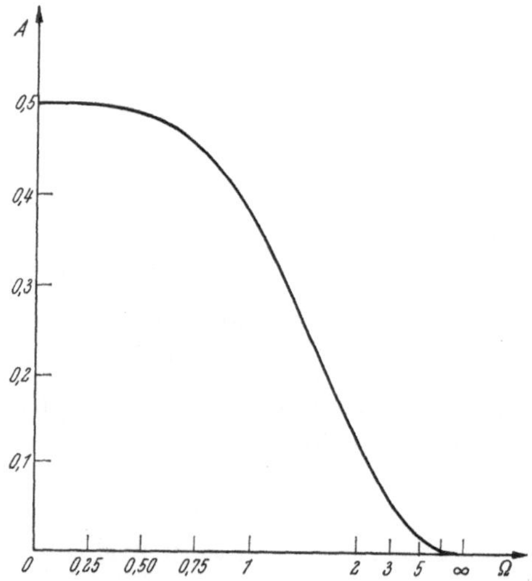

Fig. 4. Amplitude attenuation factor A, defined by eq. (7)

At low frequencies ω, corresponding to $\Omega \ll 1$, the states of temperature and stress differ very little from those in the static case $\omega = 0$, corresponding to $\Omega = 0$. If, in particular

$$\omega < \omega_\infty = \varkappa/h^2 \tag{16}$$

i.e.

$$\Omega < 1/2\sqrt{2}$$

then H' differs by less than 0.5% from its static value. Characteristic data for a standard carbon steel vessel would be

$$\varkappa = 5 \cdot 10^{-2}\,(\mathrm{m^2\,h^{-1}});\quad h = 2 \cdot 10^{-2}\,(\mathrm{m})$$

Table 1

Ω		0	0.250	0.500	1	2	3	5	10	∞
A		0.500	0.499	0.490	0.386	0.137	$5.0 \cdot 10^{-2}$	$6 \cdot 10^{-3}$	$5 \cdot 10^{-5}$	0
Φ		0	−0.063	−0.247	−0.871	−2.013	−3.005	−5.000	−10.000	$-\infty$
H'	$\theta = 0$	1.500	1.502	1.530	1.776	2.103	2.049	2.008	2.000	2
	$\theta = 0.1$	1.163	1.164	1.193	1.449	1.788	1.723	1.677	1.673	1.673
	$\theta = 0.2$	0.671	0.673	0.715	1.027	1.398	1.318	1.270	1.265	1.265
H''		0	0.057	0.244	0.776	1.409	1.098	0.996	1.000	1
H'''		1.500	1.499	1.480	1.283	1.034	1.049	1.002	1.000	1
Ψ'		$-\pi$	−3.121	−3.063	−2.974	−2.600	−3.133	−3.138	−3.141	$-\pi$
Ψ''		$\frac{1}{2}\pi$	−0.021	−0.081	−0.232	−0.130	−0.007	0.006	0.000	0
Ψ'''		0	1.561	1.365	0.841	0.212	0.012	−0.013	0.000	0

i.e. from eq. (16)

$$\omega_\infty = 125\ \mathrm{h}^{-1}$$

which is considerably larger than the lower limit ω_0 found previously. The fluctuation time corresponding to ω_∞ is ca. 3 min.

Hence, if ω satisfies the conditions

$$\max\,[\dot{\varepsilon}_0\, n\beta/\alpha,\ \dot{\varepsilon}_0\, n/\varepsilon_0] \ll \omega \ll \varkappa/h^2, \tag{17}$$

then the state of stress may be taken equal to that corresponding to a linearly thermoelastic material and a quasistatic state of temperature. Normal numerical data indicate that for a standard carbon steel vessel of 2 cm wall thickness the conditions (17) are fulfilled if the temperature fluctuation time is in the region from a few minutes up to an hour.

4. Thinwalled Spherical Shell

The thermoviscoelastic state of stress in a thinwalled spherical pressure vessel will now be analyzed. The loading consists of a constant internal pressure p, a harmonically fluctuating internal wall temperature $T_m + T_a \sin \omega t$, and a constant external wall temperature T_m. For simplicity all temperatures will be related to the constant mean temperature T_m, which may be taken as zero. With the wall thickness rather small, the condition (16) will be satisfied, and the temperature distribution may be taken as quasistatic. Hence with the notation of Fig. 5

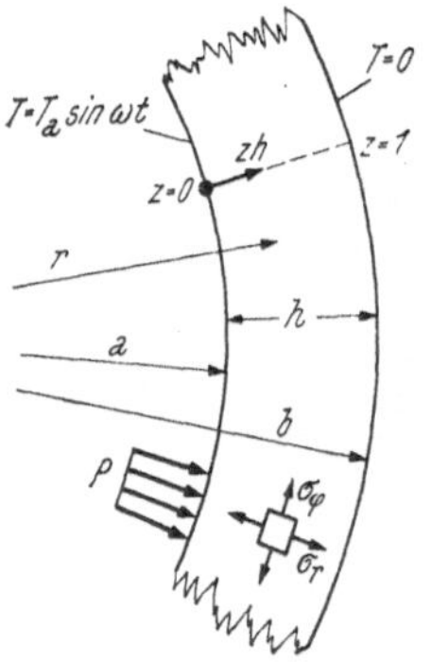

Fig. 5. Notation at thinwalled spherical shell

$$T(r, t) = [a(b - r)/(b - a)r]\,T_a \sin \omega t. \tag{18}$$

Assuming elastic incompressibility, the dilatation equation here takes the form

$$\partial u/\partial r + 2u/r = 3\alpha T(r, t). \tag{19}$$

From eqs. (18) and (19) follows after integration

$$u = F_0(t)\, a^2/r^2 + 3\alpha a\,[(b/2 - r/3)/(b - a)]\,T_a \sin \omega t$$

and hence the radial strain is

$$\varepsilon_r = \partial u/\partial r = -2F_0 a^2/r^3 - [\alpha a/(b - a)]\,T_a \sin \omega t.$$

Introducing here z and Δ as defined in Section 1, the following power series expansion is obtained

$$\varepsilon_r = -(2/a)\,[1 - 3\Delta z + 6\Delta^2 z^2 + O(\Delta^3)]\,F_0 - (\alpha T_a/\Delta)\sin \omega t. \quad (20)$$

Previous studies of creep in spherical shells, HULT [6] and SONNERUP [7], indicate that a suitable form for a similar power series expansion of σ_r would be

$$\sigma_r = -p[1 - z - \Delta z(1-z)\,F_1(t) - \Delta^2 z^2(1-z)\,F_2(t) + O_3],$$

where O_3 is an abbreviated form for $O(\Delta^3)$. Dividing through here by the nominal hoop stress

$$s = pa/2h = p/2\Delta$$

this takes the form

$$\sigma_r/s = -2\Delta(1-z) + 2\Delta^2 z(1-z)\,F_1 + 2\Delta^3 z^2(1-z)\,F_2 + O_4. \quad (21)$$

Radial equilibrium then requires for the hoop stress

$$\sigma_\varphi/s = 1 + \Delta(-2 + F_1 + 3z - 2F_1 z) + \Delta^2(3F_1 z - 4F_1 z^2 + \\ + 2F_2 z - 3F_2 z^2) + O_3. \quad (22)$$

In this spherically symmetric case the effective stress, defined by $\sigma_e^2 = (3/2)\,s_{ij}\,s_{ij}$ becomes

$$\sigma_e = \sigma_\varphi - \sigma_r \quad (23)$$

and hence, from eqs. (21) and (22)

$$\sigma_e/s = 1 + \Delta(F_1 + z - 2F_1 z) + \Delta^2(F_1 z - 2F_1 z^2 + 2F_2 z - 3F_2 z^2) + O_3. \quad (24)$$

This expression will later be inserted into the constitutive equation of the material.

From Hooke's law and the assumption of elastic incompressibility follows the radial elastic strain

$$\varepsilon_r^{(e)} = (\sigma_r - \sigma_\varphi)/E = -\sigma_e/E = -(s/E)\,[1 + \Delta(F_1 + z - 2F_1 z) + \\ + \Delta^2(F_1 z - 2F_1 z^2 + 2F_2 z - 3F_2 z^2) + O_3] \quad (25)$$

and from eq. (18) follows the radial thermal strain

$$\varepsilon_r^{(T)} = \alpha T(r,t) = [1 - z - \Delta(z - z^2) + \Delta^2(z^2 - z^3) + O_3]\,\alpha T_a \sin \omega t. \quad (26)$$

The effective creep strain, defined by $\varepsilon_e^{(c)2} = (2/3)\,\varepsilon_{ij}^{(c)}\,\varepsilon_{ij}^{(c)}$, may now be determined from eqs. (20), (25), and (26)

$$\begin{aligned}\varepsilon_e^{(c)} = -\varepsilon_r^{(c)} = -\,\varepsilon_r + \varepsilon_r^{(e)} + \varepsilon_r^{(T)} = (2/a)\,[1 - 3\Delta z + 6\Delta^2 z^2 + O_3]\,F_0 - \\ -(s/E)\,[1 + \Delta\,(F_1 + z - 2F_1 z) + \Delta^2 (F_1 z - 2F_1 z^2 + 2F_2 z - 3F_2 z^2) + \\ + O_3] + (\alpha T_a/\Delta)\,[1 + \Delta\,(1 - z) - \Delta^2 (z - z^2) + O_3]\sin\omega t. \quad (27)\end{aligned}$$

The relation connecting the effective stress according to eq. (23) and the effective creep strain according to eq. (27) is the Norton creep law in its scalar form

$$\dot{\varepsilon}_e^{(c)} = (\sigma_e/\sigma_n)^n/\tau, \tag{28}$$

where now σ_n is assumed to be temperature independent.

Introducing here the dimensionless quantities t', ε_0, M, and ν, defined in Section 1, and discarding O_3-terms, the following relation is obtained from eqs. (23), (27), and (28)

$$\begin{aligned}&(1 - 3\Delta z + 6\Delta^2 z^2)\,2F_0'/a - \\ &\quad - [\Delta\,(F_1' - 2F_1' z) + \Delta^2 (F_1' z - 2F_1' z^2 + 2F_2' z - 3F_2' z^2]\,2\varepsilon_0 + \\ &\quad + [1 + \Delta\,(1 - z) - \Delta^2 (z - z^2)]\,\omega M \cos\nu t' = \\ &= 1 + n\Delta\,(F_1 + z - 2F_1 z) + n\Delta^2 (F_1 z - 2F_1 z^2 + 2F_2 z - 3F_2 z^2) + \\ &\quad + \frac{1}{2}\,n\,(n - 1)\,\Delta^2 (F_1 + z - 2F_1 z)^2.\end{aligned}$$

Equating here equal powers of z, three differential equations are obtained for $F_0(t')$, $F_1(t')$, and $F_2(t')$.

With $\Delta \to 0$ they take the forms

$$\begin{cases} 2F_0'/a = 1 - \omega M \cos\nu t' \\ 4\varepsilon_0 F_1' + 2nF_1 = 6F_0'/a + n + \omega M \cos\nu t' \\ 6\varepsilon_0 F_2' + 3nF_2 = -12F_0'/a - 2\varepsilon_m F_1' - 2nF_1 + \\ \qquad + \dfrac{1}{2}\,n\,(n - 1)\,(1 - 2F_1)^2 - \omega M \cos\nu t' \end{cases}$$

which may be solved in succession. The stationary solutions are

$$\begin{cases} F_0 = F_0(0) + at'/2 - (a\omega M/2\nu)\sin\nu t' \\ F_1 = (n + 3)/2n - \left[\omega M/\sqrt{n^2 + 4\nu^2\varepsilon_0^2}\right]\sin(\nu t' + \varphi_1) \\ F_2 = -(2n^2 + 9n + 9)/6n^2 + \dfrac{\omega M\sqrt{(4n + 3)^2 + 196\nu^2\varepsilon_0^2}}{3(n^2 + 4\nu^2\varepsilon_0^2)}\sin(\nu t' + \varphi_2), \end{cases}$$

where

$$\begin{cases} \tan \varphi_1 = n/2\nu\varepsilon_0 \\ \tan \varphi_2 = (n^3 + 2n^2 + 52n\nu^2\varepsilon_0^2 - 24\nu^2\varepsilon_0^2)/(n^2 + 6n + 28\nu^2\varepsilon_0^2)\, 2\nu\varepsilon_0. \end{cases}$$

The following expressions are now obtained for the effective stress at the inner, middle and outer surfaces of the shell

$$\sigma_{e,i} = s[1 + \Delta F_1 + 0(\Delta^3)]$$

$$\sigma_{e,m} = s[1 + \Delta/2 + \Delta^2 F_2/4 + 0(\Delta^3)]$$

$$\sigma_{e,o} = s[1 + \Delta(1 - F_1) - \Delta^2(F_1 + F_2) + 0(\Delta^3)].$$

Hence in this first approximation the effective stress amplitude in the middle of the shell wall is negligible compared to the effective stress amplitude in the inner and outer surfaces, which is of magnitude

$$\sigma_{ea} = s\Delta F_{1a} = E\alpha T_a/\sqrt{1 + g^2}, \tag{29}$$

where

$$g = n/2\nu\varepsilon_0 = \dot{\varepsilon}_0 n/\omega\varepsilon_0.$$

Here

$$\dot{\varepsilon}_0 = (s/\sigma_n)^n/2\tau$$

denotes the rate of creep strain in the middle surface, and ε_0 as defined in Section 1, denotes the elastic strain in that surface.

If no creep deformation were present, i.e. if $\dot{\varepsilon}_0 = 0$, the effective stress amplitude in the shell surfaces, given by eq. (29) would reduce to

$$\sigma_{ea}^{(e)} = E\alpha T_a$$

and hence

$$\sigma_{ea} = \sigma_{ea}^{(e)} \cdot K_s,$$

where

$$K_s = 1/\sqrt{1 + \dot{\varepsilon}_0^2 n^2/\omega^2\varepsilon_0^2}.$$

This stress reduction factor is identical with the one found for the bar model, eq. (14), in case of temperature independent creep parameter ($\beta = 0$). Hence the same conclusions hold for the thinwalled spherical shell as for the bar model with respect to influence of frequency etc. A similar observation pertaining to a cylinder under uniaxial tension was made by MELLGREN [8].

If the pressure were zero, the stress amplitude would be equal to $\sigma_{ea}^{(e)}$ for all frequencies. If $n = 1$ (linearly thermoviscoelastic material), K_s is found to be independent of p, which reflects the fact that superposition is then permitted.

In conclusion, if

$$\dot{\varepsilon}_0 n/\varepsilon_0 \ll \omega \ll \varkappa/h^2$$

then the state of stress in a nonlinearly thermoviscoelastic, thinwalled pressure vessel may be taken equal to that corresponding to a linearly thermoelastic material and a quasistatic state of temperature.

5. Thickwalled Shells

The multi-membrane model, introduced earlier, will now be examined in some detail. For a spherical shell, the following relations hold, taking all membrane thicknesses equal ($h_i = h/N$)

a) equilibrium

$$\sum_1^N 2\pi a_i (h/N)\, \sigma_{\varphi i} = \pi a_1^2 p,$$

b) compatibility

$$4\pi a_{i+1}^2 u_{i+1} - 4\pi a_i^2 u_i = \frac{4}{3}\pi (a_{i+1}^3 - a_i^3)\, 3\alpha T_{i+1/2} \quad (i = 1, 2, \ldots, N-1),$$

c) constitutive

$$\dot{u}_i/a_i = \dot{\sigma}_{\varphi i}/6G + (\sigma_{\varphi i}/\sigma_n)^n/2\tau + \alpha \dot{T}_i \qquad (i = 1, 2, \ldots, N)$$

These are in all $2N$ equations relating the $2N$ unknowns $\sigma_{\varphi 1}, \ldots, \sigma_{\varphi N}$ and $u_1, \ldots, u_N$ to the given pressure p and temperatures $T_1, \ldots, T_N$. They have the same structure as eqs. (1)—(3) above with θ in Fig. 2 corresponding to $h/2a_1$, and hence the same conclusions are arrived at as for the multi-bar model.

For a cylindrical shell with closed ends the following relations hold

a) equilibrium

$$\sum_1^N 2(h/N)\, \sigma_{\varphi i} = 2a_1 p; \qquad \sum_1^N 2\pi a_i (h/N)\, \sigma_{xi} = \pi a_1^2 p,$$

b) compatibility

$$2\pi a_{i+1} u_{i+1} (1 + w/L) - 2\pi a_i u_i = 2\pi (a_{i+1} - a_i)\, 3\alpha T_{1+1/2}$$
$$(i = 1, 2, \ldots, N-1),$$

c) constitutive

$$\begin{cases} \dot{u}_i/a_i = (2\dot{\sigma}_{\varphi i} - \dot{\sigma}_{xi})/6G + (2\sigma_{\varphi i} - \sigma_{xi})^n/2\sigma_n^n \tau + \alpha \dot{T}_i \\ \dot{w}/L = (2\dot{\sigma}_{xi} - \dot{\sigma}_{\varphi i})/6G + (2\sigma_{xi} - \sigma_{\varphi i})^n/2\sigma_n^n \tau + \alpha \dot{T}_i \quad (i = 1,2, \ldots, N). \end{cases}$$

These are in all $3N + 1$ equations relating the $3N + 1$ unknowns $\sigma_{\varphi 1}, \ldots, \sigma_{\varphi N}; \sigma_{x1}, \ldots, \sigma_{xN}; u_1, \ldots, u_N$, and w to the given pressure p and temperatures $T_1, \ldots, T_N$. Again similar conclusions hold as before.

6. Spherical Cavities

The case of a spherical shell with infinite wall thickness, i.e. the spherical cavity in an infinitely extended medium will now be considered. For the linearly thermoviscoelastic medium such problems have been studied at length by STERNBERG, loc. cit. The multi-membrane model may be used here also, but the case is more accessible to direct analytical solution than shells with finite wall thickness. Two limiting cases will be studied, corresponding to very low and very high frequency of temperature variation.

I. Quasistatic Temperature Field

The following relations hold, assuming incompressible nonlinearly thermoelastic material

a) equilibrium

$$\partial \sigma_r / \partial r + 2(\sigma_r - \sigma_\varphi)/r = 0,$$

b) compatibility

$$\partial \varepsilon_\varphi / \partial r + (\varepsilon_\varphi - \varepsilon_r)/r = 0,$$

c) constitutive

$$\begin{cases} \varepsilon_r = (\sigma_r - \sigma_\varphi)^n / \sigma_n^n + \alpha T \\ \varepsilon_\varphi = -(\sigma_r - \sigma_\varphi)^n / 2\sigma_n^n + \alpha T, \end{cases} \tag{30}$$

where

$$T = (a/r)\, T_a \sin \omega t.$$

In absence of internal pressure the boundary conditions are

$$\sigma_r(a, t) = \sigma_r(\infty, t) = 0.$$

If n is an odd integer, eqs. (30) hold for arbitrary values of σ_r and σ_φ, and integration is then possible. The following equations result

$$\begin{cases} \sigma_r = \sigma_n (\alpha T_a)^{1/n} \cdot 2 I_r \sin \omega t \\ \sigma_\varphi = \sigma_n (\alpha T_a)^{1/n} \left[2 I_r + (r/a)^{-3/n} (r^2/a^2 - C)^{1/n}\right] \sin \omega t, \end{cases}$$

where

$$I_r = \int_1^{r/a} x^{-1-3/n} (x^2 - C)^{1/n} dx$$

and where C is determined by the condition at infinity

$$I_\infty = 0.$$

The effective stress is largest at the cavity wall and is found to be

$$\sigma_e(a, t) = 2.52\,\sigma_n(\alpha T_a)^{1/n}\sin(\omega t + \pi)$$

for all $n \geq 3$.

II. Rapid Temperature Fluctuation

If the frequency ω is very high, the thermal stresses will be identical with those in a linearly thermoelastic medium, cf. Sections 3 and 4 above. The circumferential stress is then

$$\sigma_\varphi = \left(r^{-3}\int_a^r T\varrho^2 d\varrho - T\right) E\alpha/(1-\nu)$$

and hence

$$(\partial\sigma_\varphi/\partial r)_a/\sigma_{\varphi a} = (\partial T/\partial r)_a/T_a - 1/a$$

or, referring to Fig. 6

$$\delta_\sigma < \delta_T$$

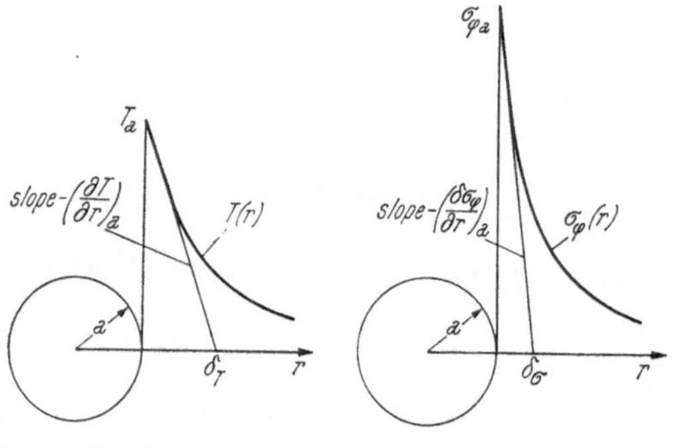

Fig. 6. Skin-effect for temperature and stress at spherical cavity

i.e. the stress distribution $\sigma_\varphi(r)$ will be steeper than the temperature distribution close to the cavity wall. Since at high frequensies the temperature distribution itself will be steep at the cavity wall (analogy to electrodynamical skin-effect), only a very narrow band will be subject to high thermal stresses. A similar effect in cylinders subject to tensile loading was previously observed by MELLGREN, loc. cit.

Acknowledgement

The numerical calculations for this paper were performed by B. Sundstrom.

References

[1] Sternberg, E.: Transient Thermal Stresses in an Infinite Medium with a Spherical Cavity. Kon. Nederl. Akad. Wetensch. Proc. B. **60** (5), 396–408 (1957).
[2] Sternberg, E., and R. Muki: On Transient Thermal Stresses in Viscoelastic Materials with Temperature-Dependent Properties. J. Appl. Mech. **28** (2), 193–207 (1961).
[3] Sternberg, E., and J. G. Chakravorty: Thermal Shock in an Elastic Body with a Spherical Cavity. Quart. Appl. Math. **17** (2), 205–218 (1959).
[4] Sternberg, E.: On the Analysis of Thermal Stresses in Viscoelastic Solids. Third Symposium on Naval Structural Mechanics 1963, Proc. 348–382, London: Pergamon Press. 1963.
[5] Carslaw, H. S., and J. C. Jaeger: Conduction of Heat in Solids, 2nd ed. Oxford: University Press. 1959.
[6] Hult, J.: Primary Creep in Thickwalled Spherical Shells. Trans. Chalmers Univ. of Techn. Göteborg 1962.
[7] Sonnerup, L.: Temperature Effects in Connection with Primary Creep in a Spherical Pressure Vessel. Conf. Thermal Loading and Creep in Structures and Components, IME Proc. 1963–64, **178**, part 3L. London 1964.
[8] Mellgren, S. A.: Measuring Accuracy in Creep Tests. WADC, TR 59-78, part II, 1959.

Discussion

Odqvist: I want to underline the result, stated by the speaker, that the skin effect of a spherical cavity heated periodically from within, is much more accentuated with respect to stress than to temperature. This agrees very well with the result found by Mr. N. G. Ohlson and myself when studying periodical thermal stress in a testbar of triangular cross section, periodically heated at a point of the edge generatrix. This tendency may thus be universal.

The Dynamic Blanking, Forging, Indenting and Upsetting of Hot Metals

By

W. Johnson and **R. A. C. Slater**

Manchester and Salford (U. K.)

Summary

General accounts of the processes described in the title are given, but special attention is paid to the force and energy requirements for effecting them. The consequences of the processes being impulsive and therefore fast as against otherwise identical, squeezing or static processes, are brought out. The role played by working temperature is demonstrated, and especially the importance attaching to the temperature which is about one half the melting temperature.

Thermal discontinuities in the processing of solids are described and other features encountered when investigating the specific processes are discussed.

Notation

A	current cross-sectional area during dynamic compression
A_0	original cross-sectional area of specimen
c	specific heat
d	permanent indentation crater diameter
H	current height of specimen during dynamic compression
H_0	original height of specimen
$e = (H_0 - H)/H_0$	compressive engineering strain at any instant
e_F	final compressive engineering strain
$\dot{e} = de/dt$	compressive engineering strain-rate at any instant
$\dot{e}_0$	initial compressive engineering strain-rate
F	dynamic uni-axial force exerted on specimen
J	mechanical equivalent of heat
k	yield shear stress
M	mass of tup
M_p	mass of projectile
m_p	mass of plunger
$M' = (M + m_p)$	total mass of system after impact
$\overline{P}_d$	mean effective dynamic indentation pressure
P_s	quasi-static indentation pressure

P_0	extrapolated value of quasi-static indentation pressure at absolute zero temperature
s	length of a line of tangential velocity discontinuity
T	absolute temperature °K
T_M	absolute melting temperature °K
T/T_M	homologous temperature
t	time from instant of impact
t_F	time from instant of impact to effect final strain, e_F
$u = Mv/M'$	velocity at commencement of compression
v	impact velocity
v_N	component velocity in a direction normal to a line of tangential velocity discontinuity
V^*	magnitude of tangential velocity discontinuity
Y	uni-axial yield stress
σ_D	dynamic true compressive stress
$\bar{\sigma}_D$	mean dynamic yield stress
σ_S	quasi-static true compressive stress
$B, C = B/\sigma_S, K$	various constants
m, n	various indices
α	$M'H_0/A_0\sigma_S$
γ, β, a not α	are constants depending on frictional behaviour between the projectile and the indented material
λ	temperature coefficient for the material
η	$2K/\alpha\dot{e}_0^2$
$\Delta\theta$	magnitude of temperature jump at a line of thermal discontinuity

Introduction

Interest in the hot working of metals has grown considerably in recent years because of the need to work unusual metals, often at relatively high temperatures, and through the introduction of new, fast deformation processes. On the one hand, newer metals have been exploited by the aerospace industry and on the other there have been the development of explosive forming techniques [1] and attempts to introduce high speed forging machines such as the Dynapak [2] and the Petroforge [3]. Because speed is so important a factor in these latter processes, attention has been directed towards the effects of strain rate on the yield stress of a metal and hence on the forces and energy necessary to secure a given degree of deformation. It has become clear that when working "hot" metals the effect of strain rate (which is related to the speed of working) on yield or flow stress is very marked; this phenomenon does not appear to be widely appreciated in the field of engineering plasticity. To increase awareness and to add to knowledge on the topic, the authors have been involved in several metal processing investigations which have aimed to show just how important is this interdependence between stress-strain rate and temperature.

We have summarised below much of the work carried out at Manchester in recent years and we try to demonstrate,

(i) The importance of recognising that changes in strain rate, at temperatures which exceed about one half the melting temperature (on the absolute scale) of the metal being worked, can alter the flow stress of a material significantly—often by several hundred per cent and sometimes by an order of magnitude; this fact is most clearly emphasised by comparing the "static" yield stress with a dynamic yield stress at a given temperature.

(ii) That fairly sharply defined regions may exist during the fast working of a metal in which the temperature may reach (with possibly deleterious consequences) a level much higher than may otherwise have been expected; this is due to the mode of plastic deformation.

This paper is divided into five sections of unequal length and the two phenomena referred to above are discussed mainly in the first two, on Blanking and Forging. In the remaining three sections we describe, principally, the determination of yield or flow stresses in "hot" impact operations though the topics themselves are discussed in wider terms, e.g. reference is included to the modes of metal deformation encountered and specific results are presented.

It is to be noted that whilst impact operations are discussed it is only in Section 5 on Projectile Impact that wave effects are thought to become important—though even there it was necessary to adopt a quasi-static treatment. We discuss mainly "mechanical" problems, but it should be remembered that in recent years much attention has been given to analogous problems in the field of structural plasticity [4], especially in the application of constitutive equations to low speed impact on hot structural members. The importance of the two focal points referred to above is still, however, not generally recognised even in the latter area of research, and thus our observations may have value to structural engineers. It should also be added that when referring to "static" operations below, the order of strain rate is 10^{-3}/sec, and the time in which processing occurs is so short that ordinary creep effects are negligible.

It is fitting to point out that some of the first work in the effect of strain rate and temperature on flow stress is that of Nadai and Manjoine [5] and some of the results we describe are clearly discernible in their early results. Our contribution has been to explore these in more detail, to point to specific facts and to have sought corroboration of this basic knowledge in actual forming processes.

1. Blanking

The Process

Modern blanking practice aims at producing (a) the finished product with no subsequent machining operations being required, or (b) components for ultimate fabrication as in "stacked stamping" which may be regarded in certain cases as an alternative to casting or otherwise lengthy and more expensive machining operations, or (c) blanks for further processing such as deep drawing.

The blanking process consists essentially of shearing a part of the desired geometry from a sheet material using a punch and a die assembly which are machined so as to produce the required shape for the blanked part. To enable this operation to be performed an operating clearance must be established between the punch and the die assembly. Industrial blanking operations are normally carried out in hydraulic presses or crank presses at punch speeds up to about 15 ft/s depending upon the size of the part to be blanked and the material used. In the blanking operation the main interest with regard to product quality is obviously focussed on the blanked part. However, in a piercing operation the shaped hole which

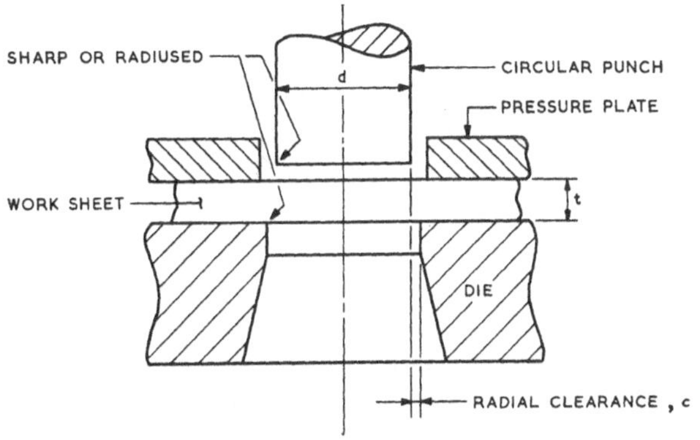

Fig. 1. Diagrammatic arrangement for axi-symmetric blanking

remains in the work sheet after the punch penetration is of prime importance. The difference between these two associated processes is governed mainly by the tool geometry. In numerous manufacturing processes, parts are simultaneously blanked and pierced to produce the desired components.

Many presses used for blanking operations are designed as multi-action presses, incorporating in addition to a punch and die assembly, an ejector for removing the blanked part free from the punch and a pressure plate or pad to securely clamp the work sheet during the process. However, in an elementary consideration of the mechanics of the blanking process it will be sufficient to refer to Fig. 1 which shows a diagrammatic arrangement for the axi-symmetric blanking of a circular disc from a sheet of metal using a cylindrical punch and a die of the circular orifice type. The blanking of circular discs from metal strip or sheet is a complex non-steady state process involving a tri-axial stress system, in some zones being compressive and in others tensile. It involves the strain hardening properties of the material, the tool geometry and such factors as speed, temperature, the punch-die clearance and the extent of punch penetration. The process of producing a blank also involves both plastic deformation and fracture phenomena.

Punch Force—Punch Displacement Characteristic

As the blanking operation proceeds a proportional increase in punch force with punch displacement occurs during the initial elastic phase. If the punch penetration were stopped at the limit of this phase, some elastic recovery would occur and no detectable penetration would be observed upon microscopical examination of the blank. Further penetration of the punch results in plastic shear deformation and the magnitude of the punch force changes with the punch displacement as affected by the strain-hardening characteristics of the material. The maximum blanking force is attained at some fraction of the thickness of the work sheet and although further strain-hardening of the material may occur, the punch force decreases as the shear area is reduced. At this stage of the process, cracks are formed which are initiated at the die edge and propagate towards the punch; similar cracks are propagated from the punch edge. As the punch does not follow the cracks it is probable that secondary cracks are formed as the punch is further displaced. No further plastic deformation occurs and the punch force decreases rapidly. Up to this stage, the blanking force is thus determined by the current shear stress of the material and the frictional resistances present. Beyond this stage, frictional conditions are the determining factor.

The process is further complicated by the effect of strain-rate upon the flow stress of the material depending upon the punch speed and also by temperature. If blanking operations are performed at very high speed, then elastic and plastic stress wave propagation through the blanked material and tools assumes greater importance.

Nature of the Sheared Edge

Three zones can usually be identified as making up the sheared edge of the blank. These are

1. The periphery of the bottom of the blank which is usually slightly rounded. This zone is usually very small in extent and of little interest except when large punch-die clearances are employed. Large bending stresses in relation to the shear stress will result in a kind of drawing operation. High tensile stresses occur at the punch and die radii.

2. The next zone, is a smooth or burnished one which comprises a considerable fraction of the blank thickness and results from plastic flow.

3. A fracture zone which usually tapers and is irregular.

The process of "Fine Blanking" introduced a few years ago, aims at producing a smooth edge profile to the blanked part and involves using very small punch-die clearances. Higher compressive hydrostatic stresses are produced than in the conventional blanking process so that a critical tensile energy level is not reached and the onset of fracture is suppressed.

Axi-Symmetric Quasi-Static Blanking at Elevated Temperatures

The authors [6] have recently studied the effects of temperature and speed on the force and energy requirements for blanking. Nominal 1 in. diameter circular discs were blanked quasi-statically at a mean shear strain rate of about $10^{-3}\,s^{-1}$ using an axi-symmetric punch and die assembly. Each blanking operation was performed as a compression test in an universal testing machine equipped with an autographic recorder specially arranged to obtain punch load—punch displacement diagrams. The materials investigated were 1/4 in. nominal thickness commercially pure aluminium (B. S. 1470 SIC) and copper (B. S. 1432) and 3/16 in nominal thickness black mild steel (En 2) at temperatures of up to 773 °K, 1073 °K and 1373 °K respectively.

The Ratio $F_{\max}/A_c\,\sigma_{(\varepsilon=0.5)}$

The experimental values of the maximum quasi-static blanking force, $F_{\max}$, corresponding to various values of percentage radial clearance[1] were compared with the product of the nominal sheared circumferential area A_c, of the blank and the true compressive stress of the material

---

[1] The non-dimensional percentage radial clearance is defined as

$$\frac{\text{radial clearance between punch and die}}{\text{thickness of specimen sheet}} \times 100$$

$\sigma_{(\varepsilon=0.5)}$ at a constant natural compressive strain, $\varepsilon = 0.5$. The ratios so obtained corresponded to different homologous temperatures (T/T_M) for the three materials and in nearly all cases were found to be approximately 0.6.

The results thus revealed that the maximum quasi-static blanking force at elevated temperatures ($0.16 \leq T/T_M \leq 0.79$) for the three materials can be estimated to within ± 15 per cent by an empirical equation:

$$F_{\max} = 0.6 A_c \sigma_{(\varepsilon=0.5)}, \qquad (2 < d/t < 6) \tag{1}$$

However, low values of the coefficient in eq. (1) were actually obtained for the aluminium and copper at the highest temperatures. The coefficients in these cases were of about the same value, and also corresponded to, about the same value of homologous temperature. It is therefore possible that these low values of the coefficient are related to visco-plastic effects which may become significant at homologous temperatures exceeding about 0.8.

Dependence of Maximum Quasi-static Blanking Force on Percentage Radial Clearance and Temperature

The maximum quasi-static blanking force determined from autographic punch load—punch displacement diagrams was found to be nearly independent of percentage radial clearance for the three materials at the various temperatures and to decrease with increase in temperature.

The relationship between the maximum quasi-static blanking force and temperature shows for aluminium, copper and the mild steel that the maximum blanking force, although substantially independent of percentage radial clearance, in each case is highly dependent on temperature. A single inflexion occurred at an approximate temperature of 200 °C for the aluminium. A similar single inflexion occurred at a temperature of about 400 °C in the comparable relationship for the other relatively pure metal, copper. However, in the case of the mild steel, two inflexions occurred in the curve at approximate temperatures of 500 °C and 800 °C. The single inflexions in the curves for both the aluminium and copper are associated with the recrystallization of the aluminium and copper respectively. The two inflexions in the similar curve for the mild steel are associated with (a) the recrystallization of α iron and (b) the A_3 phase transformation when α iron (ferrite) having a body centred cubic structure becomes γ iron (austenite) having a face-centred cubic structure. For the black mild steel En 2, (0.15—0.2% C), this transformation would be expected to occur at a temperature of between 830–850 °C as determined from the iron-carbon thermal equilibrium diagram.

The temperatures at which these inflexions occur become more evident in a logarithmic relationship as shown in Fig. 2 for the aluminium in which the recrystallization temperature is found to be 235 °C or 0.55 T/T_M.

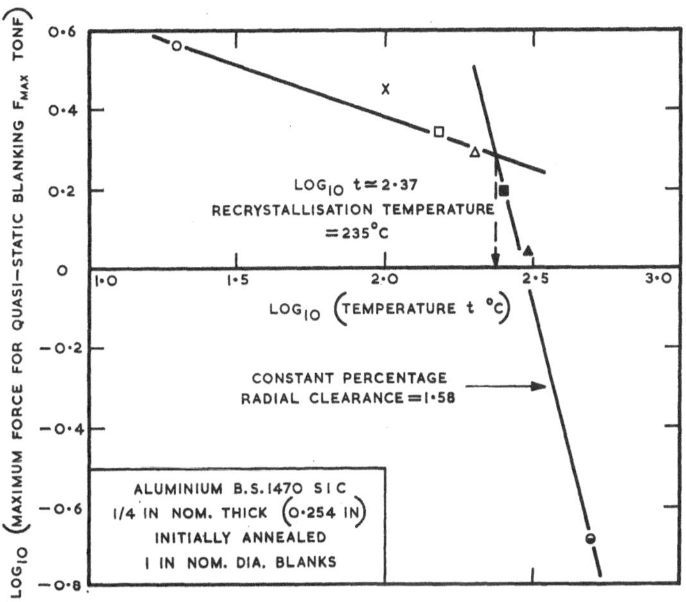

Fig. 2. Logarithmic relation between maximum force for quasi-static blanking and temperature a constant percentage radial clearance (Aluminium B. S. 1470 SIC)

Dynamic Blanking

In addition to the quasi-static blanking already referred to above, the authors have also conducted experiments in the dynamic blanking of aluminium, copper and black mild steel at elevated temperatures. These blanking operations were effected using the same blanking tool assembly as that used for the quasi-static blaking experiments. In this dynamic process the mean plastic shear strain rate was estimated to be in the range 500–6000 s^{-1}.

To provide the necessary kinetic energy to perform the fast blanking operations, a linear induction motor was used to accelerate a mass of $21^1/_2$ lb. Impact velocities approaching 50 ft/s were attained which could be varied by adjusting the phase voltage across the motor by means of a 3 phase multi-tap auto-transformer.

Comparison Between the Energy Required for Quasi-Static and Dynamic Blanking at Elevated Temperatures

If this comparison is made for the relatively pure metals, aluminium and copper, then similar relationships are obtained to that shown in Fig. 3 which is for the aluminium. This graph shows that greater energy is required for the dynamic process and that an inflexion is exhibited in the curve for the dynamic process similar to that obtained for the quasi-static process. For both the aluminium and copper the difference between the energy required for the dynamic and quasi-static processes remains sensibly constant but reduces slightly in the region of the recrystallization temperature. From Fig. 3 for the aluminium this difference in energy is approximately 105 ft lbf at 20 °C, 87.4 ft lbf at 250 °C and 98 ft

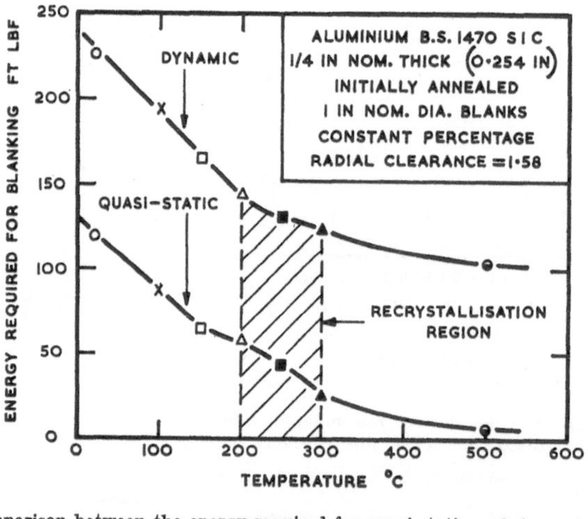

Fig. 3. Comparison between the energy required for quasi-static and dynamic blanking at elevated temperatures (Aluminium B. S. 1470 SIC)

lbf at a temperature of 500 °C. Since Fig. 3 corresponds to the least value of percentage radial clearance it can be assumed that the mean shear strain and also the mean shear strain rate during the operations were sensibly constant for the dynamic process over the entire temperature range. The dynamic yield shear stress of the material thus *appears* to have an approximately constant value during the process except as affected by recrystallization. However, above the recrystallization temperature it might have been expected that the dynamic yield shear stress would be reduced on the assumption that recrystallization proceeds. It can therefore, be inferred that the process time which was of the order of 2 to 4 ms is less than that required for the recrystallization reaction.

In the case of the aluminium it can be assumed that negligible thermal recovery takes place at high shear strain rates. Reference [6] shows that the results for the copper are similar to these in most respects except that there is evidence of partial thermal recovery at temperatures in excess of the recrystallization temperature for the metal. The mechanism is more complicated for the mild steel as it is affected not only by the recrystallization of the α iron but also by the A_3 phase transformation.

Strain-Rate Sensitivity at Elevated Temperatures

The strain-rate sensitivities of the three materials as influenced by temperature are typified in non-dimensional relationships such as Fig. 4 which is for the aluminium. For this material the ratio E_D/E_S is

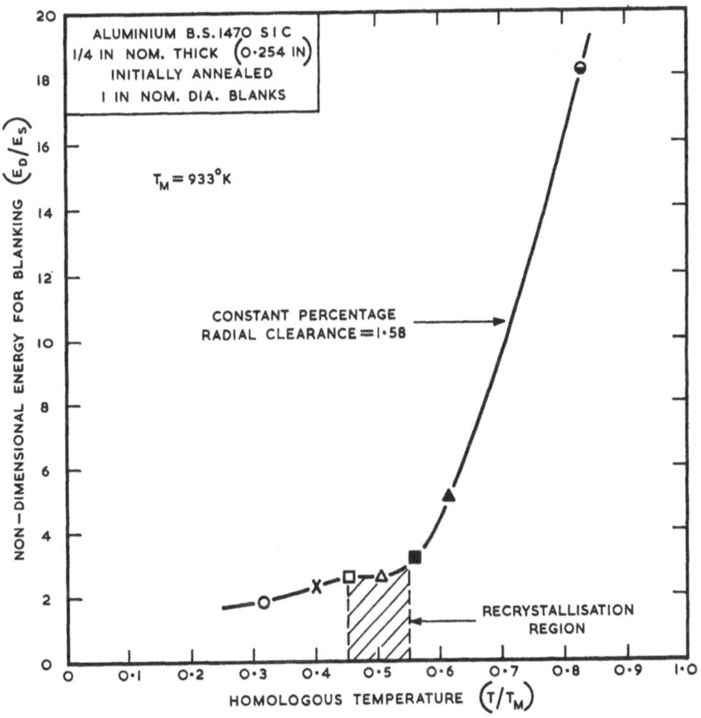

Fig. 4. Relation between non-dimensional energy for blanking (E_D/E_S) and homologous temperature (T/T_M) at constant percentage radial clearance (Aluminium B. S. 1470 SIC)

18 at the homologous temperature of 0.83, while Ref. [6] shows that the values are 8.3 for the copper at $T/T_M = 0.83$ and about 9.5 for the mild steel at $T/T_M = 0.76$. As can be seen in Fig. 4 which applies to the alu-

minium, it was found for the three materials investigated that the strain-rate sensitivities are not very significant at temperatures below the recrystallization temperature (i.e. $T/T_M \simeq 0.5$) but are considerable at temperatures above the recrystallization temperature. Inflexions associated with either recrystallization or a phase transformation, as in the case of the mild steel, were again evident in relationships such as that presented in Fig. 4.

Mode of Deformation at Very High Speed

The photograph of Fig. 5 from the investigations reported by JOHNSON and TRAVIS [7] shows 1/4 in. nominal thickness copper specimens which were blanked at room temperature and at increasing punch speeds with a 3/8 in. diameter punch at a constant radial clearance of 10 per cent, except for specimen No. 8 where the die was removed.

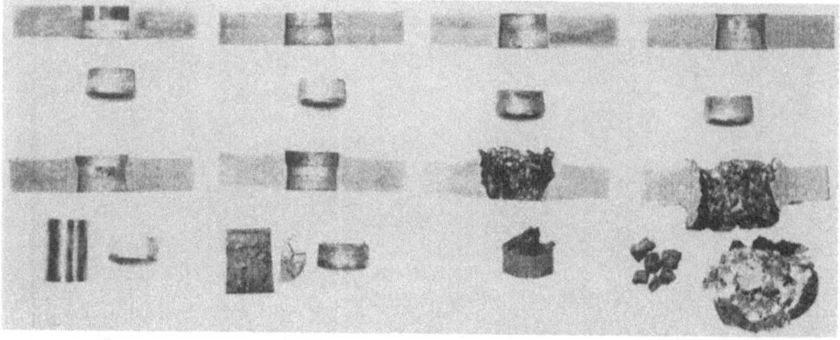

Fig. 5. Photograph of copper specimens blanked at increasing punch velocity. Velocities in ft/s, 1 : 0, 2 : 252, 3 : 438, 4 : 793, 5 : 930, 6 : 1500, 7 : 2100, 8 : 2350 (die removed)

Specimen No. 1 was blanked quasi-statically and Nos. 2, 3 and 4 are specimens blanked dynamically using least energy with punch lengths of 3, 1 and $^1/_2$ in. respectively. Specimen Nos. 5 to 8 respectively were blanked dynamically using a punch of 1/2 in. length where the speed of the punch was increasingly in excess of that required to achieve successful blanking. For the conditions prevailing during these investigations, it is evident that there was an upper limit to the impact speed, i.e. of about 600 ft/s, when the mode of deformation changed and cratering was produced which rapidly increased in severity. At about this same speed the punches were also found after impact to have suffered swelling at the impact end. Corresponding to the strain rates involved in these experiments, the plastic wave speed in copper can be shown to be slightly less than 600 ft/s. It may therefore be conjectured that if the punch moves at a speed of about this magnitude, the plastic material ahead of the punch will be displaced sideways and upwards to cause a "coronet". It will be moved sideways out of the way of the punch and in the direction of least resistance because ahead of it, in the time allowed, the whole material of what is to be the blank cannot be moved. Blanking will, in the first instance, tend not to occur since the punch will be moving faster than the speed at which the material in the shear zones can convey "the message" that it is to deform plastically.

If the ratio of the energy required to successfully blank dynamically, E_D, to that for the corresponding quasi-static operation, E_S, is plotted against the punch impact velocity, Fig. 6 is obtained [8]. This shows that

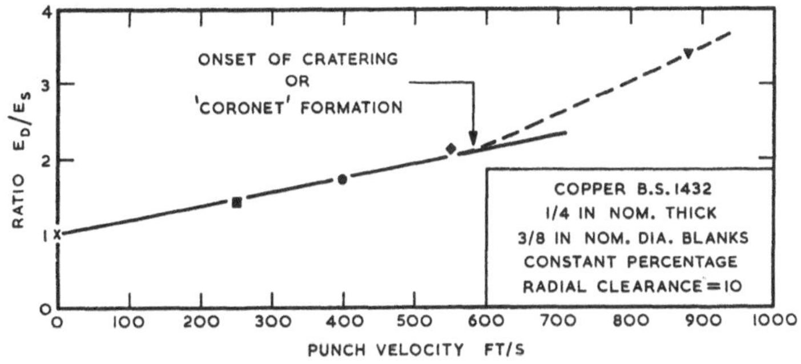

Fig. 6. Relation between the ratio E_D/E_S and punch velocity showing strain-rate effect

the ratio E_D/E_S increases from 1 to 3.38 as the velocity increases from 0 to about 900 ft/s. As a reasonable approximation it can be assumed that the ratio E_D/E_S varies in direct proportion to the punch speed until the punch attains a speed of about 600 ft/s. It can be concluded that Fig. 6 illustrates the strain-rate effect on the mean yield shear stress of the copper. Also for successful high speed blanking at room temperature the ratio E_D/E_S is bounded by $1 < E_D/E_S < 2$ for impact speeds up to about 600 ft/s when the mode of deformation is modified by the onset of cratering or "coronet" formation.

Metallurgical Examination of the Shear Zone in Blanking

Metallurgical examination of a section through the sheared surface of a blanked component using a sharp edge punch and die reveals a lenticular region of strain hardened material in the vicinity of the annular clearance between the punch and die. This region can be of appreciable width amounting to as much as half the thickness of the work sheet.

The effects of strain-rate in blanking operations were observed by Zener and Holloman [9] and further described by Zener [10]. The difference between quasi-static and dynamic blanking was demonstrated when perforating steel plate with a sharp edge punch. In the former case the plastic deformation was found to extend over a wide region but in dynamic blanking the shear zone was narrower and the blanks produced were far less distorted. This is confirmed by the photomicrographs of Fig. 7 which are for 1/4 in. thick copper partially penetrated to the same extent by a 3/8 in. diameter punch at 2 per cent radial clearance, (a) quasi-statically and (b) dynamically at a punch impact speed of 397 ft/s and show clearly the reduced width of the shear zone at high speed. However, the quantitative work of JOHNSON and SLATER [6] and JOHNSON and TRAVIS [7] does not support the claim of ZENER and HOLLOMAN or that of NOVOTNY as reported by COCKCROFT [11] that less energy is necessarily required for high speed punching than for slow speed punching.

Effect of Elastic Deformation of Tools etc.

During the course of the experimental work reported in [6] and [7] it was necessary to consider the effect of elastic deformation of the tools, relative expansion of the punches and dies and also the effects of oxidation at the highest temperatures for the copper and mild steel.

Themain concern related to effects on the punch die clearances, especially at the least values of percentage radial clearance. In the dynamic blanking experiments the peak values of force were expected to be several times the quasi-static values.

During the manufacture of the tools, the dimensional accuracy was never better than ± 0.0001 in after grinding and the co-axiality of the punches and dies was never better than ± 0.0005 in. The authors believe that the accuracy of the experimental results was limited by these factors rather than by changes in the clearances due to elastic deformation or expansion of the tools. Of course, in the dynamic blanking investigations the elastic deformation of the tools becomes more important, especially in the case of the very high speed blanking reported in [7]. However, this effect is difficult to assess and although the least punch die clearances were likely to be effected, elastic deformation of the tools was neglected.

2. Forging

Heat Lines or Lines of Thermal Discontinuity

In a virtually unknown paper presented to the Manchester Association of Engineers in 1921, H. F. MASSEY [12] reported extensive experimental investigations into the plastic flow of metals under various

(b)

(a)

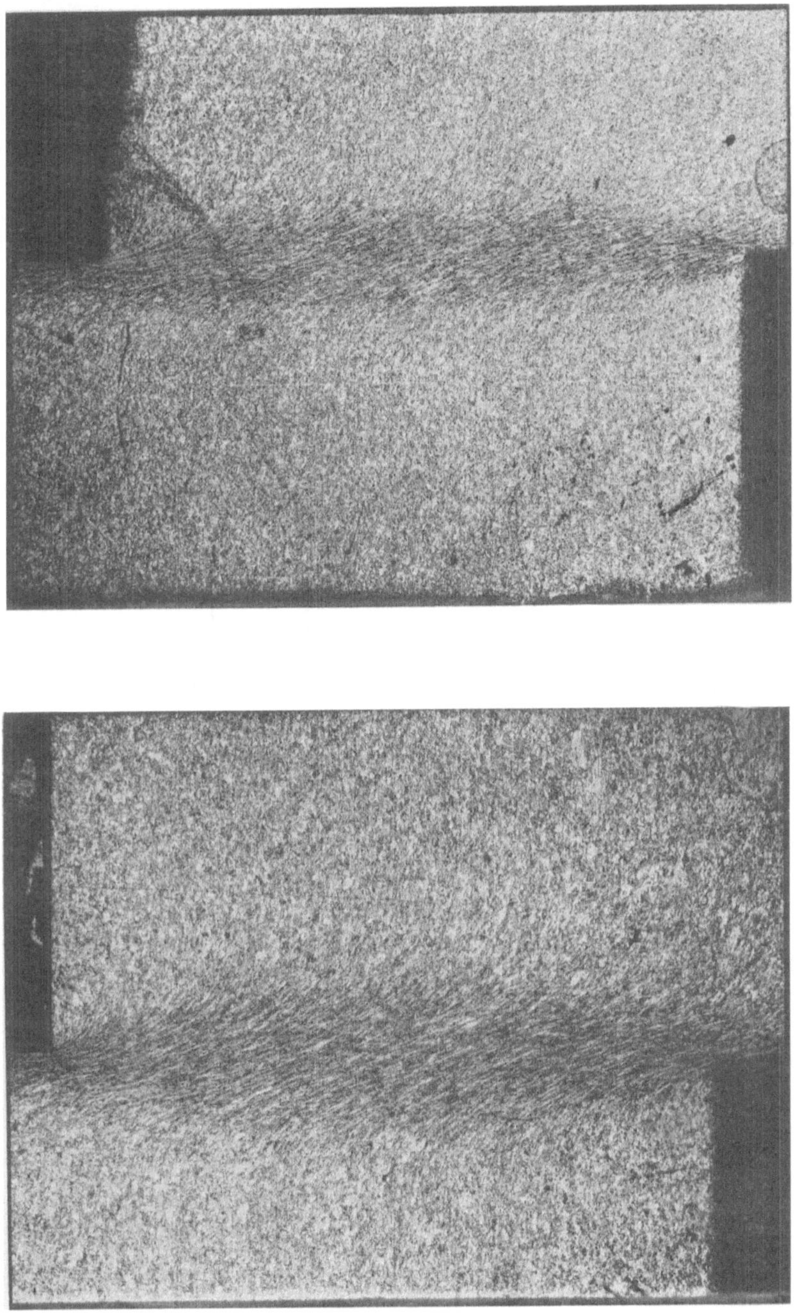

Fig. 7. Photo-micrographs of copper specimens partially blanked (a) quasi-statically and (b) dynamically at 2 per cent radial clearance

hammer forging conditions. During the forging of mild steel at a temperature of about 680 °C he observed distinct and sharply defined zones of high temperature in the form of crosses which he subsequently called "heat lines". Massey only briefly referred to this phenomenon in his paper and although he attempted to record his observations photographically the results were unsuccessful and in consequence this work seems to have been forgotten.

An account was presented by Johnson, Baraya and Slater [13] of a preliminary investigation into the phenomenon of the generation of distinct patterns of "heat lines" or lines of thermal discontinuity in the forging of mild steel at temperatures af about 700 °C. The patterns of these lines as associated with different tool geometries were found to

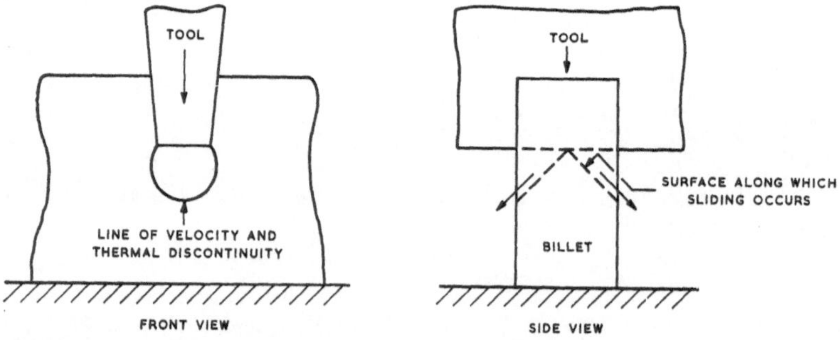

Fig. 8. Two views of the anticipated mode of deformation with velocity discontinuities in both views when using a wedge-shaped necking tool

closely resemble anticipated lines of tangential velocity discontinuity, such as are well known in theoretical plasticity. The lines are thin regions in which the temperature of the forged material is raised significantly above that of the surrounding mass and in consequence the phenomenon may have considerable metallurgical importance.

Flat bar forging and the use of necking tools including tools having semi-circular and wedge-shaped profiles were employed during the investigations. Two views of the anticipated mode of deformation using a wedge-shaped necking tool are presented in Fig. 8[1].

[1] It is regretted that in preparing this paper it has not been possibe to reproduce the coloured photographs which demonstrate clearly the patterns of thermal discontinuities obtained.

A short 16 mm colour cine-film which illustrates the phenomenon was shown at the time of presentation of the paper. The film is also available on loan from the authors.

Estimation of the Temperature Jump at a Line of Thermal Discontinuity During a Fast Metal Forming Process

From an elementary consideration of the Upper Bound Theorem as applied to a plane strain plastic deformation involving tangential velocity discontinuities it can be shown that the rate of internal energy dissipation for a velocity field consisting of a number of straight lines of tangential velocity discontinuity is given by [14].

$$d\,W/dt \leq \Sigma\, k s V^*. \tag{2}$$

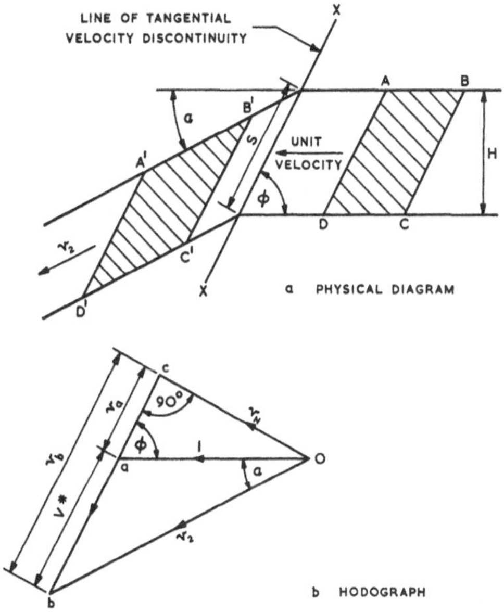

Fig. 9. Mode of plane strain deformation for determining the rate of internal energy dissipation and temperature jump for a rigid-perfectly plastic material

Assuming a plane strain mode of deformation and a fast metal forming process in which heat transfer is negligible, then an expression for the temperature jump, $\Delta\theta$, which occurs at a line of thermal discontinuity identical with a line of tangential velocity discontinuity such as XX in Fig. 9a, can be deduced as follows:

In unit time, the work done during the plastic deformation is equal to the heat dissipation,

$$\text{i.e.,} \quad k s V^* = \mathrm{J}.\,\varrho.\,\mathrm{c}.\,\mathrm{H}.\,\Delta\theta, \quad \text{and}$$

$$\therefore\ \Delta\theta = \frac{ks}{J\varrho c}\cdot\frac{V^*}{H} = \frac{ks}{J\varrho c}\cdot\frac{V^*}{s\cdot\sin\Phi}.$$

But $\sin \Phi = V_N/1$ from the hodograph[1] shown in Fig. 9b, and

$$\therefore \Delta\theta = \frac{k}{J\varrho c} \cdot \frac{V^*}{V_N}.$$

Thus if the von Mises yield criterion is assumed to be applicable then

$$k = Y/\sqrt{3},$$

so that

$$\Delta\theta = \frac{Y}{\sqrt{3}\cdot J\varrho c} \cdot \frac{V^*}{V_N}. \qquad (3)$$

The above equation for the temperature jump at a line of thermal discontinuity can be seen to consist of two terms, one peculiar to the metal being deformed, namely $\left(Y/\sqrt{3}\,J\varrho c\right)$ and the second term (V^*/V_N) peculiar to the mode of deformation. The latter term is just a measure of the shear strain undergone by material which traverses the line of discontinuity, — it can be obtained from the hodograph corresponding to the proposed velocity field for the deformation process.

Of course, in reality no precise line of thermal discontinuity can exist but thin regions can appear in which temperature jumps above that of the surrounding mass will occur[2]. The thickness of the shear zones is highly dependent on the strain-hardening characteristic of the metal at the temperature of working.

To illustrate how the temperature rise may be approximately evaluated using eq. (3), it will be assumed that a rectangular section billet of metal is forged between rigid, frictionless and parallel platens of width $2w$ which move together vertically with unit velocity. The depth of the billet at the time considered is $2h$ such that $w/h = 1$. The proposed field of tangential velocity discontinuities, in this case, constitutes a single cross with the lines intersecting as shown in Fig. 10a. The material on either side of these lines is supposed rigid so that the deformation is envisaged as the relative motion of the four separate rigid sliding blocks A, B, C and D respectively. The corresponding hodograph is then shown in Fig. 10b.

The ratio V^*/V_N for each of the four tangential velocity discontinuities is $\sqrt{2}/\left(1/\sqrt{2}\right) = 2$.

---

[1] From the Greek 'hodos', meaning way. The hodograph is a means of displaying the velocity at every point in the deforming and rigid regions throughout the body. See, for example, [14], Chapter 13.

[2] This is due to the consideration of a non-steady state problem. In a steady state process such as rolling or orthogonal metal cutting, generally two distinct zones are visible; a low temperature one before the zone of deformation and a high temperature one after it.

For mild steel preheated to say, 700 °C before the forging operation, it may be assumed that $Y = 22.5$ tonf/in^2 corresponding to an engineering shear strain rate of about 200 s^{-1}. It can also be assumed that $\varrho = 0.28$ lb/in^3, $c \simeq 0.15$ and $J = 1400$ ft lbf/C. H. U.

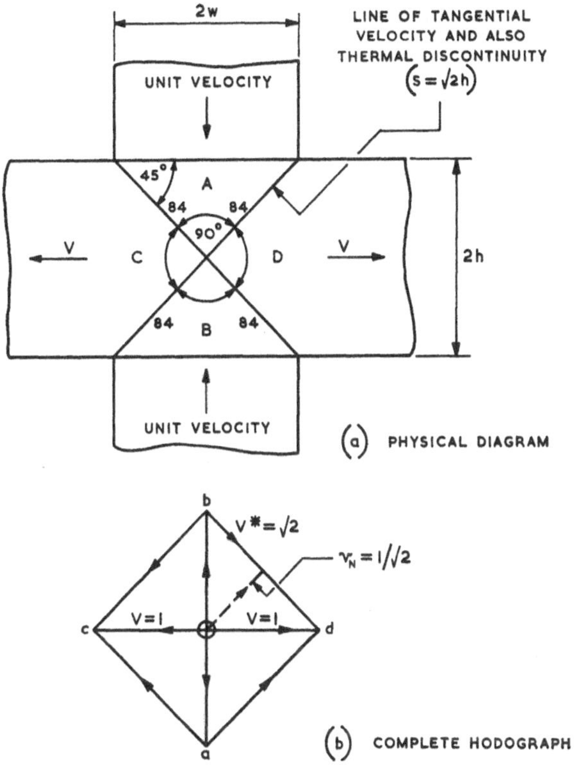

Fig. 10. Plane strain compression between frictionless parallel platens ($w/h = 1$)

From eq. (3)

$$Y/\left(\sqrt{3} \cdot J\varrho c\right) = 22.5 \times 2240/\left(12 \times \sqrt{3} \times 1400 \times 0.28 \times 0.15\right) \simeq 42\,°\mathrm{C},$$

and the temperature jump $\Delta\theta$ occurring at each of the lines of thermal discontinuity is then approximately $42 \times 2 = 84$ °C.

In the experimental work referred to previously, the temperature jump was evaluated for the above case. The temperature was measured using an optical pyrometer and was found to be about 800 °C when the temperature of the surrounding mass was about 700 °C thus indicating a temperature jump in the "heat line" zones of the order of 100 °C.

3. Upsetting

Considerable experimental work has been performed with the aim of studying material properties at elevated temperatures and high rates of strain. Only limited effort, however, has been specifically directed towards a fuller understanding of industrial forging operations. The need exists to estimate with some degree of accuracy the force and energy requirements to effect a given degree of upsetting depending upon the temperature and strain-rate. The importance of this information results from current requirement to effectively programme for automatic forging and other metal forming processes.

Dynamic Compression of Super-Pure Aluminium at Elevated Temperatures Using an Experimental Drop Hammer

In the real forging operation, the deformation of the workpiece occurs at a variable strain-rate. To simulate this situation an experimental drop hammer was employed by BARAYA, JOHNSON and SLATER [15] to effect the dynamic compression of cylindrical specimens of super-pure aluminium at temperatures up to 500 °C. In this work height/diameter ratios of 0.5, 1.0, 1.5 and 2.0 were used for the specimens.

When the mass of the tup was maintained constant at 15.7 lb but the impact velocity varied between 17.9 and 35.9 ft/s, the relation between the logarithm of the percentage reduction and temperature exhibited an inflexion in nearly all cases. These inflexions are prominent at low energy levels but not so easily located at the high energy levels. For very accurate determination of the transition temperature a greater number of tests would be required. However, it is clear that the inflexion occurs between 200 and 300 °C and approaches the higher temperature as the impact velocity increases. The results are generally in agreement with those presented by TROZERA et al. [16]. Assuming that the point of inflexion occurs at 0.55 T_M as suggested by WESTBROOK [17] then for super-pure aluminium it may be assumed that the inflexion is expected to occur at approximately 240 °C.

At a mean strain-rate of 263 s^{-1} the ratio of the mean dynamic yield stress to the quasi-static yield stress for super-pure aluminium was found to be about 11.6 at a test temperature of 500 °C corresponding to a homologous temperature $T/T_M = 0.83$.

A Tentative Analysis for the Fast Upsetting of Short Cylinders of a Strain-Hardening, Strain-Rate Sensitive Material Using a Drop Hammer

The investigation reported in Ref. [15] was continued very recently by SLATER, JOHNSON and AKU [18, 19] with a view to predicting the strain history and the impulse during the fast upsetting of short cylindrical specimens at room temperature.

Approximate equations were derived to account for the compressive engineering strain-time, dimensionless compressive engineering strain-rate—compressive engineering strain and dynamic force-time characteristics for the fast compression of prismatic blocks of a strain-hardening, strain-rate sensitive material using a drop hammer.

The theoretical characteristics were compared with experimental results obtained using pure lead and 0.55% plain carbon steel cylindrical specimens (having initial height/diameter ratio = 1) which were subjected to dynamic compression at ambient temperature (~20 °C) using an experimental drop hammer. In the case of the pure lead, the tup mass was maintained constant, compressive engineering strains of up to about 0.4 were involved and the range of compressive engineering strain-rate did not exceed 250 s^{-1}. However, during similar experimental work on the steel, the tup mass was varied, compressive engineering strains of up to 0.5 were involved and the maximum compressive engineering strain-rate was of the order of twice that for the pure lead.

It was found that at room temperature, the two materials were not significantly strain-rate sensitive in the range of strain-rate investigated. The ratio of the dynamic yield stress to the quasi-static yield stress, $\bar{\sigma}_D/\sigma_S$ was approximately 1.2 at a strain-rate of 250 s^{-1} for small strains in the case of the lead but increased slightly with increase in strain. Similarly, for the steel $\bar{\sigma}_D/\sigma_S$ was approximately 1.25 when the strain-rate was 400 s^{-1} and the compressive engineering strain was 0.25.

The analysis which follows below is described as tentative because it necessarily embodies a number of assumptions and approximations and only when it has been shown to accord with experimental results can it be regarded as adequate.

It is assumed that the deformation of the cylindrical specimen is a homogeneous uni-axial compression and hence that there is no frictional restraint at the material-platen interfaces. It is further assumed that the material is strain-hardening and, to begin with, strain-rate sensitive such that the dynamic flow stress σ_D is related to the strain-rate $\dot{e}$ and the strain e by the product type expression:

$$\sigma_D = (\sigma_S + B\dot{e}^n)\, K e^m \tag{4}$$

Now, the equation of motion for the tup impinging on the specimen is

$$A\,\sigma_D + M'\,du/dt = 0$$

and thus

$$A\,\sigma_D = (A_0 H_0/H)\,(\sigma_D + B\dot{e}^n)\,K e^m$$

since the material is assumed to be incompressible. Because

$$H_0/H = 1/(1-e),\ \dot{e} = u/H_0 \text{ and } d\dot{e}/dt = (1/H_0)\,du/dt \text{ then}$$

$$-M' H_0 \cdot d\dot{e}/dt = \{A_0/(1-e)\}\,(\sigma_S + B\dot{e}^n)\,K e^m$$

or

$$-(M' H_0/A_0\sigma_S)\,\{\dot{e}\,d\dot{e}/(1 + C\dot{e}^n)\} = \{K e^m/(1-e)\}\,de$$

since $\dot{e} = de/dt$,

$$dt = de/\dot{e} \quad \text{and} \quad C = B/\sigma_S.$$

Thus

$$-\alpha\,\{\dot{e}\,d\dot{e}/(1 + C\dot{e}^n)\} = \{K e^m/(1-e)\}\,de,$$

where

$$\alpha = M' H_0/A_0\sigma_S.$$

Integration of the above expression between the limits of strain $e = 0$ and any strain e where the corresponding strain-rates are $\dot{e}_0$ and $\dot{e}$ respectively, leads to

$$-\alpha \int_{\dot{e}_0}^{\dot{e}} \dot{e}\,d\dot{e}/(1 + C\dot{e}^n) = K \int_0^e e^m\,de/(1-e). \tag{5}$$

Each of the definite integrals in eq. (5) can be solved as a series by employing integration by parts which then produces:

$$\int_{\dot{e}_0}^{\dot{e}} \dot{e}\,d\dot{e}/(1 + C\dot{e}^n) = \frac{\dot{e}^2}{2(1 + C\dot{e}^n)} - \frac{\dot{e}_0^2}{2(1 + C\dot{e}_0^n)} +$$

$$+ \frac{n}{2(n+2)} \cdot \frac{C\dot{e}^{n+2}}{(1 + C\dot{e}^n)^2} - \frac{n}{2(n+2)} \cdot \frac{C\dot{e}_0^{n+2}}{(1 + C\dot{e}_0^n)^2} + \cdots \tag{6}$$

and

$$\int_0^e e^m\,de/(1-e) = -e^m \ln(1-e) - m e^{m-1}(1-e)\{\ln(1-e) - 1\} + \cdots \tag{7}$$

Also, exact solutions can be easily obtained for the definite integrals:

$$\int_{\dot{e}_0}^{\dot{e}} \dot{e}\,d\dot{e}/(1 + C\dot{e}^n) \quad \text{and} \quad \int_0^e e^m\,de/(1-e)$$

for values of m and n equal to an integer or the reciprocal of an integer. However, when inserting the values of these integrals into eq. (5) which is fundamental for analysing the upsetting of a strain-hardening, strain-rate sensitive material there are difficulties in deriving the dimensionless strain-rate equation for, $\dot{e}/\dot{e}_0$, without making any approximations. Similar difficulties are experienced when obtaining an equation for the dimensionless strain, e/e_F, as a function of time.

In producing a theoretical force-time curve it is necessary to calculate (a) the strain, e, at any instant and then (b) the strain-rate, $\dot{e}$, at the same instant. To demonstrate the method of analysis it will be assumed that $m = n = 1/2$. Equation (5) may then be rewritten as

$$-\frac{2\alpha}{C_4}\left[\frac{1}{3}\{(1+C\dot{e}^{1/2})^3-(1+C\dot{e}_0^{1/2})^3\}-\frac{3}{2}\{(1+C\dot{e}^{1/2})^2-(1+C\dot{e}_0^{1/2})^2\}+\right.$$
$$\left.+3\{(1+C\dot{e}^{1/2})-(1+C\dot{e}_0^{1/2})\}-\ln\frac{1+C\dot{e}^{1/2}}{1+C\dot{e}_0^{1/2}}\right]$$
$$=K\left(-2e^{1/2}+\ln\frac{1+e^{1/2}}{1-e^{1/2}}\right). \tag{8}$$

Expanding the logarithmic term on the left hand side of eq. (8) as a series to the fourth degree term in $(C\dot{e}^{1/2})$ and simplifying, leads to

$$\frac{\alpha}{2}(\dot{e}_0^2-\dot{e}^2)\approx K\left(-2e^{1/2}+\ln\frac{1+e^{1/2}}{1-e^{1/2}}\right) \tag{9}$$

$$\therefore \dot{e}/\dot{e}_0\approx\left\{1+\eta\left(-2e^{1/2}+\ln\frac{1+e^{1/2}}{1-e^{1/2}}\right)\right\}^{1/2}, \tag{10}$$

where

$$\eta = 2K/\alpha\dot{e}_0^2.$$

When the deformation is complete, the strain-rate $\dot{e} = 0$ and the strain e is the final strain e_F. Substituting these values in eq. (10) gives

$$\eta = \frac{1}{2e_F^{1/2}-\ln\dfrac{1+e_F^{1/2}}{1-e_F^{1/2}}}$$

$$\therefore \dot{e}/\dot{e}_0\approx\left[1-\frac{2e^{1/2}-\ln\dfrac{1+e^{1/2}}{1-e^{1/2}}}{2e_F^{1/2}-\ln\dfrac{1+e_F^{1/2}}{1-e_F^{1/2}}}\right]^{1/2} \tag{11}$$

If the logarithmic terms are now expanded as a series to the second term in each case, the following approximate equation is obtained:

$$\dot{e}/\dot{e}_0\approx[1-(e/e_F)^{3/2}]^{1/2}. \tag{12}$$

Using the approximate eq. (12),

$$\dot{e} = d\dot{e}/dt \approx \dot{e}_0 [1 - (e/e_F)^{3/2}]^{1/2}$$

and hence,

$$\int de/[1 - (e/e_F)^{3/2}]^{1/2} \approx \dot{e}_0 t + \psi,$$

where ψ is a constant of integration.

To solve this indefinite integral it is proposed to use a graphical solution by assuming that

$$I = \int_0^{e/e_F} de/[1 - (e/e_F)^{3/2}]^{1/2} = a + b \ln [1/\{1 - (e/e_F)\}],$$

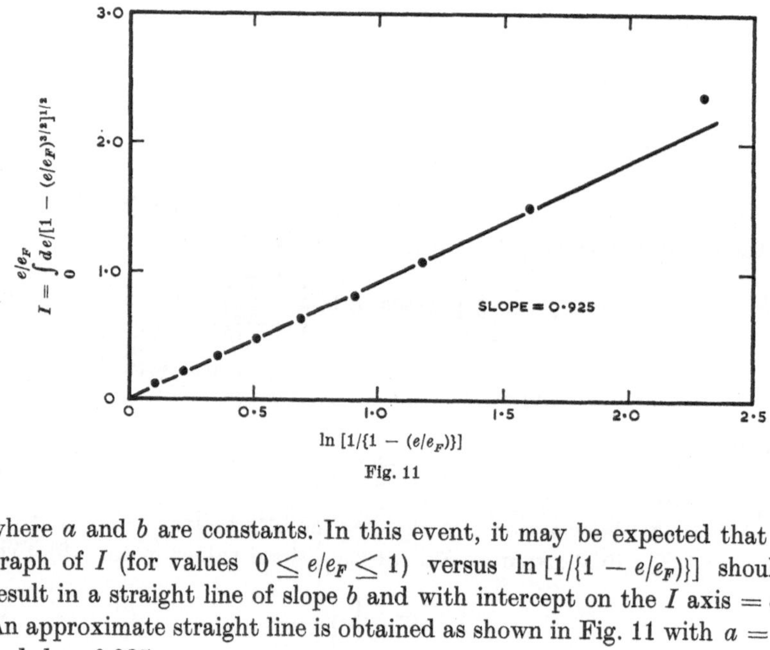

Fig. 11

where a and b are constants. In this event, it may be expected that a graph of I (for values $0 \leq e/e_F \leq 1$) versus $\ln [1/\{1 - e/e_F)\}]$ should result in a straight line of slope b and with intercept on the I axis $= a$. An approximate straight line is obtained as shown in Fig. 11 with $a = 0$ and $b = 0.925$.

$$\therefore \; I \approx 0.925 \ln [1/\{1 - e/e_F)\}]$$

$$\text{and} \int de/[e/e_F)^{3/2}]^{1/2} \approx 0.925 e_F \ln [1/\{1 - (e/e_F)\}] + \text{constant};$$

hence,

$$0.925\, e_F \ln [1/\{1 - (e/e_F)\}] \approx \dot{e}_0 t + \psi$$

and when

$$t = 0, \quad e = 0 \quad \text{so that} \quad \psi = 0.$$

Thus

$$\ln [1/\{1 - (e/e_F)\}] \approx 1.08\, \dot{e}_0 t/e_F$$

and finally,

$$e/e_F \approx [1 - \exp(-1.08\,\dot{e}_0 t/e_F)]. \tag{13}$$

From eq. (4), the uni-axial compressive force exerted on the specimen at any instant is given by

$$F = A\,\sigma_D = (A_0 H_0/H)\,(\sigma_S + B\dot{e}^n)\,K e^m$$

or

$$F = A_0 \sigma_S (1 + C\dot{e}^n)\,K e^m/(1 - e) \tag{14}$$

from which the dimensionless force is

$$F/A_0\sigma_S = (1 + C\dot{e}^n)\,K e^m/(1 - e). \tag{15}$$

Comparison Between the Theoretical and Experimental Dimensionless Strain-Rate ($\dot{e}/\dot{e}_0$) and the Dimensionless Compressive Engineering Strain (e/e_F) Characteristics

The theoretical curve of Fig. 12 presents eq. (12) on a dimensionless basis and assumes $m = n = \frac{1}{2}$ in eq. (15). For purposes of direct

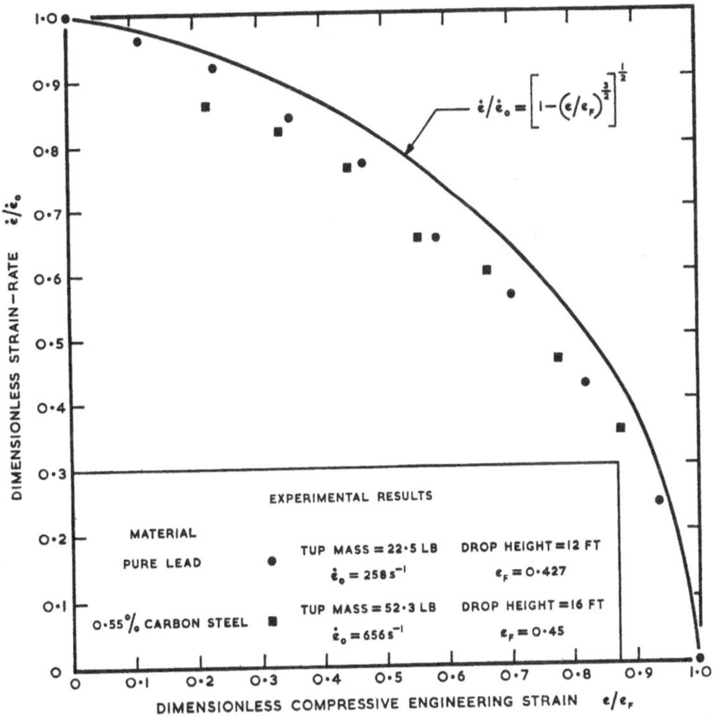

Fig. 12. Comparison between the theoretical and experimental dimensionless strain-rate ($\dot{e}/\dot{e}_0$) – dimensionless compressive engineering strain (e/e_F) characteristics

comparison the experimental results for the pure lead and 0.55% plain carbon steel are also given in Fig. 12.

Fairly good agreement is obtained between the theoretical and experimental results which indicate clearly that the process does not occur at a constant strain-rate nor does it proceed at a linearly decreasing strain-rate.

Comparison Between the Theoretical and Experimental Compressive Engineering Strain-Time Characteristics

A theoretical curve showing the relation between the compressive engineering strain and time, i.e. the strain history as calculated using the approximate eq. (13) is shown for the pure lead in Fig. 13 which also

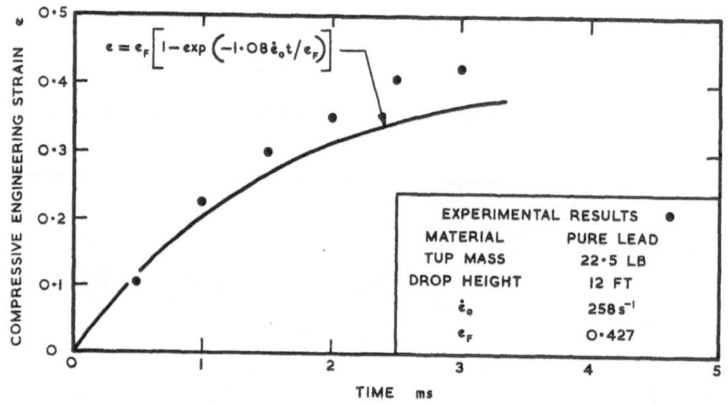

Fig. 13. Comparison between the theoretical and experimental compressive engineering strain-time characteristics for pure lead at ambient temperature

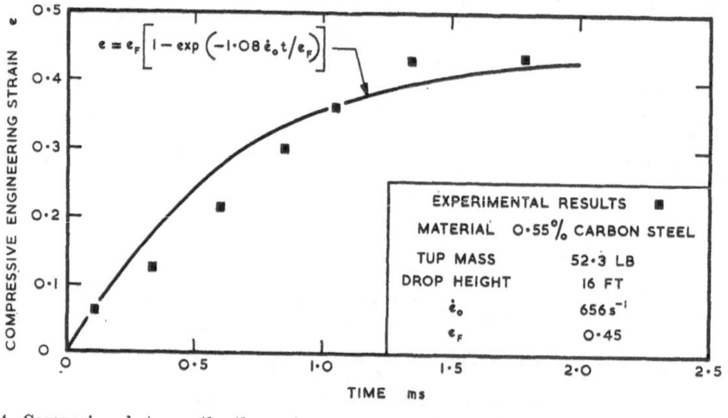

Fig. 14. Comparison between the theoretical and experimental compressive engineering strain-time characteristics for 0.55% carbon steel at ambient temperature

includes the corresponding experimental results for comparison. A similar comparison between the theoretical and experimental results for the 0.55% plain carbon steel is presented in Fig. 14. Close agreement is obtained between the predicted and experimental results considering that eq. (13) can only be regarded as approximate and depends on $m = n = \frac{1}{2}$ being used in eq. (5) which in turn is derived by assuming that the functional form of eq. (4) is applicable.

Comparison Between the Theoretical and Experimental Force-Time Characteristics for the Dynamic Compression of Pure Lead and 0.55% Plain Carbon Steel at Room Temperature

The relation between the dimensionless force, $F/A_0\sigma_S$, as given by eq. (15) and the dimensionless time, t/t_F, is illustrated in Fig. 15. This theoretical curve is applicable to the case for pure lead and was deter-

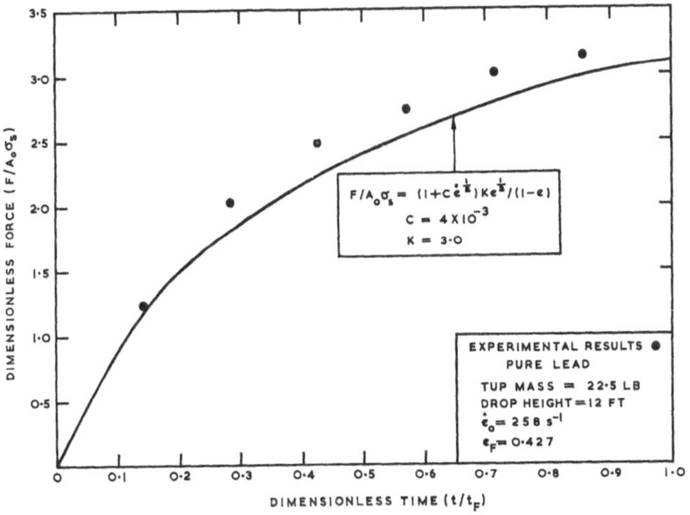

Fig. 15. Comparison between the theoretical and experimental force ($F/A_0\sigma_S$) – dimensionless time (t/t_F) characteristics for pure lead at ambient temperature

mined using $m = n = \frac{1}{2}$ and assuming appropriate values for the constants C and K. To enable the validity of the theoretical prediction to be studied, the experimental values for the pure lead are also included in Fig. 15. A similar comparison is made in Fig. 16 between the predicted and experimental results for the 0.55% plain carbon steel.

Figs. 15 and 16 show clearly that eq. (14) and hence eq. (15) are capable of predicting the uni-axial compressive force whilst the specimen deforms plastically. However, the theoretical equations are unable to predict the unloading phase of the operation once the maximum value of the force is attained.

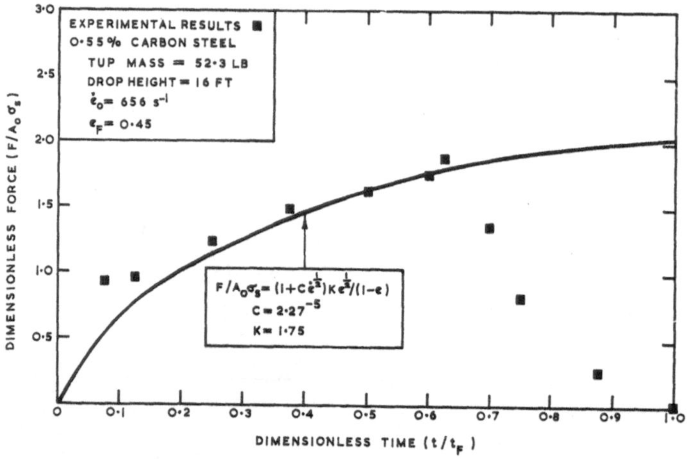

Fig. 16. Comparison between the theoretical and experimental dimensionless force $(F/A_0\sigma_S)$ – dimensionless time (t/t_F) characteristics for 0.55% plain carbon steel at ambient temperature

The Use of Plasticine to Simulate the Dynamic Compression of Prismatic Blocks of Hot Metal

White plasticine was used as a model material at room temperature by AKU, SLATER and JOHNSON [20] to study the dynamic compression of hot metals. This was possible because white plasticine at room temperature and metals worked at their normal forging temperatures exhibit substantially "flat" quasi-static stress-strain curves. In the latter case, this indicates that the rate of strain-hardening is nullified by the thermal recovery if frictional resistance is absent or negligible. Plasticine is also known to be a strain-rate sensitive material.

During the study, plasticine specimens of 1 in. nominal height having circular, rectangular, triangular and annular sections were subjected to dynamic compression at room temperature using a model drop hammer and the mean strain rates were up to 200 s^{-1}.

Upset axi-symmetrical specimens remained more or less axi-symmetric for all reductions, but the holes of the hollow cylinders did not close even with height reductions exceeding 70 per cent. Non-uniform lateral spread was observed in the case of the cubic, rectangular and triangular prisms, but the change in shape of the squares and rectangles was not appreciable even for reductions up to 70 per cent. The prisms of triangular section tended to become circular for reductions greater than 70 per cent. It is interesting to note that at a reduction of 87 per cent the triangular section became almost a complete circle of 2 in diameter as illustrated in Fig. 17. An analytical approach to this problem could possibly predict this phenomenon although in the opinion of the authors, the strain analysis is likely to prove complex!

A bibliographical search showed that little effort had been devoted to the subject of inhomogeneous deformation which may result during dynamic compression at high fractional reductions. In a fast metal forging operation where the process time is about 2 to 4 ms, it is exceedingly difficult to maintain efficient lubrication. With these points in mind, the authors also presented qualitative results obtained when various shapes of white plasticine were dynamically upset to a high fractional reduction.

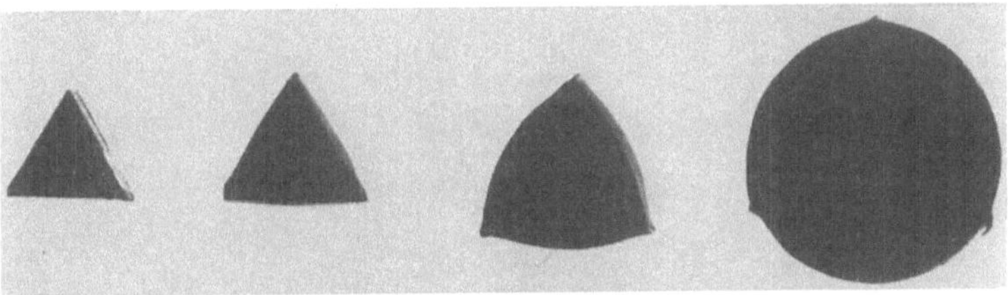

$R = 0$ $R = 0{,}32$ $R = 0{,}65$ $R = 0{,}87$

Fig. 17. Dynamically upset equilateral triangular prisms for different fractional reductions

The vertical faces of an undeformed plasticine specimen cube, a rectangular prism and an equilateral triangular prism were marked with 1/10 in. square grids using printed transfers as shown in Fig. 18. The specimens and tools were then well lubricated, (using French chalk powder,) and dynamically upset using the model drop hammer. The plan view of upset specimens of each type and the corresponding fractional reduction are also shown in Fig. 18. The authors emphasize that the plan views of these dynamically upset specimens show material which is substantially in the same horizontal plane as the ultimate material-platen interface.

Even though the specimens were well lubricated over their entire surfaces before dynamic compression, Fig. 18 shows clearly that when the fractional reduction is high, say $R > 0.5$, it is extremely difficult to ensure that the deformation is homogeneous.

All the cases considered indicate that elements which were originally on the vertical surfaces of the specimens have been displaced, deformed and rotated so that material is subsequently rolled up or down to make contact with the upper or lower platens respectively. Similar results were also obtained with other cases such as a cube having a co-axial square section hole, a rectangular prism with a square re-entrant slot and hollow circular cylinders. These observations are generally in agreement with the earlier work of SHAW [21] and the report of the more recently completed research programme at the I. I. T. Research Institute in Chicago [22] concerned with the upsetting of steel circular cylinders.

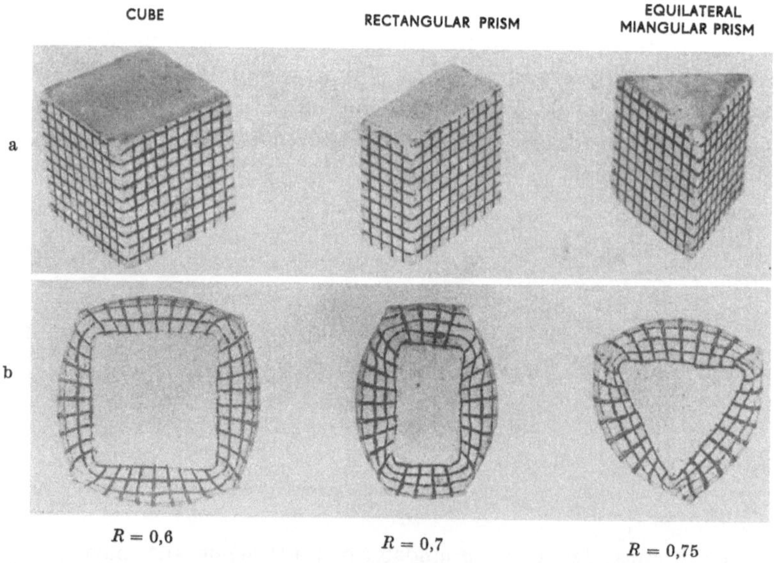

Fig. 18. (a) Undeformed plasticine cube, rectangular prism and equilateral triangular prism. (b) Corresponding dynamically upset specimens showing displacement, deformation and rotation of elements.

These plan views show material which is substantially in the same horizontal plane as the ultimate material-platen interface. (R denotes fractional reduction or compressive engineering strain)

Throughout this work it will be observed that the accurate incorporation of thermal effects into any analysis of the mechanics of the operation must be exceedingly difficult. Not merely are there different modes of deformation, strain rate and lubrication effects and highly complex strain histories, but the material properties must vary considerably when there is a large input of energy in the form of plastic work done.

4. Indentation

Dynamic Indentation at Elevated Temperatures

The dynamic indentation of copper (B. S. 1433) and an aluminium alloy (B. S. 1476 HE 10) was investigated by MAHTAB, JOHNSON and SLATER [23] using $^1/_2$ in. diameter cylindro-conical projectiles which were fired from an air-actuated gun. These projectiles which had a conical nose having a semi-angle of 45° were made from a hardened tool steel and the mass of each projectile was 19.2 gm.

Experiments were performed with impact velocities varying between 1000 and 2500 in/s and at elevated temperatures of up to 600 °C for the copper and 550 °C for the aluminium alloy. The magnitude of the corresponding range of mean strain-rate was of the order of 10^3 to 10^4 s^{-1} depending upon the material, impact velocity and temperature.

Quasi-static indentation tests were also performed on both materials at different temperatures using a conical indenter with a semi-angle of 45° and which had a similar surface finish to that of the projectiles. The tests were performed by pressing the indenter into the specimen with a specific load which was maintained constant for 10 min. after which the diameter of the impression was measured at room temperature. The quasi-static indentation pressures based on impression diameter at the test temperature were then determined.

The relationship between the static indentation pressure and temperature for the copper is presented in Fig. 19 on a semi-logarithmic scale. This graph shows that there is a change in temperature coefficient at a temperature of 330 °C and further suggests that this is the recrystallisation temperature for copper (B. S. 1433) under the test conditions and corresponds to a homologous temperature $T/T_M = 0.45$.

In previous work by MAHTAB, JOHNSON and SLATER [24] concerned with dynamic indentation using conical projectiles at ambient temperature, the results were analysed using dimensional analysis. It was shown that some of the parameters of dynamic indentation can be represented by the following equations:

$$d = \beta(\varrho v^2/P_S)^{-a} \cdot (M_P/\varrho)^{1/3} \tag{16}$$

and

$$\overline{P}_d/P_S = \gamma/\beta^3 \, (\varrho v^2/P_S)^{1+3a}. \tag{17}$$

The quasi-static indentation pressure P_S can be expressed as

$$P_S = P_0 \exp(-\lambda T). \tag{18}$$

Substituting for P_S in eqs. (16) and (17) gives

$$d = \beta\, [\varrho v^2/P_0 \exp(-\lambda T)]^{-a} \,.\, (M_P/\varrho)^{1/3} \tag{19}$$

and

$$\overline{P}_d = \gamma/\beta^3\, [\varrho v^2/P_0 \exp(-\lambda T)]^{1+3a} \,.\, P_0 \exp(-\lambda T). \tag{20}$$

In the dynamic indentation tests at elevated temperatures, the diameter of the permanent crater was measured at room temperature

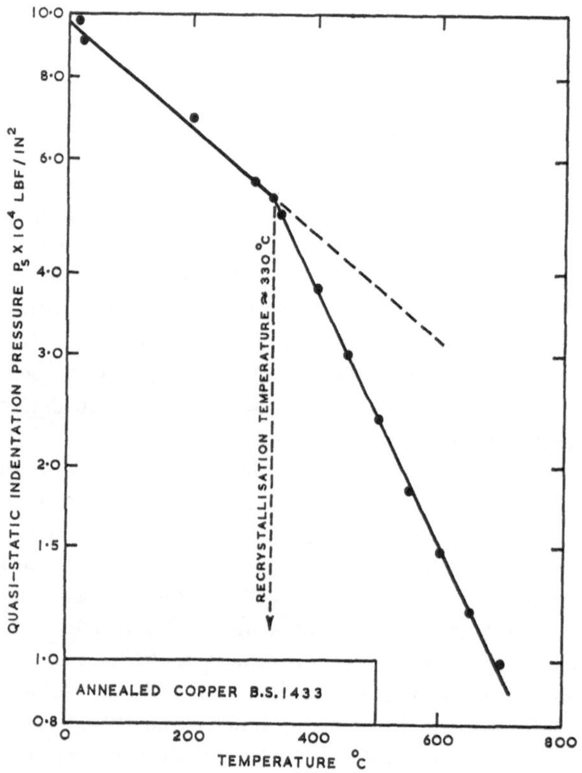

Fig. 19. Relation between quasi-static indentation pressure P_S and temperature for annealed copper (B. S. 1433) on a semi-logarithmic scale

after allowing the specimen to cool. The mean effective dynamic indentation pressure, $\overline{P}_d$, was then calculated following eq. (20). Referring to Fig. 20 in which the relation between the mean effective dynamic indentation pressure, $\overline{P}_d$ and temperature is shown for both the copper and the aluminium alloy, it can be seen that the dynamic pressure decreases with rise in temperature in a non-linear manner. However, Fig. 21 shows clearly that the ratio, $\overline{P}_d/P_S$ increases with temperature particularly when the homologous temperature $T/T_M > 0.5$. The difference between the mean effective dynamic indentation pressure and the quasi-static indentation pressure can again be attributed mainly to the effect of strain-rate upon the current yield stress of the two materials. The ratio. $\overline{P}_d/P_S$, may thus be regarded as indicative of the ratio of the dynamic yield

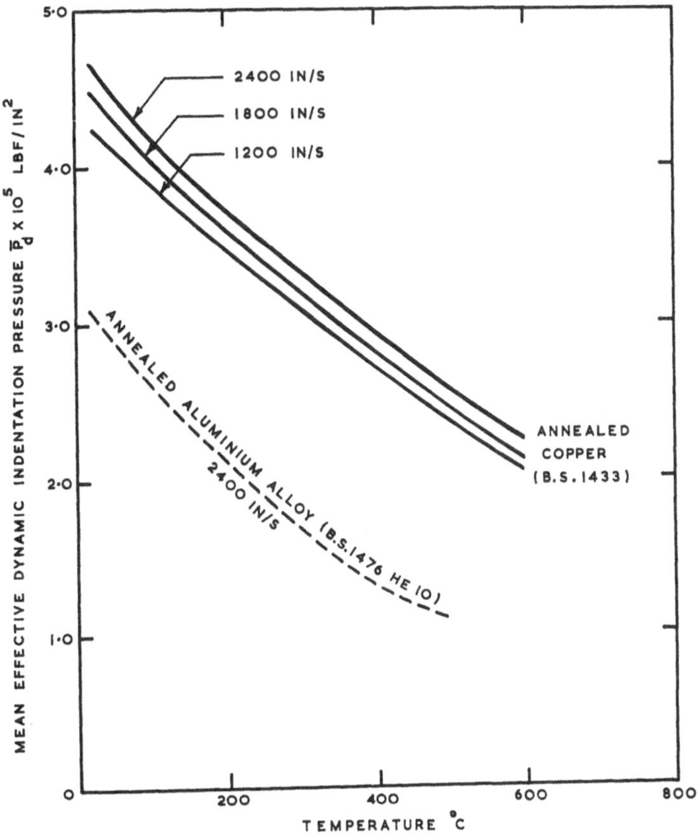

Fig. 20. Relation between the mean effective dynamic indentation pressure $\overline{P}_d$ and temperature

stress to the quasi-static yield stress of the material, Y_D/Y_S. At room temperature, Fig. 21 shows that the ratio, $\overline{P}_d/P_S$ is about 5 for annealed cooper and about 4 for the aluminium alloy corresponding to an impact velocity of 2400 in/s. At comparatively low temperatures the strain-rate is therefore not very significant but Fig. 21 also illustrates that for temperatures in excess of the recrystallisation temperature for the respective materials, the strain-rate effect is of considerable importance. At a temperature of 600 °C or an homologous temperature, $T/T_M = 0.65$, the ratio $\bar{P}_d/P_S$ for the copper is about 16 corresponding to an impact velocity of 2400 in/s or mean strain-rate of approximately 7000 s^{-1}. In the case of the aluminium alloy the ratio $\bar{P}_d/P_S$ is as high as 32 when the homologous temperature, $T/T_M \simeq 0.75$ and the mean strain-rate is approximately 7000 s^{-1}.

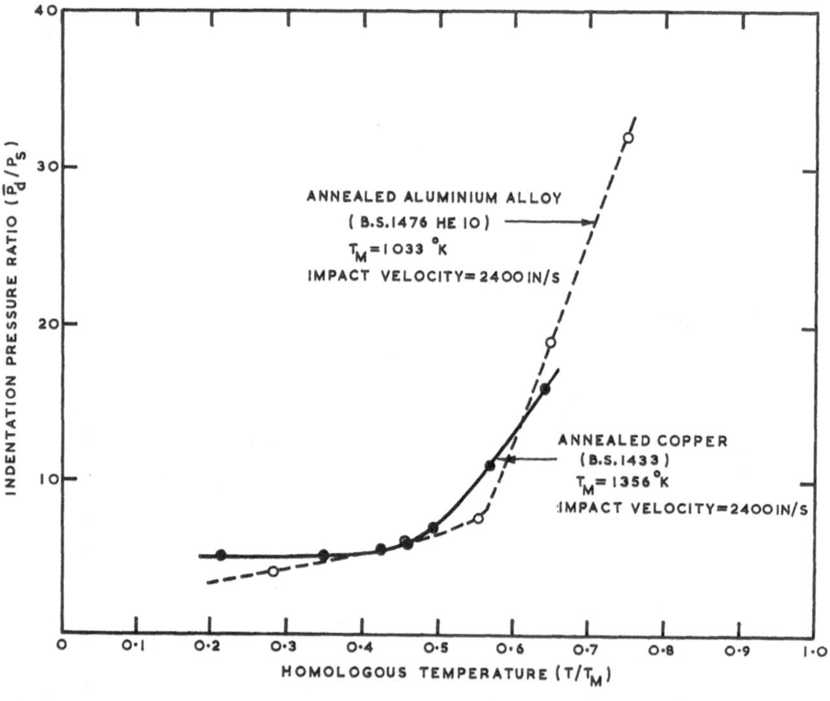

Fig. 21. Relation between the indentation pressure ratio ($\bar{P}_d/P_S$) and the homologous temperature (T/T_M)

5. Dynamic Yield Stress by High Speed Impact Against a Flat Rigid Target

A method of determining the dynamic yield stress of a metal suggested by TAYLOR [25] is that in which a flat ended cylindrical projectile of the metal is fired at normal incidence at a flat rigid target; the projectile speed for most metals is usually about 1,000 ft/s. On impact, projectiles "mushroom" and from measurements taken from the projectile, with the help of a theory of plastic wave propagation, the dynamic yield stress may be calculated. The strain rate for this process is about 10^3/sec. The method has been used by WHIFFIN [26] and LEE and TUPPER [27] to give experimental results. Recently, HAWKYARD, EATON and JOHNSON [28] have reported work in which the same method was used to determine values of mean dynamic yield stress at elevated temperatures. "Hot" projectiles of annealed mild steel, bright drawn mild steel and copper at temperatures between 15 °C and 700 °C, (i.e. between approximate homologous temperatures, T/T_M, of 0.16 and 0.64), were fired at nominally rigid flat targets.

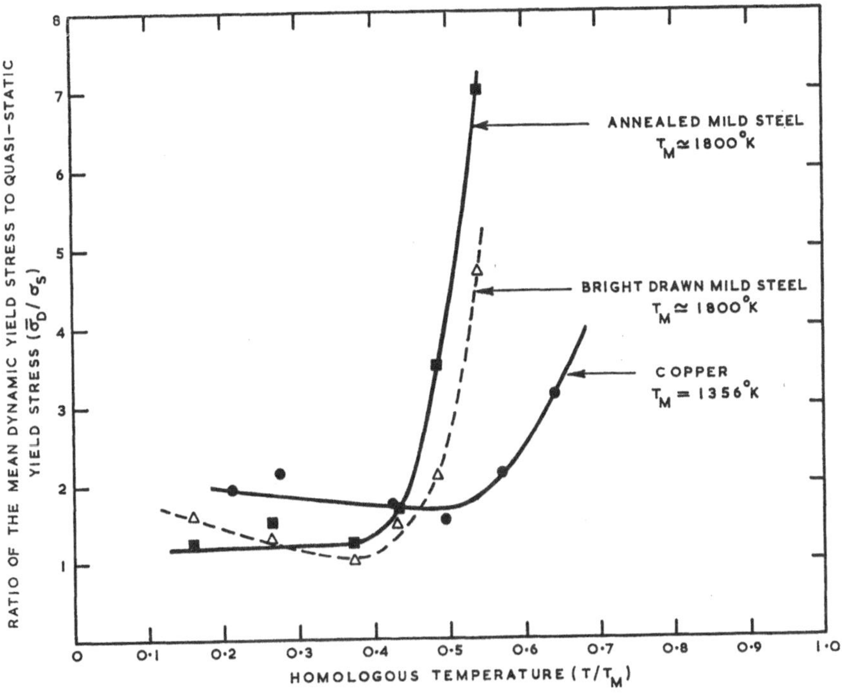

Fig. 22. Relation between the yield stress ratio ($\bar{\sigma}_D/\sigma_S$) and homologous temperature (T/T_M) for high speed impact of flat ended cylindrical projectiles

The mode of deformation after impact which they found was dissimilar to that encountered when using "room" temperature projectiles and accordingly a method of analysis different from that employed by TAYLOR was used. The kinetic energy of the projectile was equated to the plastic work of deformation (assessed from measurements of the shape of the deformed projectile) and the mean yield or flow stress deduced; thus a quasi-static type of analysis was used.

Some of these results are given in the form of the ratio of the dynamic to the static yield stress (at the same level of mean strain) at strain rates of 10^3/sec and 10^{-3}/sec respectively, and are shown in Fig. 22.

Acknowledgement

The authors gratefully acknowledge the financial support given by the British Iron and Steel Research Association, Sheffield which enabled the various investigations to be undertaken.

Thanks are due to Mr. A. BOWERS for preparing the illustrations and the assistance given by Miss JOAN HARWOOD and Miss CAROLYNNE DELVES during the preparation of the manuscript is also sincerely appreciated.

References

[1] JOHNSON, W., E. DOEGE, and F. W. TRAVIS: The Explosive Expansion of Unrestrained Tubes. Paper No. 4, High-rate Forming, Proc. Instn. Mech. Engrs., 179, Part 1, Number 7 (1964–5).

[2] Dynapak for High Energy-rate Forming. Sales Publication, General Dynamics Corporation (1963).

[3] CHAN, L. T., F. BAKHTER, and S. A. TOBIAS: Design and Development of Petro-Forge High Energy-rate Forming Machines. Proc. Instn. Mech. Engrs. **180**, Part 1, 689 (1965–66).

[4] Behaviour of Materials under Dynamic Loading. Ed. by N. J. HUFFINGTON. Amer. Soc. Mech. Engrs. (1965).

[5] NADAI, A., and M. J. MANJOINE: High Speed Tension Tests at Elevated Temperatures Parts II and III. J. Appl. Mech., Trans. Amer. Soc. Mech. Engrs. **63**, A-77 (1941).

[6] SLATER, R. A. C., and W. JOHNSON: The Effects of Temperature, Speed and Strain-rate on the Force and Energy Required in Blanking. Int. J. Mech. Sci. **9**, 271 (1967).

[7] JOHNSON, W., and F. W. TRAVIS: High Speed Blanking of Copper, Paper 16 Applied Mechanics Convention (Cambridge), Instn. Mech. Engrs., London (1966).

[8] JOHNSON, W., and R. A. C. SLATER: A Survey of the Slow and Fast Blanking of Metals at Ambient and High Temperatures. Proc. Int. Conf. on Manufacturing Technology, C. I. R. P./A. S. T. M. E., University of Michigan, Ann Arbor, Michigan, p. 848, September (1967).

[9] ZENER, C., and C. HOLLOMAN: Effect of Strain-rate upon Plastic Flow of Steel. J. Appl. Phys. **15**, 22 (1944).

[10] ZENER, C.: Fracturing of Metals, 3. New York: Amer. Soc. Metals. 1948.

[11] COCKCROFT, M. G.: The Metal Forming Symposium in Czechoslovakia, September 1960, N. E. L. Report No. 6 (1961).

[12] MASSEY, H. F.: The Flow of Metal during Forging. Trans. Manchester Assn. of Engrs., Nov. (1921).

[13] JOHNSON, W., G. L. BARAYA, and R. A. C. SLATER: On Heat Lines or Lines of Thermal Discontinuity. Int. J. Mech. Sci. **6**, 409 (1964).

[14] JOHNSON, W., and P. B. MELLOR: Plasticity for Mechanical Engineers. New York: Van Nostrand. 1962.

[15] BARAYA, G. L., W. JOHNSON, and R. A. C. SLATER: The Dynamic Compression of Circular Cylinders of Super-pure Aluminium at Elevated Temperatures. Int. J. Mech. Sci. **7**, 621 (1965).

[16] TROZERA, T. A., O. D. SHERBY, and J. E. DORN: Effect of Strain-rate and Temperature on Plastic Deformation of High Purity Aluminium. Trans. Amer. Soc. Metals **49**, 173 (1957).

[17] WESTBROOK, J. H.: Temperature Dependence of the Hardness of Pure Metals. Trans. Amer. Soc. Metals **45**, 221 (1953).
[18] SLATER, R. A. C., W. JOHNSON, and S. Y. AKU: Experiments in the Fast Upsetting of Short Pure Lead Cylinders and a Tentative Analysis. Int. J. Mech. Sci. **10**, 169 (1968).
[19] SLATER, R. A. C., W. JOHNSON, and S. Y. AKU: Fast Upsetting of Short Circular Cylinders of Plain Carbon (0.55%) Steel at Room Temperature. Presented at the 9th Machine Tool Design and Research Conference, University of Birmingham, September (1968).
[20] AKU, S. Y., R. A. C. SLATER, and W. JOHNSON: The Use of Plasticine to Simulate the Dynamic Compression of Prismatic Blocks of Hot Metal. Int. J. Mech. Sci. **9**, 495 (1967).
[21] SHAW, M. C.: Surface Conditions in Deformation Processes. Proc. 9th Sagamore Army Materials Research Conference, 107 (1964).
[22] Fundamental Study of Metal Flow in Steel Forging, The Engineer **332**, Feb. 24 (1967).
[23] MAHTAB, F. U., W. JOHNSON, and R. A. C. SLATER: Dynamic Indentation of Copper and an Aluminium Alloy with a Conical Projectile at Elevated Temperatures. Proc. Instn. Mech. Engrs. **180**, Part 1, No. 11, 285 (1965—66).
[24] MAHTAB, F. U., W. JOHNSON, and R. A. C. SLATER: Dynamic Indentation of Copper, an Aluminium Alloy and Mild Steel with Conical Projectiles and Dynamic Tip Flattening of Conical Projectiles at Ambient Temperature. Int. J. Mech. Sci. **8**, 685 (1965).
[25] TAYLOR, G. I.: The Use of Flat-ended Projectiles for Determining Dynamic Yield Stress. Part I. Proc. Roy. Soc. (London), Ser. A, **194**, 289 (1948).
[26] WHIFFIN, A. C.: The Use of Flat-ended Projectiles for Determining Dynamic Yield Stress, Part II. Proc. Roy. Soc. (London), Ser. A, **194**, 300 (1948).
[27] LEE, E. H., and S. J. TUPPER: Analysis of Plastic Deformation in Steel Cylinders Striking Rigid Targets. Trans. Amer. Soc. Engrs., J. Appl. Mech., **22**, 131 (1955).
[28] HAWKYARD, J. B., D. E. EATON, and W. JOHNSON: The Mean Dynamic Yield Strength of Copper and Low Carbon Steel at Elevated Temperatures from Measurements of the "Mushrooming" of Flat-ended Projectiles. Int. J. Mech. Sci. **10**, 929 (1968).

Thermo-Elastic-Plastic Analysis at Finite Strain

By

E. H. Lee

Stanford, Calif. (U.S.A.)

Summary

For high pressure loading situations, classical plasticity theory may not be satisfactory since the usual assumption of the total strain being the sum of elastic and plastic components may not apply for non-infinitesimal strain, and thermo-mechanical coupling influences become important. Kinematics to deal with the finite deformation case are introduced and also a two component thermodynamic model to represent elastic-plastic deformation. One component is reversible thermo-elasticity, and the other plastic flow which absorbs some of the plastic work in the generation of crystal imperfections and the rest is dissipated into heat.

Introduction

Fig. 1 shows a stress-strain curve for a ductile metal such as aluminum. Plastic flow sets in at A. On unloading from B after some plastic flow, the elastic strain at B, CD, is recovered, and the specimen under zero stress retains the plastic strain OC. The elastic strain is given by the yield stress divided by the modulus and is of the order 10^{-3} for most metals in simple tension or shear. For a ductile metal, plastic flow without fracture can continue up to strains of the order unity. For such large total strains an elastic component of the order 10^{-3} can be considered negligible, and plasticity theory for the analysis of such situations has commonly neglected the elastic component of strain, thus introducing plastic-rigid theory. Since plasticity is an incremental strain or flow

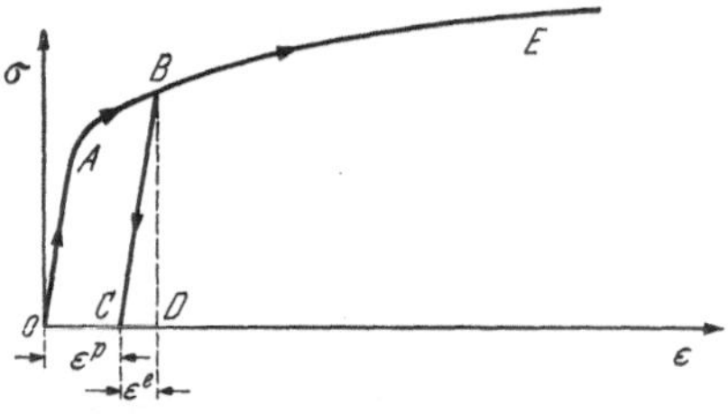

Fig. 1. Stess-strain curve for a ductile metal

type phenomenon, plastic-rigid theory is a fluid type theory in plastic regions, which can be treated in terms of the velocity field at each instant, from which deformations are obtained by integration. On the other hand, elastic-plastic theory is commonly considered applicable when total strains are of the order of elastic strains in a tension or shear test ($\sim 10^{-3}$), so that infinitesimal strain theory is appropriate, and classical elastic-plastic theory is based on this assumption. There are, however, situations which fall between these limits. For plastic flow under high pressure, as for example, that generated by explosive forming, the elastic strains can be finite and of the same order as the plastic strains [1] and for a satisfactory analysis it is necessary to develop elastic-plastic theory for which both components comprise finite strain. Such a development is clearly also needed to provide a uniform theory encompassing the two limits described above.

The consideration of finite strains demands appropriate kinematics to replace linear infinitesimal strain theory. Since finite elastic strains usually involve finite volume compression (the yield limit in shear prevents finite elastic shear components) thermo-mechanical coupling becomes significant and temperature variations must be included in the analysis. For elastic strain components such a theory is well established [2], but must be supplemented by thermodynamic representation of plastic flow. This must include the irreversibility and transfer of the part of the plastic work dissipated into the thermo-elastic system.

Kinematics

It is clear from Fig. 1 that the total strain in a simple one component stress test is equal to the sum of elastic and plastic strain components. As will be shown below, this is in general not true for combined stress

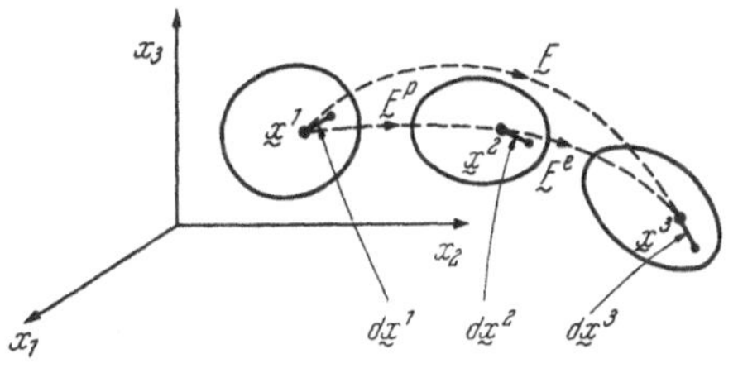

Fig. 2. Plastically and elastically deformed configurations of a body

Note: Letters with wavy line are indicated by bold-face in the text

loading and finite strain components. The kinematics of such general deformation has been discussed in [3]. Consider rectangular Cartesian coordinates (x_1, x_2, x_3) as shown in Fig. 2, and the initial undisturbed configuration of a body to be represented by the particle coordinates

(x_1^1, x_2^1, x_3^1) expressed as a column vector $\boldsymbol{x}^1$. Following elastic-plastic deformation, at time t the particle $\boldsymbol{x}^1$ is displaced to the new position $\boldsymbol{x}^3$ with respect to the same axes. The history of deformation of the body is then given by the function

$$\boldsymbol{x}^3 = \boldsymbol{x}^3(\boldsymbol{x}^1, t). \tag{1}$$

If at time t the elements of the body in $\boldsymbol{x}^3$ are unstressed and have their temperatures reduced to the initial base temperature T_0, the thermo-elastic strains will be released, and only the plastic deformation will remain. Let this configuration of plastic deformation be represented by the particle position $\boldsymbol{x}^2$. The history of plastic deformation is thus represented by the mapping:

$$\boldsymbol{x}^2 = \boldsymbol{x}^2(\boldsymbol{x}^1, t). \tag{2}$$

As discussed in [3], unloading the surface of a body which has been subjected to elastic-plastic deformation commonly does not unstress the body, but leaves a residual stress distribution, and this can be eliminated only by cutting the body into small elements. The corresponding mapping (2) will not then be continuous, but as shown in [3], this circumstance does not alter essentially the following discussion.

In the total deformation (1) a small element $d\boldsymbol{x}^1$ of the undisturbed body is deformed at time t into the element $d\boldsymbol{x}^3$ given by the total derivative of (1), the dx_1^3 component being:

$$dx_1^3 = \frac{\partial x_1^3}{\partial x_1^1}\, dx_1^1 + \frac{\partial x_1^3}{\partial x_2^1}\, dx_2^1 + \frac{\partial x_1^3}{\partial x_3^1}\, dx_3^1.$$

In suffix notation with summation convention the total derivative reads:

$$dx_i^3 = \frac{\partial x_i^3}{\partial x_j^1}\, dx_j^i, \tag{3}$$

$\partial x_i^3/\partial x_j^1$ is called the deformation gradient from configuration $\boldsymbol{x}^1$ to $\boldsymbol{x}^3$, which can be expressed as a square matrix $\boldsymbol{F}$ in which i gives the row position and j the column position of an element. Relation (3) can then be written in matrix product form:

$$d\boldsymbol{x}^3 = \boldsymbol{F}\, d\boldsymbol{x}^1. \tag{4}$$

The deformation gradient matrix $\boldsymbol{F}$ is a convenient entity for representing the geometry of the deformation from $\boldsymbol{x}^1$ to $\boldsymbol{x}^3$. For example, the initial squared length of the element is given by:

$$(ds^1)^2 = d\boldsymbol{x}^{1^T}\, d\boldsymbol{x}^1 = (dx_1^1)^2 + (dx_2^1)^2 + (dx_3^1)^2, \tag{5}$$

where the superscript T stands for the transpose of a matrix, and changes the column vector $d\boldsymbol{x}^1$ into a row vector. Its deformed length in the configuration $\boldsymbol{x}^3$ is

$$(ds^3)^2 = d\boldsymbol{x}^{3^T} d\boldsymbol{x}^3 = d\boldsymbol{x}^{1^T} \boldsymbol{F}^T\boldsymbol{F}\, d\boldsymbol{x}^1 \tag{6}$$

using (4) and the rules of matrix manipulation. Thus the common measure of finite strain, the Lagrange strain, E_{ij} or $\boldsymbol{E}$, is given by

$$(ds^3)^2 - (ds^1)^2 = 2E_{ij}\, dx_i^1\, dx_j^1 = d\boldsymbol{x}^{1^T} 2\boldsymbol{E}\, d\boldsymbol{x}^1 = d\boldsymbol{x}^{1^T} (\boldsymbol{F}^T\boldsymbol{F} - \boldsymbol{I})\, d\boldsymbol{x}^1 \tag{7}$$

where $\boldsymbol{I}$ is the unit matrix.

The deformation (2) from $\boldsymbol{x}^1$ to $\boldsymbol{x}^2$ comprises the plastic deformation, and the corresponding deformation gradient is:

$$\frac{\partial x_i^2}{\partial x_j^1} = \boldsymbol{F}^p. \tag{8}$$

Similarly the thermo-elastic deformation gradient is:

$$\frac{\partial x_i^3}{\partial x_j^2} = \boldsymbol{F}^e. \tag{9}$$

The chain rule for partial differentiation gives:

$$\boldsymbol{F} = \frac{\partial x_i^3}{\partial x_j^1} = \frac{\partial x_i^3}{\partial x_k^2}\frac{\partial x_k^2}{\partial x_j^1} = \boldsymbol{F}^e\boldsymbol{F}^p. \tag{10}$$

As explained in [3], when the configuration $\boldsymbol{x}^2$ is discontinuous the deformation gradient matrices $\boldsymbol{F}^e$, $\boldsymbol{F}^p$ become transformation matrices, and their product $\boldsymbol{F}$ is a deformation gradient matrix between the continuous configurations $\boldsymbol{x}^1$ and $\boldsymbol{x}^3$.

Relation (10), $\boldsymbol{F} = \boldsymbol{F}^e\boldsymbol{F}^p$, is the finite strain analogue of the usual infinitesimal strain relation:

$$\varepsilon_{ij} = \varepsilon_{ij}^e + \varepsilon_{ij}^p. \tag{11}$$

That this latter equation is not valid for the Lagrange strains defined in (7) is clear from the expressions:

$$2\boldsymbol{E} = \boldsymbol{F}^T\boldsymbol{F} - \boldsymbol{I} = \boldsymbol{F}^{p^T}\boldsymbol{F}^{e^T}\boldsymbol{F}^e\boldsymbol{F}^p - \boldsymbol{I} \tag{12a}$$

$$2\boldsymbol{E}^p = \boldsymbol{F}^{p^T}\boldsymbol{F}^p - \boldsymbol{I} \tag{12b}$$

$$2\boldsymbol{E}^e = \boldsymbol{F}^{e^T}\boldsymbol{F}^e - \boldsymbol{I} \tag{12c}$$

where (10) has been substituted in (12a).

Another way of looking at this situation is that, for the special case when the mappings are continuous, the plastic displacement

$$\boldsymbol{u}^p = \boldsymbol{x}^2 - \boldsymbol{x}^1 \tag{13a}$$

the elastic displacement

$$\boldsymbol{u}^e = \boldsymbol{x}^3 - \boldsymbol{x}^2 \tag{13b}$$

and the total displacement

$$\boldsymbol{u} = \boldsymbol{x}^3 - \boldsymbol{x}^1 \tag{13c}$$

so that

$$\boldsymbol{u} = \boldsymbol{u}^e + \boldsymbol{u}^p. \tag{14}$$

Since for finite strain, the strain components are non-linear functions of the displacement components, summation will not apply for strains. For infinitesimal displacements, however, the strain components are linear in the displacements and (14) leads to the common summation condition (11), which also applies generally in the one-dimensional case as illustrated in Fig. 1.

Non-linearity, however, is not the aspect of major significance in replacing (11) by (10). The matrix product $\boldsymbol{F}^e \boldsymbol{F}^p$ is in general not commutative, and the fact that $\boldsymbol{F}^e$ pre-multiplies $\boldsymbol{F}^p$ indicates that $\boldsymbol{F}^p$ represents the plastic deformation which has already occurred, and $\boldsymbol{F}^e$ represents an additional deformation of this already plastically deformed state. In such a matrix product each component represents a particular deformation, for example simple stretch or rotation, which retains its geometrical properties irrespective of the deformation it pre-multiplies. Addition of matrices for finite strain would not have this necessary property.

It is perhaps worth pointing out that the configuration $\boldsymbol{x}^2$ is not unique since destressing an element can involve arbitrary rotation of the element. This causes no difficulty as far as the kinematics is concerned, since the reverse rotation must then be added to the plastic deformation so that the total deformation from $\boldsymbol{x}^1$ to $\boldsymbol{x}^3$ is unchanged.

Work Hardening

As has been described more fully in the paper [11], we shall develop a more general theory of isotropic work-hardening analogous to that described by HILL [4] for isothermal plastic flow during an increment of strain assuming linear elasticity. The considerations which lead to a single infinity of non-intersecting yield surfaces, each given uniquely by a single stress state at the yield limit, as discussed by HILL, apply also in the present case. The insensitivity of the yield surfaces and of the resulting plastic flow and work-hardening to the hydrostatic pressure component of the stress state, and the consequent lack of volume change in plastic strain, also apply. These follow from measurements of plastic flow under high hydrostatic pressure. Since we are concerned with finite elastic strains beyond the linear range, thermo-mechanical coupling becomes significant, and finite strain thermo-elastic theory must be used to represent the elastic behavior. This then demands consideration of the influence of temperature change on plastic flow. For consistency with the thermo-elastic analysis, thermo-mechanical coupling for the plastic component of deformation is also included.

In order to develop a theory of work hardening, the work expended in plastic flow must be separated from the total work, since the work done against elastic strain is simply stored as strain energy and is available on release of the strain. In the infinitesimal strain case for which (11) applies, the rate of working expended in deforming a unit volume of material subject to a homogeneous stress σ_{ij} is

$$\dot{W} = \sigma_{ij}\dot{\varepsilon}_{ij} = \sigma_{ij}(\dot{\varepsilon}^e_{ij} + \dot{\varepsilon}^p_{ij}) \tag{15}$$

which obviously separates into an elastic rate of work term, $\sigma_{ij}\dot{\varepsilon}^e_{ij}$, and a plastic rate of work term, $\sigma_{ij}\dot{\varepsilon}^p_{ij}$. The integral of the latter, W^p, constitutes the independent variable which determines the work-hardening. For finite strain the matrix product relation (10) replaces the simple summation of strain components (11), and a breakdown of the total work rate into elastic and plastic components must be achieved to develop a corresponding work-hardening theory. In [3] this was achieved only for the special case of stressing without rotation for which principal axes of stress and strain remain fixed in the body.

Consider the deforming body at time t. The rate at which mechanical work is being expended in deforming the body, that is the total rate of work less the rate of production of kinetic energy, is given by (see for example [5]):

$$\dot{W} = \int_{V^3} \sigma_{ij} \frac{\partial v^3_i}{\partial x^3_j} \, d V^3 \tag{16}$$

where v^3_i is the particle velocity in the contiguration x^3, and V^3 is the volume in this state. It will prove convenient to use matrix notation and write:

$$\sigma_{ij} = \boldsymbol{T} \tag{17a}$$

$$\frac{\partial v^3_i}{\partial x^3_j} = \frac{\partial v^3_i}{\partial x^1_k} \frac{\partial x^1_k}{\partial x^3_j} = \dot{\boldsymbol{F}}\boldsymbol{F}^{-1} = \boldsymbol{L} \tag{17b}$$

$\boldsymbol{F}$ being $\partial x_i^3/\partial x_j^1$ as in the previous section, and $\dot{\boldsymbol{F}}$ representing time differentiation at constant $\boldsymbol{x}^1$, and thus differentiation relative to a fixed particle since the body is at rest in the initial state $\boldsymbol{x}^1$. Moreover, in order to determine the work expended on a specified element of material, it is convenient to change the variables to integration over V^1, the body in the configuration $\boldsymbol{x}^1$. (16) then becomes:

$$\dot{W} = \int_{V^1} \mathrm{tr}(\boldsymbol{T}\boldsymbol{L}) \det(\boldsymbol{F})\, d\,V^1 \tag{18}$$

where "tr" stands for the trace of a matrix (the sum of diagonal terms) and "det" for determinant. Det $(\boldsymbol{F})$ is the Jacobean of the transformation of variables for the volume integral. Substituting (10) for $\boldsymbol{F}$ in (18) and (17b), and using the product formula for differentiation of a matrix product, (18) becomes

$$\dot{W} = \int_{V^1} \mathrm{tr}[\boldsymbol{T}(\dot{\boldsymbol{F}}^e\boldsymbol{F}^p + \boldsymbol{F}^e\dot{\boldsymbol{F}}^p)\,\boldsymbol{F}^{p^{-1}}\boldsymbol{F}^{e^{-1}}] \det(\boldsymbol{F}^e) \det(\boldsymbol{F}^p)\, d\,V^1 \tag{19}$$

Zero volume change in plastic flow determines

$$\det(\boldsymbol{F}^p) = 1 \tag{20}$$

and (19) then reduces to

$$\dot{W} = \int_{V^1} \mathrm{tr}\left[\boldsymbol{T}\dot{\boldsymbol{F}}^e\boldsymbol{F}^{e^{-1}}\right] \det(\boldsymbol{F}^e)\, d\,V^1 + \int_{V^1} \mathrm{tr}\left[\boldsymbol{T}\boldsymbol{F}^e\dot{\boldsymbol{F}}^p\boldsymbol{F}^{p^{-1}}\boldsymbol{F}^{e^{-1}}\right] \det(\boldsymbol{F}^e)\, d\,V^1 \tag{21}$$

which divides into two vomponents

$$\dot{W}^e = \int_{V^1} \mathrm{tr}\left(\boldsymbol{T}\dot{\boldsymbol{F}}^e\boldsymbol{F}^{e^{-1}}\right) \det(\boldsymbol{F}^e)\, d\,V^1 \tag{22a}$$

$$\dot{W}^p = \int_{V_1} \mathrm{tr}\left(\boldsymbol{T}\boldsymbol{F}^e\dot{\boldsymbol{F}}^p\boldsymbol{F}^{p^{-1}}\boldsymbol{F}^{e^{-1}}\right) \det(\boldsymbol{F}^e)\, d\,V^1 . \tag{22b}$$

According to the kinematics discussed in connection with Fig. 2, $\boldsymbol{F}^e$ represents the deformation gradient associated with elastic strain of the plastically deformed body at each instant. Thus (22a) might simply be considered as the rate of storage of elastic strain energy when the body is taken through the deformation history $\boldsymbol{F}^e(\boldsymbol{x}^2, t)$. However, care must be exercised since the situation differs essentially from the usual problem of deforming an elastic body, since normally the unstressed reference state is a fixed undistorting configuration of the body. In the present problem, the unstressed reference state $\boldsymbol{x}^2$, is continuously deforming plastically.

It is known that moderate plastic flow has only a minor influence on subsequent elastic characteristics, for example, that the generation of elastic anisotropy in rolling is negligible up to 30% reduction ([4] p. 24).

Indeed, stress analysis is usually based on standard elastic constants even though the components may have been cold-worked. We shall therefore consider the elastic properties relating the stress $\boldsymbol{T}$ in $\boldsymbol{x}^3$ to the elastic deformation gradient $\boldsymbol{F}^e$ to be constant and to represent isotropic material response. Since the material in the unstressed configuration $\boldsymbol{x}^2$ has the same uniform density as in the initial state $\boldsymbol{x}^1$, response from $\boldsymbol{x}^2$ is not influenced by its deformation or rotation for isotropic elasticity with prescribed characteristics. Thus the relationship between stress $\boldsymbol{T}$ and deformation $\boldsymbol{F}^e$ will be the same as that for deformation of an elastic body from a fixed reference configuration, and (22a) will represent the elastic energy invested in the body mechanically. The elastic properties can be conveniently expressed in terms of the Helmholtz free energy function per unit mass $\psi(\boldsymbol{C}^e, \theta)$, where $\boldsymbol{C}^e = \boldsymbol{F}^{e^T} \boldsymbol{F}^e$ and θ is the temperature, and the stress and entropy in the thermo-elastic system are then given by ([6] p. 302 and 309):

$$\boldsymbol{T} = 2\varrho_0 \boldsymbol{F}^e \frac{\partial \psi}{\partial \boldsymbol{C}^e} \boldsymbol{F}^{e^T} / \det(\boldsymbol{F}^e) \tag{23a}$$

$$S^e = -\frac{\partial \psi}{\partial \theta} \tag{23b}$$

where ϱ_0 is the initial density, and S^e the entropy. Examples of forms of the free energy function ψ are given in [3] and [7].

Since (22a) gives the elastic work, (22b) must represent the mechanical work expended in plastic flow. This appears to exhibit involved coupling with the elastic deformation $\boldsymbol{F}^e$, but this can be largely eliminated as follows.

Consider the term

$$\operatorname{tr}\left(\boldsymbol{T} \boldsymbol{F}^e \dot{\boldsymbol{F}}^p \boldsymbol{F}^{p^{-1}} \boldsymbol{F}^{e^{-1}}\right). \tag{24}$$

It was pointed out in the discussion on kinematics that the state $\boldsymbol{x}^2$ is not uniquely defined since de-stressing can involve an arbitrary rotation. Let us consider the special case of de-stressing without rotation so that the principal directions of stress remain fixed in the body. For an isotropic elastic material the principal directions of elastic strain will coincide with those of stress, and this special elastic strain deformation gradient $\overline{\boldsymbol{F}}^e$ will be symmetric, and represent pure stretch in the directions of the principal axes without rotation or shear. The availability of this choice for $\boldsymbol{F}^e$ can also be seen analytically since the polar decomposition theorem ([6] p. 52) permits any matrix of the form $\boldsymbol{F}^e$ to be expressed as

$$\boldsymbol{F}^e = \boldsymbol{V}\boldsymbol{R} \tag{25}$$

where $\boldsymbol{V}$ is symmetric and $\boldsymbol{R}$ a rotation matrix. Thus (10) can be written:

$$\boldsymbol{F} = \boldsymbol{F}^e\boldsymbol{F}^p = \boldsymbol{V}\boldsymbol{R}\boldsymbol{F}^p = \boldsymbol{V}(\boldsymbol{R}\boldsymbol{F}^p) = \overline{\boldsymbol{F}}^e\overline{\boldsymbol{F}}^p. \tag{26}$$

The trace of a product of matrices is not influenced by cyclic permutation of the elements as is clear from the subscript representation, so that (24) can be written for the special choice of $\boldsymbol{x}^2$:

$$\operatorname{tr}\left(\overline{\boldsymbol{F}}^{e^{-1}}\boldsymbol{T}\overline{\boldsymbol{F}}^e\dot{\overline{\boldsymbol{F}}}^p\overline{\boldsymbol{F}}^{p^{-1}}\right). \tag{27}$$

The trace is a scalar invariant of the tensor arguments, so that change of axes does not modify its value. Selecting axes parallel to the principal stress directions $\overline{\boldsymbol{F}}^e$ and $\boldsymbol{T}$ become diagonal matrices which are commutative, so that $\overline{\boldsymbol{F}}^e$ and $\overline{\boldsymbol{F}}^{e-1}$ cancel, and (27) reduces to

$$\operatorname{tr}\left(\boldsymbol{T}\dot{\overline{\boldsymbol{F}}}^p\overline{\boldsymbol{F}}^{p^{-1}}\right) \tag{28}$$

which retains this form on transformation back to the original coordinates. Thus the plastic rate of work term (22b) becomes

$$\dot{W}^p = \int\limits_{V^1} \operatorname{tr}\left(\boldsymbol{T}\dot{\overline{\boldsymbol{F}}}^p\overline{\boldsymbol{F}}^{p^{-1}}\right)\det(\boldsymbol{F}^e)\,dV^1 = \int\limits_{V^1} \sigma_{ij}\frac{\partial v_i^2}{\partial x_j^2}\det(\boldsymbol{F}^e)\,dV^1 \tag{29}$$

where v_i^2 is the velocity of plastic straining in the configuration $\boldsymbol{x}^2$. Note that the right hand equality is written for illustrative purposes and applies only when the configuration $\boldsymbol{x}^2$ happens to be continuous and differentiable. Thus the plastic rate of work term takes on the simple form of work done by the stress $\boldsymbol{T}$ on the plastic rate of deformation with a scalar coupling term to the elastic deformation: $\det(\boldsymbol{F}^e)$. It is now seen that the result is identical with that discussed in [3] in which the analysis was limited to total deformation with principal directions fixed in the body, in which all matrices considered were diagonal so that the matrix contribution of $\boldsymbol{F}^e$ cancelled without the above special consideration. It is clear physically why this special choice of $\overline{\boldsymbol{F}}^e$ is needed, since when an increment of strain takes place the plastic component will have principal axes parallel to that of stress, and positive work will be done during the flow. If the de-stressing includes rotation, the plastic strain increment could have a quite independent orientation and would then not lead to plastic work as it is actually taking place.

The stress rate-of-deformation relation for plastic flow now follows by direct generalization of the analysis in [3] and [8]. Relation (29) can be written in the form

$$\dot{W}^p = \int\limits_{V^1} \operatorname{tr}\left[\boldsymbol{T}(\det\boldsymbol{F}^e)\,\dot{\overline{\boldsymbol{F}}}^p\overline{\boldsymbol{F}}^{p^{-1}}\right]dV^1 \tag{30}$$

We have to devise a yield law and work hardening analysis which is consistent with the experimental facts that superposed hydrostatic pressure does not appreciably modify the stress strain relation, and that the development of hardening by the generation of dislocations will require as much work expended under pressure as at ambient pressure, as discussed in [3]. The ($\det \boldsymbol{F}^e$) term in (30), which equals ϱ_0/ϱ, is less than unity with hydrostatic pressure and hence tends to reduce the plastic work for a given strain increment. To eliminate this contradictory circumstance, we suggest the yield condition:

$$f[\boldsymbol{T}(\det \boldsymbol{F}^e)] = c \tag{31}$$

where f takes on one of the usual forms, such as Mises or Tresca, but could be a more general function of J_2 and J_3, the second and third invariants of the argument deviator, as discussed in [4]. Note that $\det(\boldsymbol{F}^e)$ is a scalar multiplier which does not alter the orientation of the tensor argument. In static experiments carried out to date, from which the inclusion of ($\det \boldsymbol{F}^e$) in (31) could be tested, its value has been too close to unity for this form to be denied or confirmed.

For a work hardening material subject to temperature changes, c in (31) has been considered to be a function of W^p and temperature θ, where W^p now corresponds to homogeneous stress distribution in a unit initial volume of material. However, it is known that increasing the temperature increases the mobility of dislocations and hence permits the generation of increased dislocation density and so of hardening with a smaller expenditure of plastic work. Hence a functional of W^p is needed, rather than a function, and the form

$$c = c\left[\int_0^t \alpha(\theta)\, \dot{W}^p\, d\tau,\ \theta\right] \tag{32}$$

suggests itself. $\alpha(\theta)$ will be an increasing function of θ, to achieve the desired temperature-dislocation mobility influence. Relation (31) with changing c according to (32) will yield the desired set of similar yield surfaces when f is an isotropic homogeneous function of its argument components, as is the Mises condition.

Following PRAGER [8] it is revealing to write the yield condition (31), (32) in the form

$$f(\boldsymbol{Q}) - c[\varphi, \theta) \leq 0 \tag{33}$$

where $\boldsymbol{Q}$ stands for:

$$\boldsymbol{Q} = \boldsymbol{T}(\det \boldsymbol{F}^e) \tag{34}$$

and φ represents the functional:

$$\varphi = \int_0^t \alpha(\theta)\,\dot{W}^p\,d\tau. \tag{35}$$

The equality sign in (33) denotes plastic flow or neutral loading, and the inequality sign, an elastic state. For continuing plastic flow, differentiation of (33) gives

$$\operatorname{tr}\frac{\partial f}{\partial Q}\dot{Q} - \frac{\partial c}{\partial \varphi}\dot{\varphi} - \frac{\partial c}{\partial \theta}\dot{\theta} = 0 \tag{36}$$

since the equality sign continues to hold. Since the plastic work and hence φ is positive, continued plastic flow is associated with

$$\operatorname{tr}\frac{\partial f}{\partial \boldsymbol{Q}}\dot{\boldsymbol{Q}} - \frac{\partial c}{\partial \theta}\dot{\theta} > 0 \tag{37}$$

and this becomes the loading condition. Note that this corresponds to the yield criterion f increasing in isothermal plasticity, but for the temperature influence on the yield stress which is included here.

In order to define a law for the rate of plastic deformation, we must introduce the velocity strain in plastic flow often loosely called the rate of strain. This is defined as the symmetric part of the velocity gradient in the continuous case:

$$\bar{\boldsymbol{D}}^p = (\bar{\boldsymbol{L}}^p + \bar{\boldsymbol{L}}^{p^T})/2 = \left(\frac{\partial v_i^2}{\partial x_j^2} + \frac{\partial v_j^2}{\partial x_i^2}\right)/2; \quad \bar{\boldsymbol{L}}^p = \dot{\bar{\boldsymbol{F}}}^p \bar{\boldsymbol{F}}^{p^{-1}}. \tag{38}$$

Because $\boldsymbol{T}$ is symmetric (30) can be written

$$\dot{W}^p = \int_{V^1} \operatorname{tr}[\boldsymbol{T}(\det \boldsymbol{F}^e)\,\bar{\boldsymbol{D}}^p]\,dV^1 \tag{39}$$

Using the concept of plastic potential, as utilized, for example, by PRAGER [8], we write

$$\bar{\boldsymbol{D}}^p = k\frac{\partial f}{\partial \boldsymbol{Q}}\left(\operatorname{tr}\frac{\partial f}{\partial \boldsymbol{Q}}\dot{\boldsymbol{Q}} - \frac{\partial c}{\partial \theta}\dot{\theta}\right) \tag{40}$$

where k is a function of the current state of be determined. Since the equality sign applies in (33) we can substitute (40) into (36), using (39) and obtain:

$$\operatorname{tr}\frac{\partial f}{\partial \boldsymbol{Q}}\dot{\boldsymbol{Q}} - \frac{\partial c}{\partial \theta}\dot{\theta} = \frac{\partial c}{\partial \varphi}\alpha(\theta)\operatorname{tr}\left[\boldsymbol{Q}\,k\frac{\partial f}{\partial \boldsymbol{Q}}\left(\operatorname{tr}\frac{\partial f}{\partial \boldsymbol{Q}}\dot{\boldsymbol{Q}} - \frac{\partial c}{\partial \theta}\dot{\theta}\right)\right]. \tag{41}$$

The scalar quantities k and $\left(\mathrm{tr}\,\frac{\partial f}{\partial \boldsymbol{Q}}\,\dot{\boldsymbol{Q}} - \frac{\partial c}{\partial \theta}\,\dot{\theta}\right)$ factorize from the trace expression, the latter simplifies by use of Euler's theorem since f is homogeneous in $\boldsymbol{Q}$, and (41) reduces to

$$l = \frac{\partial c}{\partial \varphi}\, k \alpha n f \tag{42}$$

where n is the order of $\boldsymbol{Q}$ in $f(\boldsymbol{Q})$, and (40) becomes

$$\bar{\boldsymbol{D}}^p = \frac{\partial f}{\partial \boldsymbol{Q}}\left(\mathrm{tr}\,\frac{\partial f}{\partial \boldsymbol{Q}}\,\dot{\boldsymbol{Q}} - \frac{\partial c}{\partial \theta}\,\dot{\theta}\right) \Big/ \alpha n f\, \frac{\partial c}{\partial \varphi}. \tag{43}$$

This relation is equivalent to that for isothermal work hardening given by Hill ([4], p. 38) and is seen not to be appreciably complicated by the inclusion of finite strain, non-linear elasticity and thermo-mechanical coupling. In the previous analysis [3] for application to waves of one-dimensional strain, only one independent strain component is sought and the plastic work equation equivalent to (39) is sufficient to determine this, and the plastic potential concept is not needed. For the analysis of general deformation, however, this additional requirement is necessary, and incidentally will lead to a maximum plastic work theorem and a uniqueness of solution argument as in the isothermal, linear elastic case.

The Thermodynamic System

The thermo-mechanical coupling associated with plastic flow can be introduced as in the fixed principal directions case [3]. The heat capacity of the body is considered associated with the thermo-elastic system, which corresponds to the basic crystal structure of the metal. Plastic flow is associated with the generation of dislocations, which cause work hardening, but do not influence sufficient atoms to essentially modify the crystal form. Thus the elastic properties and specific heat are not appreciably influenced by plastic flow. The migration of dislocations is an irreversible process in which mechanical energy is dissipated into heat and the randomness associated with their distribution can be associated with an increase in entropy. Moreover a strain energy field surrounds each dislocation which absorbs some of the work of plastic flow as the dislocation density grows. For the specific thermo-mechanical coupling process generated by plastic flow, we assume that measured by Taylor and Farren [9] and Taylor and Quinney [10], in which a temperature rise was observed in elastic-plastic tests which exceeded that expected from thermo-elastic theory by an amount equivalent to a heat source equal to $\gamma \dot{W}^p$, where γ is a factor which varies slowly between about 0.9 and unity with increasing plastic flow. The rest of the plastic work rate, $(1 - \gamma)\dot{W}^p$, is considered to be stored in the developing dislocation system. Thus the internal energy per unit mass associated with thermo-elastic deformation E^e, and that stored in the dislocation distribution E^p, are, for adiabatic deformation, determined by

$$\varrho_0 \dot{E}^e = \dot{W}^e + \gamma \dot{W}^p \tag{44}$$

$$\varrho_0 \dot{E}^p = (1 - \gamma)\dot{W}^p \tag{45}$$

where eqs. (22) are specialized for a unit initial volume uniformly strained. Growth of elastic entropy S^e arises due to the mechanical energy dissipated in plastic flow, and is given by:

$$\theta \varrho_0 \dot{S}^e = \gamma \dot{W}^p \tag{46}$$

These relations assume continuous variations of deformation. In the case of shock-wave loading, involving discontinuous changes of stress and strain, additional increase in entropy occurs over and above that given by (46).

Combination of the components developed above determines the total stress-deformation relation. Equation (23) for elastic strain, (33) and (43) for plastic flow and (44) and (46) for the coupling between these components.

Conclusions

Plastic flow is an extremely complicated process, and a theory including its various aspects in detail would be almost prohibitively complex. That presented above inlcudes several aspects necessary for the adequate study of plastic flow under high pressure which are not included in a satisfactory manner in classical theory. Others, such as rate effects, lack of isotropy of work hardening including the Bauschinger effect, may be important in certain applications, and the present theory would need to be extended for these cases.

Acknowledgement

The major part of this development was carried out under the sponsorship of the Ballistic Research Laboratory, Aberdeen Proving Ground under a contract with Stanford University. This support is appreaciated by the author. Thanks also are due to Professor T. C. T. Ting, currently visiting Stanford from the University of Illinois, Chicago Circle, for stimulating discussions which helped to clarify aspects of the theory. In addition to the IUTAM Symposium on Thermo-inelasticity, it is also planned to present this material at the Army Symposium on Solid Mechanics, to be held at Johns Hopkins University, Baltimore, Maryland in September 1968.

References

[1] LEE, E. H., and D. T. LIU: Finite Strain Elastic-Plastic Theory. Proc. IUTAM Symp. on Irreversible Aspects of Continuum Mechanics. Wien-New York: Springer. 1968.

[2] PEARSON, C. E.: Theoretical Elasticity, p. 198–203. Cambridge: Harvard University Press. 1959.

[3] LEE, E. H., and D. T. LIU: Finite-Strain Elastic-Plastic Theory with Application to Plane Wave Analysis. J. Appl. Phys. **38**, 19–27 (1967).

[4] HILL, R.: The Mathematical Theory of Plasticity, p. 23–45. Oxford: Clarendon Press. 1950.

[5] PRAGER, W.: Introduction to Mechanics of Continua, p. 87. Boston: Ginn and Co. 1961.

[6] TRUESDELL, C., and W. NOLL: The Non-linear Field Theories of Mechanics. Handbuch der Physik, Vol. III/3. Berlin-Göttingen-Heidelberg: Springer. 1965.

[7] LEE, E. H., and T. WIERZBICKI: Analysis of the Propagation of Plane Elastic-Plastic Waves at Finite Strain. J. Appl. Mech. **34**, 931–936 (1967).

[8] PRAGER, W.: Non-isothermal Plastic Deformation. Koninkl. Ned. Akad. Wetenschap., Proc. **61**, 176 (1958).

[9] FARREN, W. S., and G. I. TAYLOR: The Heat Developed during Plastic Deformation of Metals. Proc. Roy. Soc. **107**, 422–51 (1925).

[10] QUINNEY, H., and G. I. TAYLOR: The Latent Energy Remaining in a Metal after Cold Working. Proc. Roy. Soc. **143**, 307–26 (1934).

[11] LEE, E. H.: Elastic-Plastic Deformation at Finite Strains. J. Appl. Mech. **36**, 1–6 (1969).

Discussion

ODQVIST: You referred to the paper by W. PRAGER of 1958 where he generalizes DRUCKER's postulate on the yield surface to varying temperature. In that paper PRAGER takes into account anisotropy induced by deformation. The strain itself is considered to be small as well as the strain rate of the yield surface. Could you give a more precise relation of your theory as compared with that of PRAGER?

LEE: Prof. PRAGER's paper deals with plastic-rigid analysis, and with the influence of both temperature, strain and work hardening on plastic flow. The present paper describes a theory of plasticity which includes finite deformation, both elastic and plastic, and thermomechanical coupling. Once the elastic strain has been separated, the plasticity theory required takes on a similar form to that for plastic-rigid theory. Since temperature changes must be taken into account in the finite strain circumstance, a plastic potential analogous to that considered by PRAGER is used. However, in view of the finite elastic strain, some coupling with elasticity associated with "geometrical nonlinearity" is needed, and this modifies the yield condition. In the present theory only work hardening was included, and is spite of the finite strain analysis, the general structure developed by PRAGER can be applied.

Sur la Normalité de la Fonction de Fluage Lors d'Essais de Traction d'un Dural à 200°C

Par

R. Mazet

Châtillon (France)

Résumé

Des essais de traction à vitesse constante, effectués à 200 °C sur des éprouvettes en duralumin, sont interprétés à partir des lois de Hooke et de Norton et d'une théorie simplifiée des dislocations. Dans un large intervalle d'allongements, la distribution statistique des seuils de fluage obéit à une loi normale du type de Rayleigh. La condition que l'étendue de cet intervalle soit maximale est utilisée pour déterminer la contrainte en fluage établi.

Introduction

Les relations de causalité qui commandent la résistance des solides à l'échelle macroscopique ne peuvent se traduire en formules mathématiques valables que si elles expriment — sous une forme globale donc nécessairement approchée — les phénoménes complexes dont le solide est le siége à l'échelle microscopique. Il est bien connu que les matériaux du commerce sont des milieux essentiellement hétérogénes où l'abondance et la mobilité des dislocations jouent un rôle essentiel. Il est donc à prévoir que les relations contraintes-déformations appelées à servir de base au calcul des structures hors du domaine élastique — calcul de plus en plus répandu dans la construction moderne — devront tirer leur origine de considérations statistiques. Le présent travail est une tentative en vue de savoir s'il est possible de définir une forme de loi statistique « spécifique » d'un matériau donné.

Pose du problème

Le matériau utilisé est un duralumin AU_4G trempé et vieilli à 200 °C pendant 24 heures. Trois éprouvettes de dimensions identiques ont été taillées dans la même tôle, dans la même direction, et soumises à un essai d'écrouissage en traction simple à température et vitesse constantes, la température étant maintenue à 200 °C et la vitesse variant d'une éprouvette à l'autre dans un large intervalle (rapports : 1, 20, 60). Les

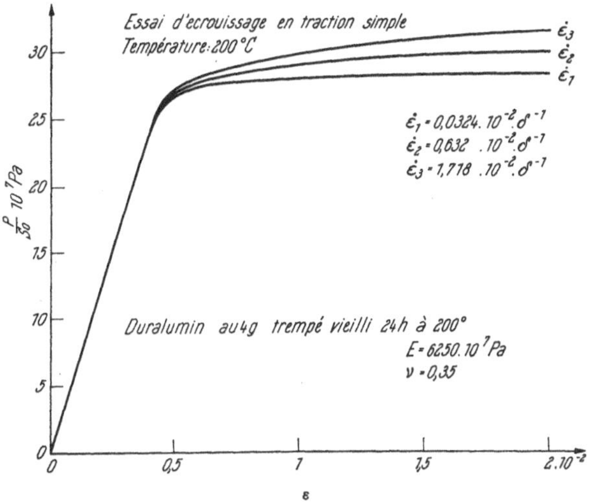

Fig. 1. Essais de trois éprouvettes de dural entraction simple à vitesse constante à 200 °C

résultats sont portés sur la Fig. 1. Le problème consiste à chercher une relation entre la contrainte $\frac{P}{S}$, l'allongement ε et la vitesse d'allongement $\dot{\varepsilon}$ qui rende compte aussi exactement que possible des résultats d'expériences à partir d'une répartition statistique des propriétés des éléments constitutifs du matériau (monocristaux ou „micrograins") et, notamment, de leur limite élastique.

Hypothèses de base

Ces hypothèses concernent, d'une part, le comportement du micrograin (marqué de l'indice i), d'autre part la loi statistique. Elles sont au nombre de six:

a) L'allongement global ε, compté parallèlement à l'axe de l'éprouvette, est le même pour tous les micrograins; il est la somme d'un allongement élastique ε_i^e et d'un allongement plastique ε_i^p variables d'un micrograin à l'autre (*hypothese de partition*):

$$\varepsilon = \varepsilon_i^e + \varepsilon_i^p. \tag{1}$$

b) La contrainte σ_i subie par un micrograin dans la direction de la charge peut se définir indifféremment comme fonctionnelle temporelle de la déformation élastique ε_i^e ou de la déformation plastique ε_i^p (*hypothèse de l'indifférence des contraintes*)

$$\sigma_i = \mathfrak{F}(\varepsilon_i^e) = \mathfrak{G}(\varepsilon_i^p). \tag{2}$$

Cette hypothèse revient à admette que, la déformation plastique étant due au mouvement des dislocations, on peut, la connaissant ainsi que son évolution dans le temps, calculer les contraintes dans le milieu élastique qui l'enrobe (cf. le modèle d'inclusion d'Eshelby) et, de là, déduire la déformation élastique dont le rôle est d'assurer la compatibilité des déformations globales.

c) Il existe, pour chaque micrograin, une *limite élastique apparente* ε_{Mi} dans la direction de la sollicitation. Cette apparence peut être due soit à une orientation aléatoire des plans de glissement du monocristal au sein du polycristal ou „grain", soit à une valeur aléatoire du seuil de glissement, soit aux deux causes réunies. Cette limite ε_{Mi} dépend, en principe, de $\dot{\varepsilon}$

$$\varepsilon \leq \varepsilon_{Mi}(\dot{\varepsilon}) \Rightarrow \varepsilon_i^p = 0. \tag{3}$$

d) *Lois de comportement:* Les fonctionnelles de la formule (2) sont indépendantes de l'instant considéré. Elles présentent, en outre, le caractère newtonien [1], c'est-à-dire qu'elles ne mettent en jeu, au plus, que la déformation (élastique ou plastique) à l'instant t et la vitesse de déformation au même instant:

$$\mathfrak{F}(\varepsilon_i^e) \equiv f(\varepsilon_i^e, \dot{\varepsilon}_i^e), \quad \mathfrak{G}(\varepsilon_i^p) \equiv g(\varepsilon_i^p, \dot{\varepsilon}_i^p). \tag{4}$$

Avant d'aller plus loin, nous devons préciser l'expression des fonctions f et g. En ce qui concerne f, nous prendrons la loi de Hooke:

$$f(\varepsilon_i^e, \dot{\varepsilon}_i^e) \equiv E\,\varepsilon_i^e \qquad (E\text{: Module d'Young}). \tag{5}$$

Pour g, considérons la courbe contrainte-déformation à vitesse constante d'un monocristal de duralumin à réseau CFC (Fig. 2). Elle comprend plusieurs phases: OA phase élastique, AB phase des glissements faciles, BC phase du durcissement rapide, CD phase du relâchement. Pour éviter l'introduction d'un trop grand nombre de paramètres sujets à des variations aléatoires, nous schématiserons fortement le phénomène physique en remplaçant la courbe réelle $OABCD$ par la courbe en pointillé $OA'B'CD'$ dans laquelle $A'B'$ est verticale et la tangente en B' à $B'CD'$ est parallèle à OA'. Seule, l'abscisse de $A'B'$, ε_{Mi}, sera sujette à une variation aléatoire entre une limite inférieure et $+\infty$. Nous supposerons en outre que la déformation plastique, dans la phase $B'CD'$ qui correspond au fluage, obéit à la loi de Norton:

$$\varepsilon > \varepsilon_{Mi}(\dot{\varepsilon}) \Rightarrow g(\varepsilon_i^p, \dot{\varepsilon}_i^p) \equiv E(\tau \dot{\varepsilon}_i^p)^{1/n} \tag{6}$$

(τ, n: constantes spécifiques[1]) avec $\dot{\varepsilon}_i^p = 0$ pour $\varepsilon = \varepsilon_{Mi}$.

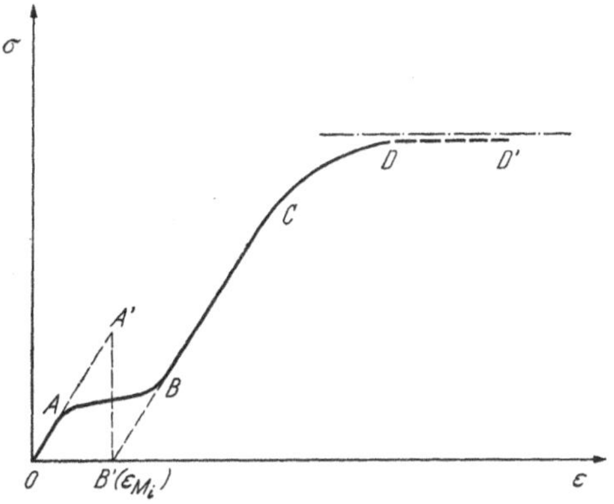

Fig. 2. Courbe contrainte-déformation d'un monocristal *CFC*

e) La distribution statistique des limites élastiques (ou seuils de fluage) ε_{Mi} est définie par la *fonction de fluage* $\theta(\varepsilon | \dot{\varepsilon})$ qui mesure, dans un volume donné, le rapport entre le nombre des micrograins subissant une déformation plastique $[\varepsilon_{Mi}(\dot{\varepsilon}) < \varepsilon]$ et le nombre total des micrograins; on peut écrire:

$$\theta(\varepsilon | \dot{\varepsilon}) = \frac{1}{S_0} \int_{S_0} (\partial s_i'')_0 \tag{7}$$

en désignant par $(\partial s_i'')_0$ l'aire d'un élément de la section initiale S_0 dont la limite élastique a été dépassée quand l'allongement atteint la valeur ε avec la vitesse $\dot{\varepsilon}$.

---

[1] La constante spécifique qui intervient dans la loi de Norton n'est pas τ, mais τ' telle que

$$\sigma_i' = \frac{E}{1+\nu} (\tau' \dot{\varepsilon}_i'^p)^{1/n} \quad (\nu\text{: coefficient de Poisson})$$

σ_i' et $\dot{\varepsilon}_i'^p$ étant les modules des déviateurs respectivement des contraintes et des vitesses de déformation. En réalité, τ dépend faiblement de ε; en effet: $\frac{\tau}{\tau'} = \left(\frac{3}{2}\right)^{(n-1)/2} \frac{1+\nu^*}{(1+\nu)^n}$, ν^* étant le coefficient défini à propos de l'hypothèse *f*. Pratiquement on peut prendre: $\frac{\tau}{\tau'} \approx \left(\frac{3}{2}\right)^{(n-1)/2} (1+\nu)^{-n}$.

La fonction $\theta(\varepsilon\,|\,\dot{\varepsilon})$, d'abord nulle, est évidemment croissante quand ε varie au-delà d'une certaine valeur ε_F, mais reste comprise entre 0 et 1. On la supposera spécifique du matériau considéré, du traitement qu'il a subi avant l'essai et de la température T.

f) La déformation élastique s'accompagne d'une contraction de coefficient (de Poisson) ν; la déformation plastique s'effectue à volume constant $\left(\nu = \frac{1}{2}\right)$. Il en résulte que la section de l'éprouvette, initialement égale à S_0, est devenue pour l'allongement ε:

$$S = S_0[1 - 2\nu\varepsilon - 2(\nu^* - \nu)\,\theta\varepsilon] \approx \frac{S_0}{1+\varepsilon} \tag{8}$$

ν^* étant un coefficient variable compris entre ν et $\frac{1}{2}$, tel que:

$$\frac{\int\limits_S \partial s_i''}{\int\limits_{S_0} (\partial s_i'')_0} = 1 - 2\nu^*\varepsilon.$$

Formule fondamentale

Soit P la charge de l'éprouvette. Posons:

$$\frac{P}{ES} = k.$$

Nous nous proposons, connaissant la fonction de fluage θ, d'exprimer k en fonction de ε et de $\dot{\varepsilon}$ au moyen des hypothèses a à e (l'hypothese f sert seulement à calculer k à partir de P et de S_0).

$$k = \frac{1}{S}\int\limits_S \varepsilon_i^e\, ds\, i \text{ avec } \varepsilon_i^e = \begin{cases} \varepsilon \;\; si \;\; \varepsilon < \varepsilon_{Mi} & \text{(9a)} \\ (\tau\dot{\varepsilon}_i^p)^{1/n} \;\; si \;\; \varepsilon > \varepsilon_{Mi} & \text{(9b)} \end{cases}$$

Pour $\varepsilon = \varepsilon_{Mi}$, ε_i^e a deux valeurs: ε_{Mi} á gauche, zéro à droite[1]. La formule (9b) se transforme, compte tenu de (1) et du fait que $\dot{\varepsilon}$ est constant, en

$$\varepsilon_i^e = \varepsilon - \varepsilon_{Mi} - \int\limits_{\varepsilon_{Mi}}^{\varepsilon} \left(\frac{\varepsilon_i^e}{\lambda}\right)^n d\varepsilon \;\; si \;\; \varepsilon > \varepsilon_{Mi} \tag{10}$$

avec $\lambda = (\tau\dot{\varepsilon})^{1/n}$ (λE est la *contrainte de fluage établi*).

---

[1] Une schématisation plus fine que celle de la Fig. 2 atténuerait cette discontinuité en la remplaçant par une chute rapide de ε_i^e.

La variation de ε_i^e en fonction de ε, définie par les formules (9a) et (10), est représentée par le tracé $OA'B'CD'$ de la Fig. 2.

Pour en déduire k, il suffit de calculer l'intégrale $\int\limits_S \varepsilon_i^e \, \partial s_i$ en utilisant la formule (7):

$$k = \frac{1}{S}\left(\int\limits_{S'} \varepsilon \, \partial s_i' + \int\limits_{S''} \varepsilon_i^e \, \partial s_i''\right). \tag{11}$$

S' désignant l'aire non fluante (au sens de Lebesgue) et S'' l'aire fluante, constituées respectivement d'éléments $\partial s_i'$, pour lesquels $\varepsilon_{Mi} \geq \varepsilon$, et $\partial s_i''$, pour lesquels $\varepsilon_{Mi} < \varepsilon$. (11) s'écrit encore:

$$k = \varepsilon \int\limits_{S'} \frac{\partial s_i'}{S} + \int\limits_{S''} \varepsilon_i^e \frac{\partial s_i''}{S}.$$

On démontre que

$$\frac{\partial s_i'}{S} = \frac{(\partial s_i')_0}{S_0} \frac{1 - 2\nu\varepsilon}{1 - 2[\nu + (\nu^* - \nu)\theta]\varepsilon} \quad [(\partial s_i')_0\text{: aire initiale de l'élément } \partial s_i']$$

$$\frac{\partial s_i''}{S} = \frac{(\partial s_i'')_0}{S_0} \frac{1 - 2\nu_i^* \varepsilon}{1 - 2[\nu + (\nu^* - \nu)\theta]\varepsilon} \quad [(\partial s_i'')_0\text{: aire initiale de l'élément } \partial s_i'']$$

ν_i^* étant, comme ν^*, compris entre ν et $\frac{1}{2}$. Ce résultat autorise à confondre

$$\frac{\partial s_i'}{S} \text{ avec } \frac{(\partial s_i')_0}{S_0} \text{ et } \frac{\partial s_i''}{S} \text{ avec } \frac{(\partial s_i'')_0}{S_0},$$

moyennant quoi

$$k = (1 - \theta)\,\varepsilon + \frac{1}{S_0} \int\limits_{S_0''} \varepsilon_i^e (\partial s_i'')_0. \tag{12}$$

Pour calculer $\int\limits_{S_0''} \varepsilon_i^e (\partial s_i'')_0$, remarquons que:

$$(\partial_s'' \, i)_0 = S_0 \, d\theta \, \{\varepsilon_{Mi}\},$$

$d\theta \, \{\varepsilon_{Mi}\}$ étant l'accroissement de θ lorsque ε varie de ε_{Mi} à $\varepsilon_{Mi} + d\varepsilon_{Mi}$[1].

Par suite:

$$\int\limits_{S_0''} \varepsilon_i^e \, (\partial s_i'')_0 = S_0 \int\limits_{S_0''} \varepsilon_i^e \, d\theta \, \{\varepsilon_{Mi}\} = S_0 \int\limits_{S_0''} \left[\varepsilon - \varepsilon_{Mi} - \int\limits_{\varepsilon_{Mi}}^{\varepsilon} \left(\frac{\varepsilon_i^e}{\lambda}\right)^n d\varepsilon\right] d\theta \, \{\varepsilon_{Mi}\} \tag{13}$$

$$\int\limits_{S_0''} \varepsilon \, d\theta \, \{\varepsilon_{Mi}\} = \theta \varepsilon, \tag{14}$$

[1] En d'autres termes, $d\theta\{\varepsilon\}$ est la probabilité pour que, si l'on prend au hasard un monocristal du matériau, sa limite élastique soit comprise entre ε et $\varepsilon + d\varepsilon$.

$$\int_{S_0''} \varepsilon_{Mi}\, d\theta\, \{\varepsilon_{Mi}\} = [\varepsilon_{Mi}\, \theta\, \{\varepsilon_{Mi}\}]_0^\varepsilon - \int_0^\varepsilon \theta\, \{\varepsilon_{Mi}\}\, d\varepsilon_{Mi} = \theta\varepsilon - \int_0^\varepsilon \theta(u)\, du, \tag{15}$$

$$\int_{S_0''} d\theta\, \{\varepsilon_{Mi}\} \int_{\varepsilon_{Mi}}^{\varepsilon} \left(\frac{\varepsilon_i^e}{\lambda}\right)^n d\varepsilon = \int_0^\varepsilon d\theta\, (u) \int_u^\varepsilon \left[\frac{\varepsilon^e\,(v\,|\,u)}{\lambda}\right]^n dv = \int_0^\varepsilon dv \int_0^v \left[\frac{\varepsilon^e\,(v\,|\,u)}{\lambda}\right]^n d\theta\,(u). \tag{16}$$

Dans les intégrales doubles des deux derniers membres, u représente la valeur courante de ε_{Mi} et v la valeur courante de ε.

Considérons maintenant:

$$I(v) = \int_0^v \left[\frac{\varepsilon^e\,(v\,|\,u)}{\lambda}\right]^n d\theta\,(u)$$

et faisons la nouvelle hypothèse:

g) *n est grand devant l'unité.* Cette hypothèse est parfaitement justifiée dans le cas concret qui nous occupe, car il résulte d'expériences faites sur le même matériau, à la même température, que n peut être pris égal à 27; les essais de fluage, de relaxation et d'écrouissage donnent sensiblement la même valeur [2].

Le fait que n soit grand nous permet de remplacer $\varepsilon^e(v\,|\,u)$ par une expression approchée qui majore légèrement la valeur vraie. En effet, l'arc $B'D'$ de la courbe de la Fig. 2 est très proche de la tangente en B' dans la phase $B'C$ du „durcissement rapide" et très proche de l'horizontale dans la phase CD' du „relâchement"; nous pouvons donc prendre:

$$Si\ v \leq \lambda, \quad \varepsilon^e \approx v - u$$

$$Si\ v \geq \lambda, \quad \varepsilon^e \approx \begin{cases} v - u & \text{pour } u \geq v - \lambda \\ \lambda & \text{pour } 0 \leq u \leq v - \lambda. \end{cases}$$

D'où:

$$I(v) \approx \int_0^{v-\lambda} d\theta + \int_{v-\lambda}^{v} \left(\frac{v-u}{\lambda}\right)^n d\theta \tag{17}$$

avec la condition que $v - \lambda$ soit remplacé par zéro si $v \leq \lambda$.

On démontre aisément que le socond terme du deuxième membre de (17) est de l'ordre de $\dfrac{\lambda}{n}$ et, par suite, négligeable devant le premier dès que v est plus grand que λ [le fait de le négliger compense en partie la majoration opérée sur $\varepsilon^e(v\,|\,u)$]. Cette circonstance nous conduit à prendre en définitive:

$$I(v) \approx \int_{0}^{v-\lambda} d\theta = \theta\,\{v-\lambda\}$$

d'où, en revenant à la formule (16):

$$\int_{S_0''} d\theta\,\{\varepsilon_{Mi}\} \int_{\varepsilon_{Mi}}^{\varepsilon} \left(\frac{\dot{\varepsilon}_i^e}{\lambda}\right)^n d\varepsilon = \int_{0}^{\varepsilon} \theta\,\{v-\lambda\}\,dv \tag{18}$$

et enfin, revenant à la formule (12) compte tenu de (13), (14), (15) et (18):

$$k = (1-\theta)\,\varepsilon + \theta\varepsilon - \theta\varepsilon + \int_{0}^{\varepsilon} \theta\,(u)\,du - \int_{0}^{\varepsilon} \theta\,\{v-\lambda\}\,dv =$$

$$= (1-\theta)\,\varepsilon + \int_{\varepsilon-\lambda}^{\varepsilon} \theta(u)\,du \qquad \text{pour } \varepsilon \geq \lambda. \tag{19}$$

Telle est la relation cherchée entre k, ε, $\dot{\varepsilon}$ (ou λ) et θ.

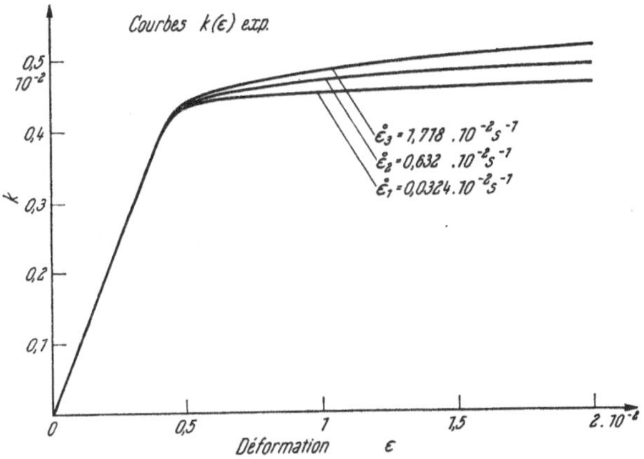

Fig. 3. Courbes $k(\varepsilon)$ expérimentales

Détermination expérimentale de la fonction θ

Si l'on se donne les valeurs de $k\big(\varepsilon\,|\,\lambda\,(o\,|\,u)\big)$ correspondant à une expérience d'écrouissage à vitesse constante [nous disposons de trois expériences de ce type; les courbes k correspondantes — obtenues, d'après (8), par la formule $k = \dfrac{P}{E S_0}\,(1+\varepsilon)$ — sont reportées sur la Fig. 3], on peut en déduire $\theta\,(\varepsilon\,|\,\lambda)$. Il suffit pour cela de calculer λ connaissant τ et n

et de résoudre l'équation intégrale (19). La méthode employée consiste à poser:

$$\varepsilon = n \frac{\lambda}{\tau},$$

τ étant un entier arbitraire suffisamment grand (dans le calcul: $\tau = 40$) et n un entier $\geq \tau$.

Désignant ensuite par i un entier compris entre 1 et n $(1 \leq i \leq n)$ et posant:

$$\theta_i = \theta \left\{ i \frac{\lambda}{\tau} \right\}$$

on remplace la relation (19) par la formule de récurrence:

$$k \left\{ n \frac{\lambda}{\tau} \right\} = (1 - \theta_n)\, n \frac{\lambda}{\tau} + \frac{\lambda}{\tau} \sum_{n-\tau}^{n-1} \theta_i \qquad (20)$$

qui s'écrit encore:

$$\theta_n = 1 - \frac{\tau}{\lambda n} k \left\{ n \frac{\lambda}{\tau} \right\} + \frac{1}{n} (\theta_{n-\tau} + \theta_{n-\tau+1} + \cdots + \theta_{n-1}). \qquad (21)$$

Les valeurs de θ se calculent de proche en proche à partir des τ premières valeurs $\theta_1, \theta_2, \ldots, \theta_\tau$ que l'on détermine en identifiant k dans l'intervalle $[o, \lambda]$ à une expression de la forme:

$$\varepsilon - m(\varepsilon - \varepsilon_F)^2, \qquad (22)$$

m et ε_F étant deux constantes (ε_F sera appelée la *limite élastique vraie* à la vitesse $\dot{\varepsilon}$). On trouve ainsi:

$$\theta_i = \begin{cases} 0 & \text{pour } i \frac{\lambda}{\tau} \leq \varepsilon_F \\ 2m \left[\varepsilon - \varepsilon_F \left(1 + \log \frac{\varepsilon}{\varepsilon_F} \right) \right] & \text{pour } \varepsilon_F \leq i \frac{\lambda}{\tau} \leq \lambda. \end{cases}$$

On notera que, dans la formule (20), le dernier terme représente une borne inférieure de l'intégrale $\int_{\varepsilon-\lambda}^{\varepsilon} \theta(u)\, du$, ce qui conduit à une sous-estimation de la valeur de $\theta \left\{ i \frac{\lambda}{\tau} \right\}$. On pourrait de la même façon prendre la borne supérieure $\frac{\lambda}{\tau} \sum_{n-\tau+1}^{n} \theta'_i$, ce qui donnerait pour $\theta \left\{ i \frac{\lambda}{\tau} \right\}$ une valeur surestimée θ'_i. Soit alors:

$$\theta'_i - \theta_i = \delta_i,$$

δ_n se calcule par la formule de récurrence:

$$\delta_n = \frac{1}{n-1}\,(\theta_n - \theta_{n-\tau} + \delta_{n-\tau+1} + \cdots + \delta_{n-1}) \text{ avec } \delta_i = 0 \text{ pour } i \leq \tau \tag{23}$$

et, si l'on prend pour $\theta\ (\varepsilon)$ la moyenne $\frac{\theta_n + \theta_n'}{2}$, c'est-à-dire:

$$\theta\left\{n\,\frac{\lambda}{\tau}\right\} = \theta_n + \frac{\delta_n}{2}.$$

l'erreur relative commise sur θ est inférieure à $\frac{\delta_n}{2\theta_n}$.

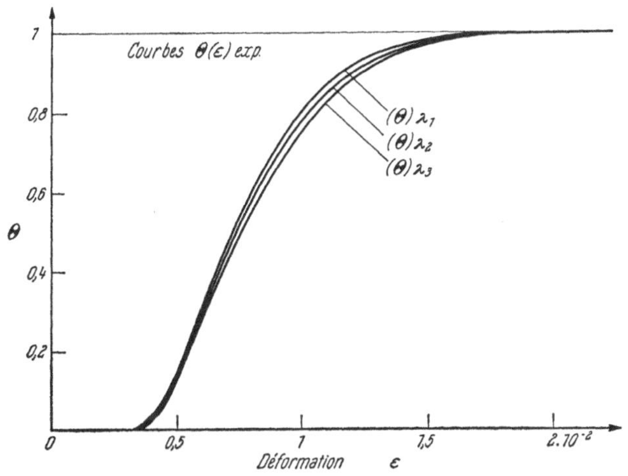

Fig. 4. Courbes $\varepsilon\,(\theta)$ expérimentales

Les courbes $\theta(\varepsilon \mid \lambda)$ ainsi obtenues, à moins de 0,7%, pour les trois valeurs:

$$\lambda_1 = 0{,}472 \cdot 10^{-2} \qquad \lambda_2 = 0{,}514 \cdot 10^{-2} \qquad \lambda_3 = 0{,}547 \cdot 10^{-2}$$

sont portées sur la Fig. 4. On voit que la limite élastique moyenne augmente légèrement avec la vitesse.

Comparaison avec une loi statistique normale

L'allure des courbes $\theta(\varepsilon)$ évoque une loi statistique se rapprochant de la loi normale de Rayleigh pour l'ensemble des réels positifs d'un intervalle $[\varepsilon_L, +\infty]$. Considérons, en effet, une fonction de distribution de la forme:

$$\theta_R(\varepsilon) = 1 - e^{-\frac{(\varepsilon-\varepsilon_L)^2}{2\Delta^2}} \quad \text{pour } \varepsilon \geq \varepsilon_L, \tag{24}$$

dans laquelle Δ représente la dispersion et $\varepsilon_L + \Delta = \varepsilon_M$ la valeur moyenne de la grandeur considérée (ici: la limite élastique ε_{Mi} d'un micrograin). Si nous introduisons la nouvelle fonction:

$$F(\varepsilon) = \sqrt{-2 \log(1 - \theta)} \tag{25}$$

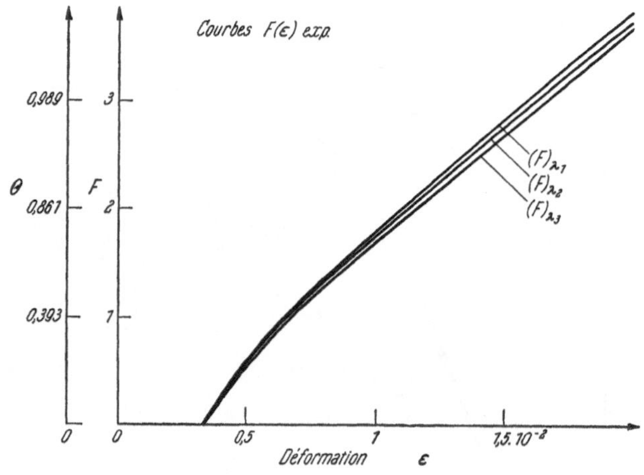

Fig. 5. Courbes $F(\varepsilon)$ expérimentales

celle-ci, pour la distribution (24), prend la forme linéaire $\frac{\varepsilon - \varepsilon_L}{\Delta}$ et est, par suite, représentable par une demi-droite. La ressemblance plus ou moins accusée entre la courbe $F(\varepsilon)$, associée à la fonction θ, et une demi-droite montre dans quelle mesure la loi physique de distribution des limites élastique ε_{Mi} est proche d'une loi normale.

Les courbes $F(\varepsilon|\lambda)$ correspondant aux trois éprouvettes essayées sont portées sur la Fig. 5. On voit qu'une partie importante de ces courbes — couvrant un dépassement d'environ trois fois la limite élastique «conventionnelle» — est presque parfaitement rectiligne; seuls, le début et la fin — qui nécessiteraient en tout état de cause une étude plus fine que celle à laquelle nous nous limitons ici — présentent un infléchissement marqué.

Relation entre une loi normale et le déplacement des dislocations

On peut se demander quelles hypothèses devraient être faites sur le mouvement des dislocations au sein du matériau pour que la loi statistique de Rayleigh soit rigoureusement vérifiée.

Désignons par N_0 le nombre des dislocations existant dans l'unité de volume du matériau avant l'essai d'écrouissage (on suppose leur répartition homogène). Pour ne pas être gênés par l'existence éventuelle de sources de FRANCK-READ, nous admettrons que le nombre N_0 — conséquence des traitements préalables subis par l'éprouvette — est assez grand pour que le nombre des dislocations nouvelles apparues au cours de l'essai puisse être négligé. Soit alors N le nombre des dislocations présentes dans l'unité de volume lorsque l'allongement de l'éprouvette est ε. Ce nombre ne peut que diminuer[1] quand ε augmente à partir d'un certain seuil ε_L en-dessous duquel la déformation est purement élastique. Soit $d\varepsilon^p$ la déformation plastique macroscopique correspondant à l'allongement $d\varepsilon$. Une théorie élémentaire des dislocations [3] établit que, dans la zone des déformations plastiques « moyennes », on a sensiblement:

$$d(N^{1/3}) = -\chi\, d\varepsilon^p, \tag{26}$$

χ étant une constante. On en déduit:

$$N^{1/3} = N_0^{1/3} - \chi\varepsilon^p,$$

$$\frac{d(N^{1/3})}{N^{1/3}} = \frac{dN}{3N} = -\frac{\chi}{N_0^{1/3} - \chi\varepsilon^p}\, d\varepsilon^p \approx -\frac{\chi}{N_0^{1/3}}\, d\varepsilon^p. \tag{27}$$

Il reste à relier $d\varepsilon^p$ à $d\varepsilon$. Pour cela, écrivons:

$$\varepsilon^p = \varphi\{\varepsilon - \varepsilon_L\}$$

et supposons la fonction φ développable selon les puissances de $\varepsilon - \varepsilon_L$. En remarquant que $\varepsilon^p = \varepsilon - k(\varepsilon)$ et considérant l'allure des courbes expérimentales $k(\varepsilon\,|\,\lambda)$, on voit que φ admet un développement de la forme:

$$\varphi\{\varepsilon - \varepsilon_L\} = \frac{(\varepsilon - \varepsilon_L)^2}{A} + \cdots$$

---

[1] Cette diminution provient du fait que certaines dislocations se libèrent de leurs « ancrages » et diffusent instantanément vers les joints du monocristal, provoquant la déformation plastique.

(A : constante), d'où:

$$d\varepsilon^p = \left(\frac{\varepsilon - \varepsilon_L}{A} + \cdots\right) d\varepsilon \tag{28}$$

Si on limite le développement à son premier terme

$$d\varepsilon^p \approx \frac{\varepsilon - \varepsilon_L}{A} d\varepsilon,$$

il vient:

$$\frac{dN}{N} \approx -\frac{3\chi}{A N_0^{1/3}} (\varepsilon - \varepsilon_L)\, d\varepsilon = -\frac{\varepsilon - \varepsilon_L}{\Delta^2} d\varepsilon$$

$\left(\Delta^2 = \dfrac{A N_0^{1/3}}{3\chi}\right)$, d'où finalement:

$$N = N_0 e^{-(\varepsilon - \varepsilon_L)^2/2\Delta^2}. \tag{29}$$

En nous reportant à la définition de θ [formule (7)] et aux considérations précédant la formule (12), nous pouvons écrire:

$$\theta \approx \frac{1}{S} \int_S ds_i'' = \frac{N_0 - N}{N_0} = 1 - e^{(\varepsilon - \varepsilon_L)^2/2\Delta^2} = \theta_R(\varepsilon).$$

Nous voyons donc que l'existence, pour la limite élastique du monocristal, d'une loi de distribution normale est liée à l'hypothese simple:

$$dN = -\frac{1}{\Delta^2} N(\varepsilon - \varepsilon_L)\, d\varepsilon. \tag{30}$$

Cette loi semble vérifiée dans un intervalle d'étendue notable au-delà de la limite élastique macroscopique.

Détermination de λ par l'essai d'écrouissage

Dans ce qui précède, nous avons calculé la valeur de λ en utilisant la formule:

$$\lambda = (\tau \dot{\varepsilon})^{1/n},$$

τ et n étant extraits de la référence [2] où ils sont déduits comme moyennes de nombreux essais de divers types: fluage, relaxation, écrouissage avec une bonne concordance entre les valeurs individuelles. A vrai dire, les résultats des essais d'écrouissage sont sujets à caution, car ils préjugent du fait que l'éprouvette a atteint, en fin d'essai, un régime de fluage établi, *ce qui n'est nullement prouvé.* Tant qu'à émettre une hypothèse sur le comportement global du matériau, il n'est pas plus téméraire d'admettre que les écarts entre la fonction θ et une fonction θ_R de Rayleigh sont dus à des effets secondaires se manifestant les uns au début, les autres à la fin du fluage et, par suite, de déterminer la constante λ figurant dans la relation (19) par la condition que *la partie rectiligne de la courbe $F(\varepsilon)$ soit maximale.* Cette valeur «théorique» du quotient par E de la contrainte de fluage établi [en effet, λ est, d'après (19), la limite de k pour $\varepsilon = +\infty$, si toutefois $(1 - \theta)\,\varepsilon \to 0$] pourra être utilement comparée à la valeur de départ extraite de la référence [2] et, si l'écart est de l'ordre de grandeur des erreurs de mesure, une telle détermination se trouvera justifiée.

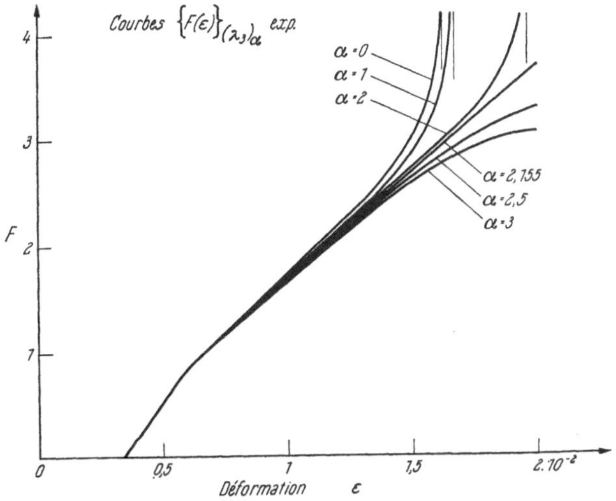

Fig. 6. Courbe $F(\varepsilon)_{\lambda_3}$ expérimentale pour diverses valeurs de λ_3 [$\lambda_3 = (0{,}547 - 0{,}0137\,\alpha)\,10^{-2}$]

Il n'est pas difficile, ayant déterminé θ pour un λ donné, de passer de là aux valeurs de θ (et, par suite, de F) correspondant à une valeur de λ peu différente, soit $\lambda + \delta\lambda$. Il suffit de prendre $\delta\lambda = \pm\,\alpha\,\dfrac{\lambda}{\tau}$ avec α entier positif. $\theta_n + \delta\theta_n$ s'obtient par le même calcul que θ_n en remplaçant τ par $\tau' = \dfrac{\tau(\lambda + \delta\lambda)}{\lambda} = \tau \pm \alpha$.

Entre deux valeurs de α, on interpole par un calcul de variations en faisant intervenir des valeurs de θ_i déjà calculées.

Des variations progressives de λ, ainsi opérées dans le sens convenable, allongent la partie rectiligne de $F(\varepsilon)$ jusqu'à ce que la courbure au point terminal change de signe; ensuite la partie rectiligne diminue (Fig. 6). La valeur séparative de λ se détermine par ce procédé avec une très grande précision. Pour les trois éprouvettes traitées, le résultat a été le suivant:

$\dot{\varepsilon}$ $(10^{-2}\,s^{-1})$	valeur de départ de λ (d'après la réf. 2) (10^{-2})	valeur finale de λ (10^{-2})	Ecart (en %)
0,0324	0,472	0,465	1,6
0,632	0,514	0,492	4,2
1,718	0,547	0,517	5,5

On voit que l'écart est, pour les trois vitesses, inférieur à 6%. Il reste dans les limites de l'approximation avec laquelle n et τ sont connus.

Ayant ainsi déterminé λ, on mesure le coéfficient angulaire $\frac{1}{\Delta}$ et l'abscisse pour $F = 1$ (c'est $\varepsilon_M = \varepsilon_L + \Delta$) de la partie rectiligne de $F(\varepsilon)$ et l'on en déduit les valeurs de Δ et de ε_L figurant dans l'expression de θ_R.

Grandeurs caractéristiques du comportement élasto-plastique d'un matériau métallique

Ce qui précède montre que le comportement en traction simple d'une éprouvette de duralumin à la température T et à la vitesse $\dot{\varepsilon}$ dépend, avec une bonne approximation, des grandeurs suivantes, fonctions de T et de $\dot{\varepsilon}$, qui, toutes, ont une signification concrète et sont directement ou indirectement accessibles à l'expérience:

a) λ, quotient par E de la contrainte de fluage établi,

b) ε_L et Δ ($\varepsilon_M = \varepsilon_L + \Delta$ est la valeur moyenne des ε_{Mi} ou *limite élastique moyenne*; Δ mesure la dispersion des ε_{Mi} autour de ε_M),

c) les coefficients m et ε_F de l'expression (22) [s'obtiennent à partir de deux données numériques dans l'intervalle $(0, \lambda)$, l'une pouvant être la limite élastique vraie ε_F, l'autre la limite élastique «conventionnelle»].

Pour reconstituer $k(\varepsilon)$ à partir de ces grandeurs, il suffit: — dans l'intervalle $(0, \lambda)$, de prendre

$$k \approx \begin{cases} \varepsilon & \text{pour } 0 < \varepsilon \leq \varepsilon_F \\ \varepsilon - m(\varepsilon - \varepsilon_F)^2 & \text{pour } \varepsilon_F \leq \varepsilon < \lambda \end{cases}$$

— pour $\varepsilon > \lambda$, d'appliquer la formule (19) avec:

$$\theta \approx \theta_R = 1 - e^{-(\varepsilon - \varepsilon_L)^2/2\Delta^2}$$

(au voisinage de $\varepsilon = \lambda$, les valeurs de k, différemment approchées à gauche et à droite, doivent être raccordées par continuité).

Comparaison avec l'expérience

Sur les Fig. 7, 8, 9, on a porté, pour les trois vitesses, les courbes représentatives de θ, de F et de $\frac{P}{S_0}\left(=\frac{Ek}{1-\varepsilon}\right)$ obtenues comme il vient

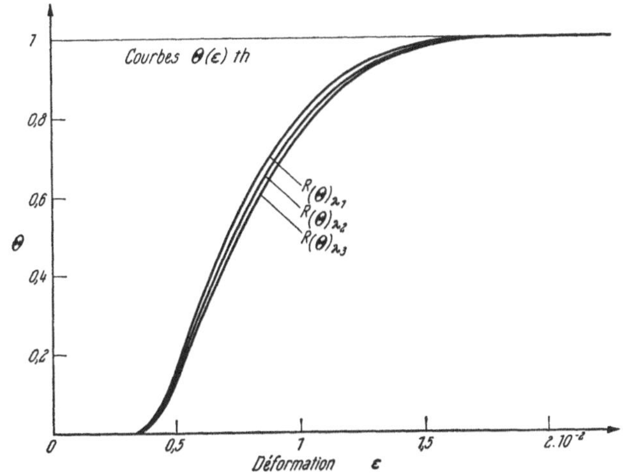

Fig. 7. Courbes $\theta_R(\varepsilon)$ théoriques

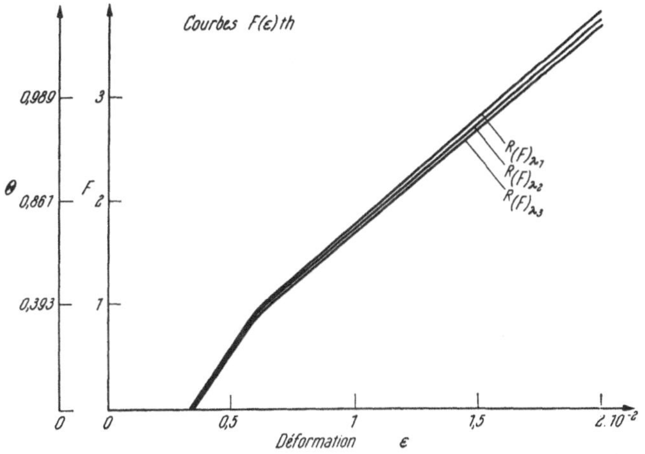

Fig. 8. Courbes $F_R(\varepsilon)$ théoriques

d'être dit, ainsi que, sur la Fig. 9, les trois courbes expérimentales initiales de la Fig. 1. Nulle part, dans l'intervalle $0 \leq \varepsilon \leq 2 \cdot 10^{-2}$ — soit cinq fois la limite élastique conventionnelle — l'écart n'atteint 1,5%.

A des allongements supérieurs, la fonction θ, dont les valeurs sont alors très voisines de l'unité (elles dépassent 0,997), croît moins vite que θ_R. Cette zone des grands allongements, dont l'intérêt pour la construction mécanique est d'ailleurs minime, bien que considérable pour la connaissance de la «résistance ultime», fera l'objet d'un autre mémoire.

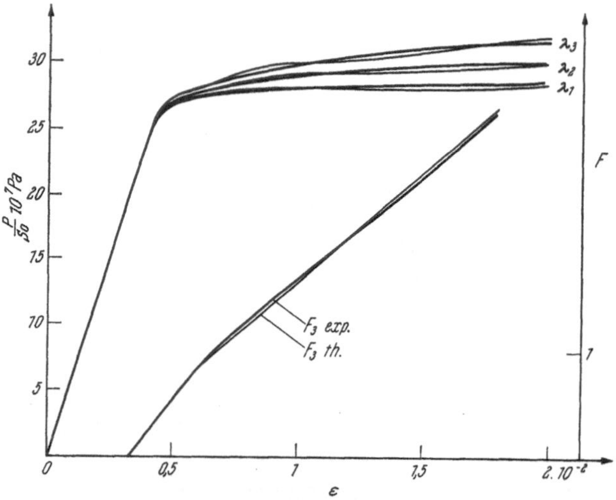

Fig. 9. Comparaison théorie – expérience pour les courbes $\frac{P}{S_0}(\varepsilon)$ et $F(\varepsilon)_{\lambda_3}$

Conclusion

De l'analyse de trois essais de traction simple à vitesse constante effectués sur des éprouvettes en duralumin à 200 °C, on a dégagé la définition de grandeurs caractéristiques qui peuvent servir à comparer les qualités de divers matériaux métalliques dans le domaine élastoplastique contigu au domaine élastique. Les résultats, parfaitement cohérents entre eux, montrent que la dispersion que l'on pouvait craindre, eu égard au nombre très limité des expériences, a été fortement réduite par l'attention apportée au choix des éprouvettes. Les phénomènes physiques à l'échelle du monocristal, complexes et encore mal connus, ont été assez grossièrement schématisés. Sous ces réserves, une certaine «normalité» de caractère statistique a été mise en évidence dans l'hétérogénéité du «grain de matière» considéré comme un amas élémentaire de monocristaux séparés par des joints. La formule (19) est assez simple pour pouvoir être appliquée dans un calcul de structure. Toutefois elle ne vaut pour l'instant que dans les limites où elle a été établie, c'est-à-dire pour $\dot{\varepsilon} = c^{te}$. Hors de ce cas, et notamment s'il y a relaxation ou charge imposée, les rôles principaux sont dévolus à ε^p et $\dot{\varepsilon}^p$ que masquent ici ε et $\dot{\varepsilon}$; de même, s'il y a retour en arrière, déformation alternée, etc . . . sans parler ici des problèmes tridimensionnels. L'étude du comportement dynamique des matériaux au delà du domaine élastique, s'il doit apporter quelque jour de profondes modifications dans le dimensionnement des pièces, n'en est encore qu'à ses débuts et nécessitera encore de longues et patientes recherches.

Remerciements

L'auteur remercie Madame S. Chopin et Monsieur R. Ohayon pour l'aide précieuse qu'il lui ont apportée dans la programmation des calculs et le dépouillement des résultats.

Références

[1] Mazet, R.: Sur un modèle apte à traduire le fluage sous charge constante des structures. «Creep in Structures», p. 327. Berlin-Göttingen-Heidelberg: Springer. 1962.

[2] Lemaître, J.: Sur la détermination des paramètres de fluage de Norton-Hoff. C. R. Acad. Sci. Paris **261**, 3731 (1965).

[3] Zarka, J.: Thèse de la Faculté des Sciences de Paris (1968).

Thermal Fatigue and Thermal Shock Investigations

By

F. K. G. Odqvist and **N. G. Ohlson**

Stockholm (Sweden)

Summary

Results are given from thermal fatigue tests of two steels, run in a fatigue testing machine. Non-uniform temperature distribution during heating and cooling in each cycle gives rise to stresses in the specimen. The number of cycles before the first crack appears is recorded and is compared with the maximum temperature as well as with the amount of plastic strain during the cycle. An estimate of the strains and stresses is given, supported with experimental data, obtained from measurements of surface strains using an optical grating technique.

I. Materials Tested

The following materials have been tested as to their resistance to thermal shock.

1. Cr-Mo-V-alloyed tools' steel, Bofors ROP 19, with the following approximate composition: 0.40% C, 1.0% Si, 5.3% Cr, 1.4% Mo and 1.0% V.

2. Nb-stabilized austenitic stainless steel, Sandvik 8 R 40, containing 0.084% C, 0.45% Si, 1.66% Mn, 0.015% P, 0.011% S, 17.4% Cr, 11.5% Ni, 0.18% Mo, 0.87% Nb and 0.04% Ta.

Different heat treatments of material 1 have been tested.

A. Formation of austenite at 1000 °C for 0.5 h, air cooling, annealing for 4 h at 625 °C to a hardness of 42 RC.

B. As A but annealed at 525 °C.

C. As A but heated to 1050 °C instead of 1000 °C for 0.5 h.

The different kind s of heat treatment used (A, B, C) proved to give little deviations in the results, the scatter being considerable.

Material 2 was tested in the as-received condition.

II. Thermal Fatigue of Beams

It is wellknown that our chief problem—to determine the stress distribution in a region from boundary values of temperature—is a very simple one in principle. We just need to solve the differential equation $\Delta^2 T = 0$ for a stationary temperature field or $\Delta^2 T = \frac{c\varrho}{k} \cdot \frac{\partial T}{\partial t}$ for a time-dependent one to obtain the function $T(x, y, z, t)$. If the theory of elasticity is valid throughout the specimen, strains and stresses can be easily calculated. In our case we deal exclusively with thermal stresses in a beam with transversely varying temperature. The only independent variables are the distance y from the edge and the time t.

As everyone knows, very few structures can be analysed as to thermal stresses without considerable simplifications of the same kind as those mentioned above. One possibility to avoid cumbersome calculations is to use model experiments, from which thermal resistance and thermal fatique life can be estimated. This method has often been utilized in industry. As some examples of test pieces may be mentioned the model of the sharp edge of a turbine blade [1], a tube (2, 3], which offers a fairly simple stress distribution, and a thin disc.

a) Bars in Uniform Uniaxial State of Strain

A detailed discussion of the parameters governing an experiment of our kind demands knowledge of the conditions during one cycle. Any would-be phase transitions should be taken into account. Changes of yield stress with temperature must also be considered. Let us consider what happens to a test piece of material 1 of the design shown in Fig. 1, placed between rigid walls, held at constant distance, and subjected to uniform temperature changes. An arbitrary point in the bar starts its stress-strain history from the origin in the diagram of Fig. 2, and using the temperature as a parameter along the curve, one fatigue cycle can be followed. The broken vertical lines correspond to specified temperatures and associated purely thermal strains. The broken line which is part of the curve, ending on $\sigma = 95$ kp/mm², indicates the path followed if the reduction of yield stress σ_s proportional to the decrease in hardness after one cycle is noticed. At 850 °C a sudden change in the coefficient of thermal expansion α due to the formation of austenite is observed. It is probable that other values of α

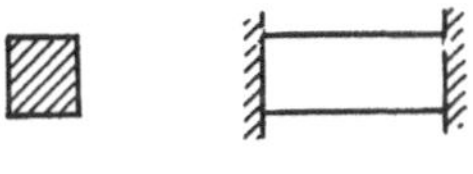

Fig. 1. Bar with rectangular cross-section

apply during cooling, as the phase composition has become different, but that possibility is not considered in Fig. 2.

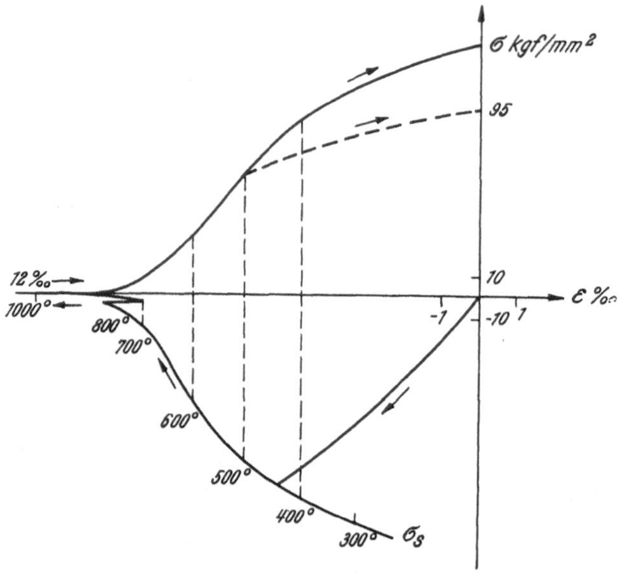

Fig. 2. Stress-strain curve for the bar in Fig. 1 subjected to uniform heating during the first cycle ($T_{max} = 1000\,°C$)

b) Evaluation of Tests of Bars in Our Experiments. Uniaxial State of Strain

Comparisons between thermal and mechanical fatigue cycling tests [4] for test pieces like that of Fig. 1, reveal similar slope of the curves in a plot of the number of cycles to failure, N, vs. the plastic strain interval, $\delta\varepsilon_p$. Our results are presented in diagrams of N vs. maximum temperature of the cycle, T_{max}. As to the martensitic steel, a log $N/\delta\varepsilon_p$-plot is shown (Fig. 12a), where $\delta\varepsilon_p$ at the edge has been found from the simplified calculation of the stresses described in Section V. Also, a diagram showing the mean value of the plastic strain within the plastic zone is added (Fig. 12b). In order to underline the similarity to mechanical straining, the thermal fatique lives of material 2 have been plotted in a conventional fatigue diagram for stainless steel, using $^1/_2\, E\alpha\, \Delta T$ as an equivalent stress amplitude (Fig. 11). Here, E denotes Young's modulus and ΔT the temperature interval run through in each cycle.

III. Test Equipment, Design of Specimens, and Discussion of Cracks

The fatigue testing machine is shown in Fig. 3. Six specimens can be tested at a time. They are fixed onto a rotor which stops in six positions per revolution with one nozzle corresponding to each position. In one of these, heating is applied while the others are intended to give cooling.

Fig. 3. Fatigue testing machine

A gas burner accounts for the heating in one nozzle and certain precautions have been taken to obtain good reproducibility in temperature. The cooling is achieved by means of compressed air through five of the nozzles.

In Fig. 4, the principal diagram of the machine is shown. Inside the dotted line is a unit admitting automatical temperature recording by means of a camera during the cycle.

The temperature as a function of time in an arbitrary testing cycle can be found in Fig. 5. The duration of the cycle as well as its maximum temperature and the velocities of heating and cooling can be changed within wide limits.

Special interest has been devoted to the heating period, as it is most easily investigated.

The specimen in the form of a prismatic bar with triangular cross-section can be seen in Fig. 6. It is similar to that used by the authors of [1] and [5]. Heating is applied approximately at a point of the edge of the

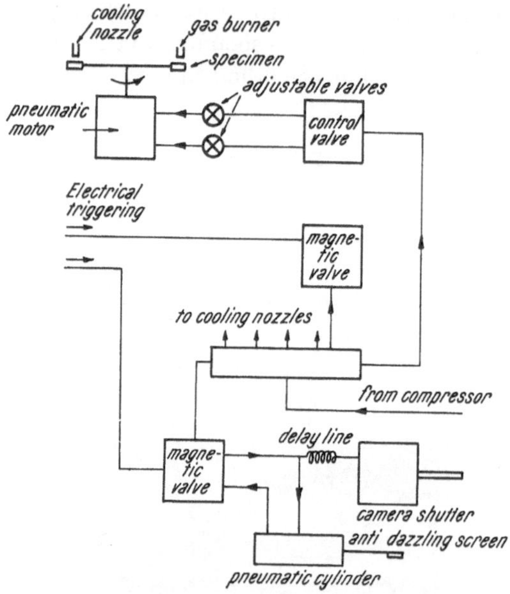

Fig. 4. Principle diagram of testing machine

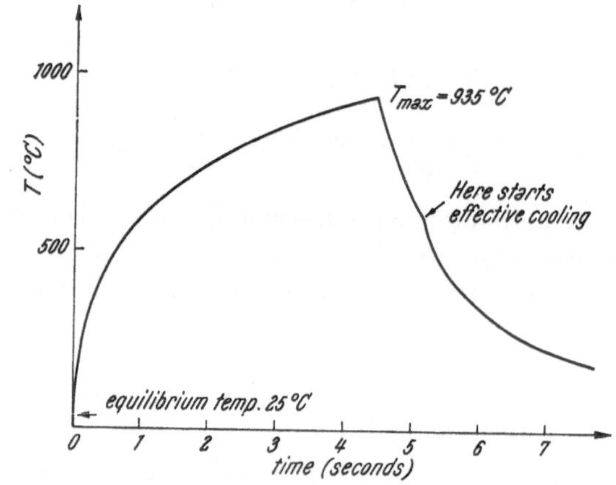

Fig. 5. Example of testing cycle, showing the maximum temperature in the shock zone versus time

specimen just as in [5]. Experiments seem to underline, however, that homogeneous temperature all along the edge gives essentially the same fatigue life.

Critical aspects concerning the design of the test pieces should be emphasized. The results of a test should be able to be described using comparatively few parameters. This is hardly the case with this type of specimen. By introducing a method of strain measurement which will be described in Section VI b, this drawback might be overcome, to some extent at least. Fatigue curves—strain vs. number of cycles to fracture—are then easily plotted and representative data deduced as a guide for the practically working designer.

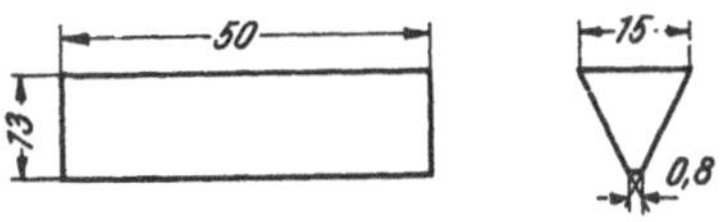

Fig. 6. Specimen

The metallurgical problems cannot be so easily discarded, unfortunately. The adjustment of the flame as to oxygen content is of great impor-

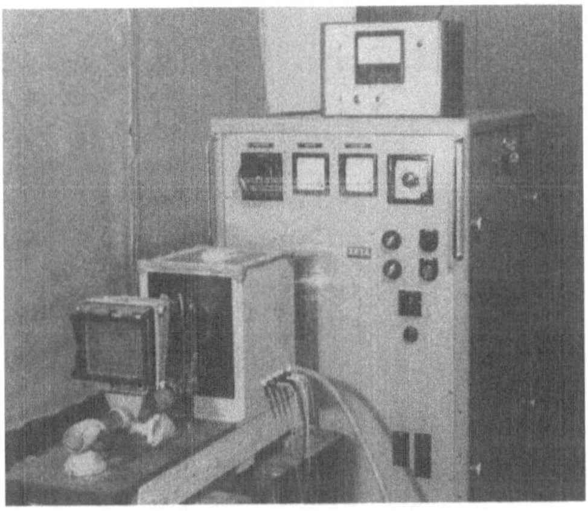

Fig. 7. Vacuum chamber used for the determination of surface strains

tance. In most tests published here an oxidizing flame has been used. A neutral atmosphere or vacuum would have been preferable.

The testing machine described above has been furnished with some extra equipment to facilitate certain measurements. Specimens which have been run in the machine for some time, can be removed and tested in a vacuum chamber—Fig. 7—to find out possible changes in the material, change of ultimate tensile stress, etc. To make these strain determinations this chamber is greatly desirable. Induction heating is used, intending to simulate the corresponding part of the cycle in the fatigue testing machine. By shaping the coil it is possible to obtain a temperature field almost identical with that used in specimens heated by the gas burner.

A chief task in fatigue tests is the detection of the first crack and the observation of the number of cycles when it starts. As a rule, such an observation offers no problem when the crack causes immediate breakdown of the specimen. This is not so in our case. We have chosen to

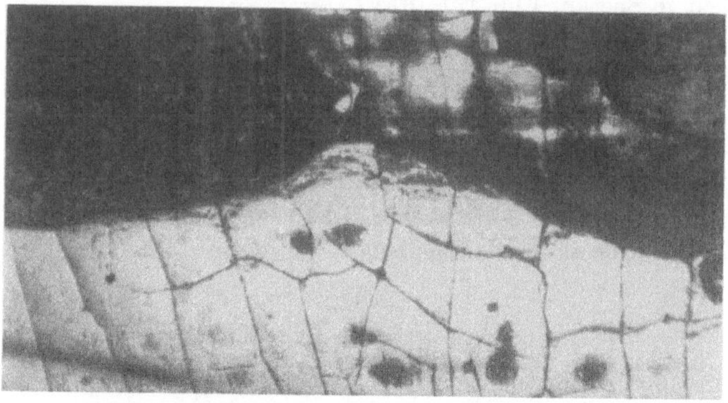

Fig. 8. Crack pattern in material 1. 1 cm corresponds to 0.1 mm

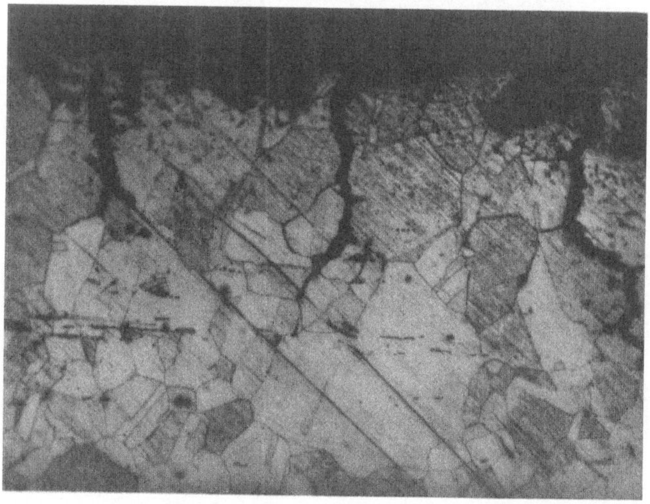

Fig. 9. Cracks in material 2. 1 cm corresponds to 0.1 mm. The edge of the specimen in Fig. 8 and Fig. 9 is to be found alongside the upper border of each photo

observe the specimen in a telescope every hundred cycles to be able to detect the gleaming of the crack surfaces in the flame. Such optical phenomena do not always accompany the formation of the cracks, however. Microscopical examinations, carried out every two hundred cycles, in those cases complete the picture.

As the cracks look rather different in different materials subjected to very much the same kind of testing, one suspects that the mechanisms of formation may differ. Let us consider the crack system in the two steels!

Material 1. A dense net of cracks is being formed within the period of a few cycles. They start at the edge, spreading on both sides and crossing the edge. Near the centre of the shock region some cracks parallel to the edge also occur. See Fig. 8! In the following, „fatigue life" is associated with the first observed crack irrespective of its appearance. Specimens which have fatigue lives of 2000 cycles or more always show the densest nets, whose cracks are very shallow, a few microns in depth. However, some cracks parallel to the edge are deeper.

Half the number of specimens—48 in all—have a rough surface (about $2.5 \cdot 10^{-3}$ mm), the rest being polished to about $0.06 \cdot 10^{-3}$ mm. As will be seen from the diagrams, a polished surface gives no better fatigue characteristics, but some fault may occur owing to increasing ease in detecting the cracks instantaneously. At any rate, this polishing implies less deviation from the mean value of fatigue life.

The fatigue life has been plotted against the maximum temperature in Fig. 10 and Fig. 11.

The cracks are about 0.08—0.10 mm apart in most unpolished pieces, while they are about 0.05 mm apart in the polished. Their length is about 1.0 mm in the most stressed region. To investigate whether these small cracks may serve as sources for more severe fracture, three cracked pieces were run at the highest maximum temperature together with three virginal pieces. The virginal specimens behaved in a normal way, whereas those already cracked apparently proved to be stronger against formation of new cracks.

Material 2, comprising 24 specimens, all of which were given a polished surface. The test conditions are a little too severe for this stainless steel. For comparison, however, it is excellent. The crack pattern is shown in Fig. 9.

IV. Thermal Fatigue Tests

These tests were carried out to give an idea of the fatigue life as a function of maximum temperature of the cycle. Measuring surface temperature, see Section VIa.

For material 1, the results have been compiled in the following table. In estimating the scatter, a normal distribution of fatigue life has been assumed for each test of sufficiently large number of specimens.

Its parameters m = mean value and s = standard deviation have been calculated.

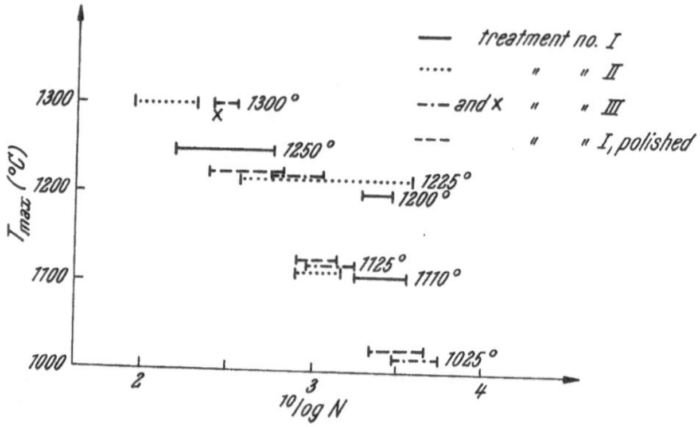

Fig. 10. Thermal fatigue life (material 1)

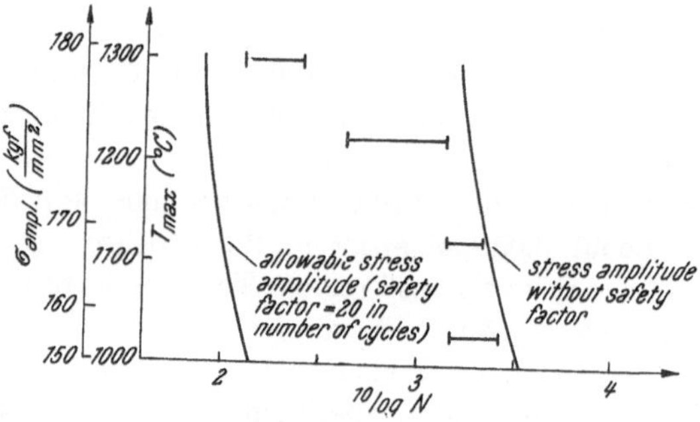

Fig. 11. Thermal fatigue life. Comparison between the present tests and conventional fatigue tests (material 2)

Table 1

Treatment	Max. temp. (°C)	Fatigue life (cycles) m	s
A	1225	385	122
		385	155
A	1125	1015	428
B	1225	980	557
B	1300	150	62
B	1125	1190	925
		1072	382
C	1225	680	124
C	1125	1230	900
		945	158
C	1025	4040	2340
		3430	1100

It will be observed that the maximum temperatures listed above are reached only in a very small region near the edge. To give an idea of the temperature distribution, it might be mentioned that for $T_{max} = 1300\,°C$, the 1000°-isotherm lies about 0.4 mm from the edge.

There is evidently no great difference in fatigue life between the three treatments. Treatment C seems to be superior in spite of the fact that it shows the largest grain size of the three. It is probable that the heat resistance which is raised with the grain size may explain this behaviour (cf. [6]).

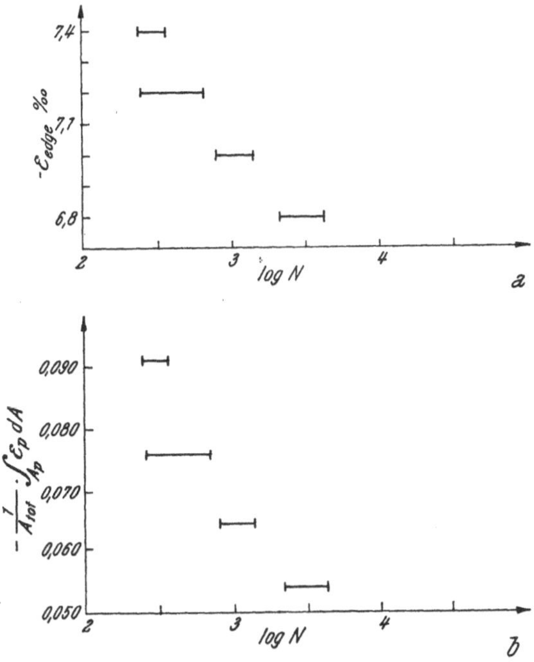

Fig. 12. Thermal fatigue life of material 1 as related to strain

In order to estimate the fatigue life one can use Coffin's eq. [7]:

$$N^{0.5} \cdot \delta\varepsilon_p = C_1,$$

where N = number of cycles to failure, $\delta\varepsilon_p$ = plastic strain during the cycle, and C_1 = constant. Slightly modified, $N^{\varkappa} \cdot \delta\varepsilon_p = C_2$, $\varkappa$ being a constant which is determined from experimental data, has been applied to our test. At 1000 °C maximum temperature one obtains in material 1 $\delta\varepsilon_p = 5 \cdot 10^{-3}$ according to data from grating measurements (see Section VII). With $\varkappa = 0.25$ and $C_2 = 0.05$ (values estimated from similar materials mentioned in literature), one finds $N = 10^4$ which is the correct order of magnitude.

To express the resistance to thermal shocks of a certain ductile alloy one might use the parameter $z = E\alpha/k$,. where E = Young's modulus, α = coefficient of thermal expansion, and k = thermal conductivity. The quantity z is temperature dependent. Table 2 below gives the values of z at the situations of our tests and shows decrease of z with fatigue life for one and the same material. No conclusions could be drawn from one material to another, as will be seen from comparison with Table 3.

Table 2. *Values of z* (s/m²)

Material	1	2
temp. (°C)	z	
100	$6.8 \cdot 10^4$	$20.6 \cdot 10^4$
800	$5.8 \cdot 10^4$	$10.9 \cdot 10^4$

Table 3. *Fatigue life (cycles)*

Material	1, treatment A, polished	2
T_{max} (°C)		
1025	3200 ± 1000	1960 ± 560
1125	1100 ± 300	1750 ± 400
1225	450 ± 200	860 ± 400
1300	300 ± 60	200 ± 60

V. Estimate of Strains and Stresses

We want to obtain an approximate expression for the deformation caused by non-uniform temperature distribution. Data from material 1 have been used. We assume that the temperature as well as stress

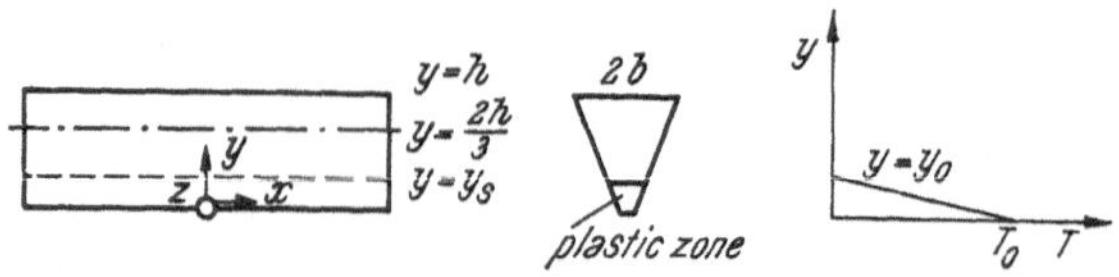

Fig. 13. Specimen, temperature distribution and coordinates used

and strain change only in the y direction. Our considerations in this section evidently generalize those of Section II. Our calculation is valid at the maximum temperature of the cycle.

Preliminary measurements of the temperature distribution were carried out by means of thermocouple probes. In order to fit these measurements, we assume the linear temperature decrease from the edge as shown in Fig. 13. Naturally, this is a very crude approximation to give an analytical expression which is fairly simple to handle.

$T = T_0 \cdot \left(1 - \frac{y}{4}\right)$, $T_0 = 1000\,°\mathrm{C}$, y is in millimeters.

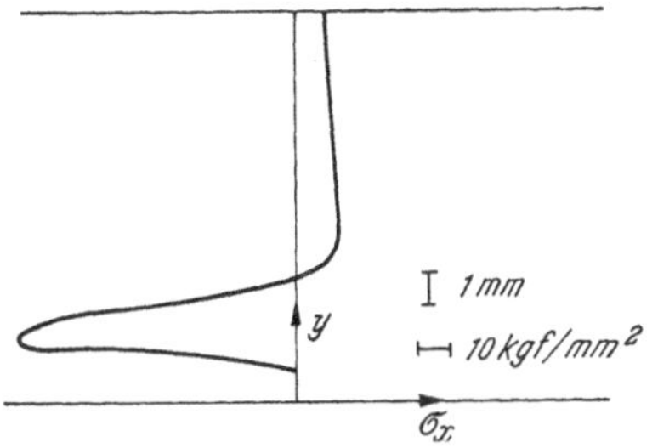

Fig. 14. Stress distribution at maximum temperature

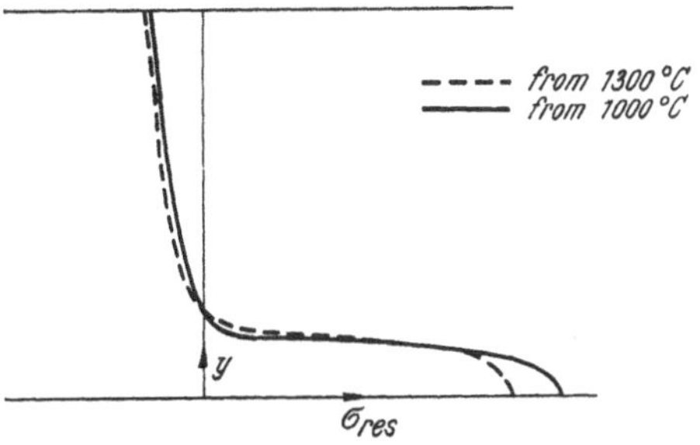

Fig. 15. Residual stresses after cooling

We further assume the variation of E with temperature to be

$E = E_0 \cdot \left(1 - \frac{T}{2T'}\right)$, $T' = 1000\,°\mathrm{C}$, $E_0 = 22 \cdot 10^3\ \mathrm{kgf/mm^2}$.

The yield stress is described by the expression $\sigma_s = 270 - \frac{9T}{25}$ for tensile stresses, in the region $500° < T < 750°$. For $T > 750°$, σ_s has been put equal to zero.

From the only remaining equation of compatibility,

$$d^2 \varepsilon_x / d y^2 = 0,$$

we obtain

$$\varepsilon_x = p + q y.$$

$$\text{Thus,}\quad \sigma_x = \begin{cases} -\sigma_s \text{ in the plastic region } 0 < y < y_s \\ -E\alpha T + pE + qEy \text{ in the elastic region } y_s < y < h. \end{cases}$$

We can now determine the three unknown variables y_s, p, and q from the following conditions.

1. σ_x continuous for $y = y_s$

$$-\sigma_s(y_s) = [-\alpha T(y_s) + p + q y_s] \cdot E(y_s).$$

2. The total normal force across the section should be zero

$$\int_0^{y_s} -\sigma_s \frac{2b}{h} y\,dy + \int_{y_s}^{y_0} (-\alpha E T + pE + qEy) \frac{2b}{h} y\,dy +$$

$$+ \int_{y_0}^{h} (pE_0 + qE_0 y) \frac{2b}{h} y\,dy = 0.$$

3. The total bending moment M_z should be zero

$$\int_0^{y_s} - \sigma_s y \left(\frac{2h}{3} - y\right) dy + \int_{y_s}^{y_0} (-\alpha E T + pE + qEy)\, y \left(\frac{2h}{3} - y\right) dy +$$

$$+ \int_{y_0}^{h} (pE_0 + qE_0 y)\, y \left(\frac{2h}{3} - y\right) dy = 0.$$

Then the elastic stress becomes

$$\sigma_x = -\alpha E T + 12.7 \cdot \frac{E}{E_0} - 0.2 \cdot \frac{E}{E_0} \cdot y,$$

$$y_s = 2.$$

The stress distribution is shown in Fig. 14.
The strain at the edge, $y = 0$, is $-6.8 \cdot 10^{-3}$.

To arrive at the residual stress after cooling to room temperature, we superpose an elastic stress

$$\sigma_e = \alpha E T + \frac{P_x}{A} + \frac{M_z}{I_z} y, \text{ where}$$

$$P_x = \frac{2b}{h} \left(\int_0^{y_s} -\sigma_s y\,dy + \int_{y_s}^{y_0} -\alpha E T\, y\,dy \right) \quad \text{and}$$

$$M_z = \frac{2b}{h} \left(\int_0^{y_s} - \sigma_s y \left(\frac{2h}{3} - y\right) dy + \int_{y_s}^{y_0} -\alpha E T\, y \left(\frac{2h}{3} - y\right) dy \right).$$

The residual stress obtained after cooling from 1000 °C edge temperature is as follows (Fig. 15). No flow occurs at the edge. The decrease in σ_s after one cycle as discussed in Fig. 2 is not considered here.

For $T_0 = 1300$ °C, we find that $y_s = 2.5$ and the edge strain is $-7.4 \cdot 10^{-3}$. The residual stress has been plotted in Fig. 15 (dotted line).

The net strain which is measured includes the thermal expansion. Therefore, at the edge the strain is $6.4 \cdot 10^{-3}$ at 1000 °C.

VI. Two Measuring Methods

a) Determination of Surface Temperature

The problem is to find the temperature distribution in a region of about 50 mm^2 where heating rates of 200 °C/s may occur. The atmosphere surrounding the region is so active that thermocouples will be destroyed almost immediately.

Method of solution. If a photographic emulsion is prepared for indication of longwave light there is a connection between surface temperature of the body pictured and the density of the film after development. It is presumed that the emissivity of the surface is independent of temperature.

Equipment. A Linhof Technica camera with a 100 mm objective and 9×12 cm plates was used. In front of the lens was placed a "black" filter with a cut-off wavelength of 760 mμ. Good results have been obtained on Gevaert Scientia 52 A 86 and Kodak IRER.

Densities S between 0.3 and 2.0 logarithmic units are especially convenient for the evaluation in densitometers.

$$S = {}^{10}\log I_0/I, \quad \text{where}$$

I_0 = intensity of in-coming light and
I = intensity of transmitted light.

The measuring spot is 0.25–1 mm^2 depending on the construction of the densitometer.

Calibration of the emulsion was performed by means of thermocouples. See diagram 16a! The camera has been kept open for 1 second

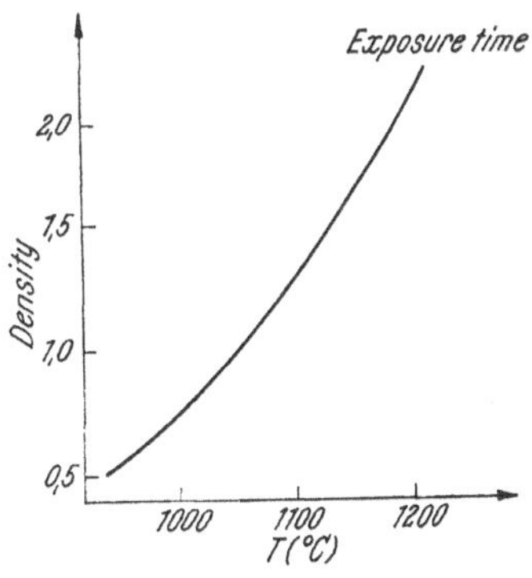

Fig. 16a. Calibration curve for Kodak IR. ER through filter Wratten 87

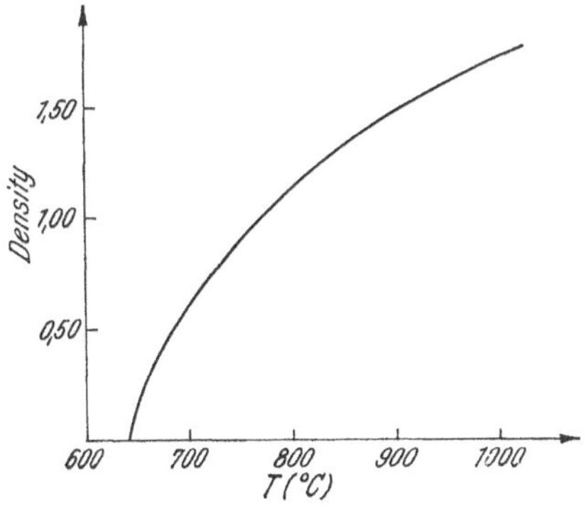

Fig. 16b. Density vs. temperature with camera open for 6 s during heating

which is possible with repeated exposures, each of short duration. The error was about 0.05 at $S = 1$, i.e. scarcely greater than the accuracy of the densitometer would permit.

More accurate determination in low-temperature regions can be obtained, if the camera is left open from the beginning of the heating till the desired moment is reached. Calibration according to these circumstances must be carried out, Fig. 16b.

Accuracy. If the thermocouples are calibrated in advance, 5°-isotherms lying 0.5 mm apart can be resolved. The absolute accuracy is $\pm$ 25° at 1200 °C between two exposures.

b) *Determination of Surface Strains*

Find the strains in the region described at the beginning of Section VIa. The temperature may be above 1000°. All strain measurements include thermal and mechanical strain. It is a main purpose of this paper to separate the two different strains.

Method of solution. Produce a grating on the surface to be examined with its lines perpendicular to the direction in which the strain is to be measured. The line density should be about 50 lines/mm. The grating parameter—i.e. the spacing between the lines—changes proportional to the strain. If the grating is properly illuminated, it can be photographed in the strained state. Assuming that the camera introduces no distortion of the picture and the emulsion of the film is unshrinkable, one obtains a transmission grating with the same changes in parameter

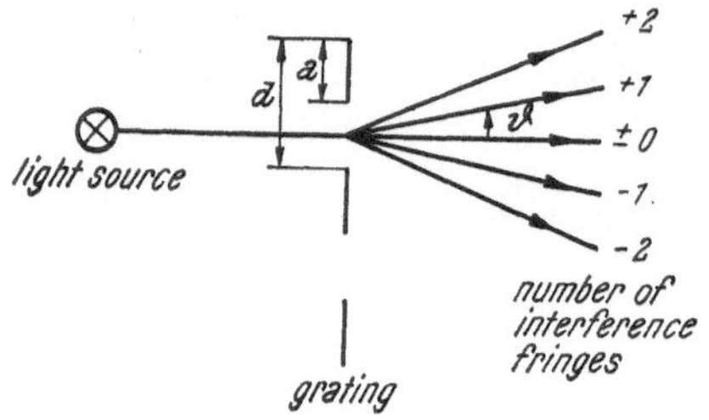

Fig. 17. Analysis of transmission grating

as the original. If the grating is inserted into a spectrometer the position of the interference fringes can be observed. That determines the parameter d and hence the strain ε, according to Bragg's relation

$$\sin \vartheta_n = n \cdot \lambda / d_0$$

$$\sin (\vartheta_n + \delta \vartheta_n) = n \cdot \lambda / d = n \cdot \lambda / d_0 \cdot (1 + \varepsilon).$$

See Fig. 17! λ denotes the wavelength of the light of the spectrometer, ϑ_n is the angle of deviation of n:th interference, d_0 is the grating parameter of a grating without distortion, and d the parameter when the strain is ε. From the equation ε may be determined.

Accuracy. With this method it is possible to measure the strains instantaneously in an area of about 2 mm² at temperatures of above 1200 °C. The error is less than 10^{-3}.

VII. Analysis of Strains Using the Principle Described in Section VIb

The state of surface strain has been determined by means of three sorts of gratings

a) grating lines parallel to the edge

b) grating lines perpendicular to the edge

c) grating lines at 45° to the edge.

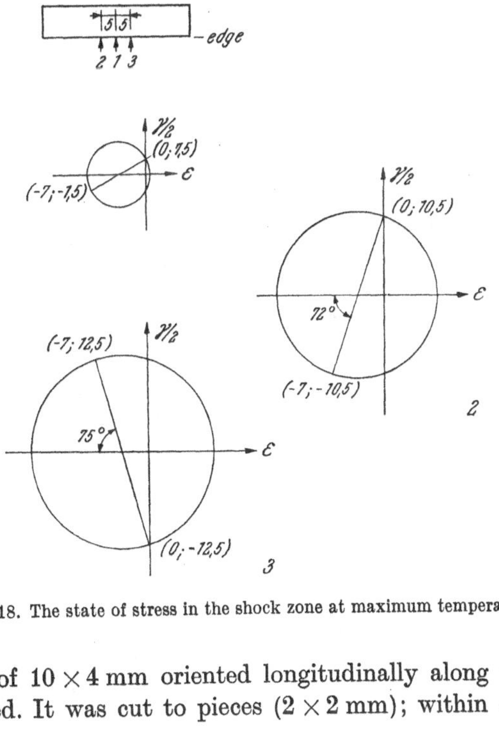

Fig. 18. The state of stress in the shock zone at maximum temperature

A region of 10 × 4 mm oriented longitudinally along the edge has been examined. It was cut to pieces (2 × 2 mm); within each of these the strains and temperature were assumed constant and their mean values noted. From the measured strains the thermal expansion has been subtracted.

Strains were measured in the first, second and tenth cycle at the end of the heating period and after cooling to establish any would-be residual strain. The surface strains are represented by means of Mohr's circle for the following points, Fig. 18.

1. on the line of temperature symmetry, near the edge, i.e. in the centre of the shock zone

2. in the outer region of the shock zone, about 5 mm from its centre, near the edge

3. in the same position as 2, but on the opposite side.

Results

Point 1: uniaxial state of strain is obviously a good approximation

Point 2: the principal strain direction is at 36° angle to the edge

Point 3: the principal strain direction is at 37.5° angle to the edge.

The cracks in the surroundings of 2 and 3 form an angle of at least 50° with the edge, according to micrograph.

Finally, we make a remark on the changes caused by repeated shocks. From microhardness investigations it is known that the hardness in the shock zone is gradually reduced. According to L. PRANDTL [8], hardness is proportional to σ_s. Hence we may conclude that

$$\sigma_s = 80 \text{ kgf/mm}^2 \text{ after } 10 \text{ cycles}$$

$$\sigma_s = 55 \text{ kgf/mm}^2 \text{ after } 300 \text{ cycles.}$$

Initially, $\sigma_s = 130$. Probably that will change the state of stress and especially the stress amplitude in each cycle. Consequently, the fatigue damage may be considerably greater during the first few cycles before the softening has set in. That can be seen from measuring the strain perpendicular to the edge in the tenth cycle: the mean value has increased by $2.3 \cdot 10^{-3}$ from the second cycle.

Acknowledgement

This work has been carried out at the Division of Strength of Materials, Royal Institute of Technology, Stockholm, sponsored by the Research Group for High-temperature Mechanics of Materials (Varmhållfasthetskommittén) of the Swedish Steelmakers' Association (Jernkontoret) and by the Swedish Technical Research Council.

List of Symbols

E	Young's modulus
I, I_0	light intensity
N	number of cycles to failure
S	density of photographic film
T	temperature, $T(x, y, z, t)$
ΔT	temperature interval in a cycle
c	specific heat
C_1, C_2	constants
d, d_0	grating parameter
k	thermal conductivity
m, s	mean value and standard deviation of a normal distribution mentioned in the text
n	$0, \pm 1, \pm 2, \ldots$
p, q	constants
t	time
z	thermal shock parameter
x, y, z	Cartesian coordinates
α	coefficient of linear thermal expansion
ε	strain
$\delta \varepsilon_p$	plastic strain interval
ϑ_n	angle of deflection, corresponding to the n:th interference
$\delta \vartheta_n$	change of ϑ_n due to ε
$\varkappa$	constant
λ	wave length of spectrometer light source
ϱ	density of material
σ	stress
σ_s	yield stress

References

[1] Baron, H. G., and B. S. Bloomfield: Resistance to Thermal Stress Fatigue of Some Steels, Heat-resisting Alloys and Cast Iron. J. Iron and Steel Inst. **177**, 223 (March 1961).

[2] Coffin, L. F., Jr., and N. Y. Schenectady: An Investigation of Thermal-Stress Fatigue as Related to High-temperature Piping Flexibility. Trans. ASME **76**, 931 (1954).

[3] Taira, S., et al.: Thermal Fatigue under Multiaxial Thermal Stresses. Proc. of the 6th Japan Congress on Testing Materials. March 1963.

[4] Swindeman, R. W., and D. A. Douglas: The Failure of Structural Metals Subjected to Strain-Cycling Conditions. J. Basic Eng. **81**, 203 (1959).

[5] Muscatell, F. L., et al.: Thermal Shock Resistance of High Temperature Alloys. Amer. Soc. for Testing Materials, Proc., **57** (1957).

[6] Tomalin, Steels in Steam Turbines, High-temperature Properties of Steel, BISRA, Eastbourne, April 1966.

[7] Coffin, L. F., Jr.: Thermal Stress Fatigue. Prod. Engineering, p. 175 (June 1957).

[8] Hill, R.: The Mathematical Theory of Plasticity, p. 260. Oxford (1956).

Thermal Effects in Viscoplasticity

By

W. Olszak and P. Perzyna

Warsaw (Poland)

1. Introduction

The basic object of the present paper is the description of the behaviour of an elastic/viscoplastic material during thermodynamic process within the framework of thermodynamics with internal state variables[1].

The essential feature of an elastic/viscoplastic material is that the material behaves before yielding elastically and, after yielding, exhibits combined elastic, viscous and plastic properties. The necessity of simultaneous consideration of viscoelastic and plastic properties of a material is indicated by the results of experimental investigations of dynamic loads. They have also shown that many materials behave differently under dynamic loading and static loading. The principal cause of these differences is the sensitivity of the material to the rate of deformation.

In our thermodynamic description of an elastic/viscoplastic material we shall assume that the deformation gradient and the temperature are the thermodynamic state variables and the inelastic deformation gradient is the internal state variable or hidden parameter[2].

The rate of internal state variable is assumed to be proportional to the excess of stress over the state corresponding to the static yield condition[3].

The constitutive equations and the thermodynamic restrictions are derived and thermal effects on relaxation phenomena are discussed. Finite deformations of a body during a general thermodynamic process

[1] Thermodynamics with internal state variables has been presented independently by Coleman and Gurtin [1967, 1] and Valanis [1967, 4]. The application of the concept of internal state parameters to the description of the properties of a rate sensitive plastic material has been proposed by Perzyna and Wojno [1968, 1].

[2] We follow here the concept presented by Perzyna and Wojno [1968, 1].

[3] This idea was discussed by Olszak and Perzyna [1967, 2].

are considered. The reduced constitutive equations obtained are invariant under a change of the reference frame. Some generalizations and particular cases are discussed.

2. Constitutive Assumptions

Let us consider a body B with particles X and assume that this body may deform and conduct heat. We shall assume that couple stresses and body couples are absent.

The thermodynamic process of a body B is described by nine functions $\{\chi, \boldsymbol{T}, \boldsymbol{b}, \psi, \boldsymbol{q}, r, \eta, \vartheta, \boldsymbol{A}\}$ of the particle X and time t. These functions have the following interpretations. The function of motion $\chi(X, t)$ determines the spatial position occupied by the material point X at time t, which in the reference configuration R occupied the position $\boldsymbol{X}$, i.e.,

$$\boldsymbol{x} = \chi(\boldsymbol{X}, t). \tag{2.1}$$

The components of the function χ are assumed to be continuously differentiable. The function $\boldsymbol{T}(X, t)$ is a symmetric Cauchy stress tensor, $\boldsymbol{b}(X, t)$ is the body force per unit mass, $\psi(X, t)$ denotes the specific free energy per unit mass, $\boldsymbol{q}(X, t)$ the heat flux vector, $r(X, t)$ the heat supply per unit mass and unit time, $\eta(X, t)$ is the specific entropy, $\vartheta(X, t)$ the local absolute temperature, and $\boldsymbol{A}(X, t)$ is the internal state tensor. We assume that $\boldsymbol{A}$ is a second order tensor.

Since we identify the material point X with its position $\boldsymbol{X}$ in the reference configuration R, the deformation gradient $\boldsymbol{F}$ is determined by

$$\boldsymbol{F} = \operatorname{Grad} \chi(\boldsymbol{X}, t), \tag{2.2}$$

where Grad is computed with respect to the material coordinates $\boldsymbol{X}$, and it is assumed that $\det \boldsymbol{F} > 0$.

The system of nine functions $\{\chi, \boldsymbol{T}, \boldsymbol{b}, \boldsymbol{q}, r, \eta, \psi, \vartheta, \boldsymbol{A}\}$ defined for every particle X in B and for any time t is called the thermodynamic process in B if, and only if, it is compatible with the condition for the balance of linear momentum (Cauchy's first law of motion)

$$\operatorname{div} \boldsymbol{T} - \varrho \ddot{\boldsymbol{x}} + \varrho \boldsymbol{b} = 0, \tag{2.3}$$

and with the balance of energy (the first law of thermodynamics)

$$\operatorname{tr}\{\boldsymbol{F}^{-1} \boldsymbol{T} \dot{\boldsymbol{F}}\} - \operatorname{div} \boldsymbol{q} - \varrho(\dot{\psi} + \vartheta \dot{\eta} + \dot{\vartheta} \eta) + \varrho r = 0, \tag{2.4}$$

where the operator div is computed with respect to space coordinates $\boldsymbol{x}$, the dot denotes the material differentiation with respect to time t.

In order to define a thermodynamic process in B it suffices to prescribe the seven functions $\{\chi, \boldsymbol{T}, \boldsymbol{q}, \eta, \psi, \vartheta, \boldsymbol{A}\}$. The two remaining functions $\boldsymbol{b}$ and r are then uniquely determined by eqs. (2.3) and (2.4).

By $\boldsymbol{\Gamma} = (\boldsymbol{F}, \vartheta)$ we denote the thermodynamic state variable pair, and similarly by $\boldsymbol{\Sigma} = \big(\boldsymbol{T}_R(t), -\eta(t)\big)$ is denoted the stress-entropy pair at time t, where $\boldsymbol{T}_R(t)$ is the first Piola-Kirchhoff stress tensor which statisfies the relation

$$\boldsymbol{T}_R = (\det \boldsymbol{F})\, \boldsymbol{T}(\boldsymbol{F}^{-1})^T, \tag{2.5}$$

if ϱ denotes the actual mass density. Let $\boldsymbol{g} = \operatorname{grad} \vartheta$ be the spatial gradient of the temperature.

In the present theory an elastic/viscoplastic material at the point X in body B is characterized by the system of constitutive equations as follows[1]

$$\begin{aligned} \psi &= \hat{\psi}(\boldsymbol{\Gamma}, \boldsymbol{g}; \boldsymbol{A}),\\ \boldsymbol{\Sigma} &= \hat{\boldsymbol{\Sigma}}(\boldsymbol{\Gamma}, \boldsymbol{g}; \boldsymbol{A}),\\ \boldsymbol{q} &= \hat{\boldsymbol{q}}(\boldsymbol{\Gamma}, \boldsymbol{g}; \boldsymbol{A}),\\ \dot{\boldsymbol{A}} &= \hat{\boldsymbol{H}}(\boldsymbol{\Gamma}, \boldsymbol{g}; \boldsymbol{A}), \end{aligned} \tag{2.6}$$

where $\hat{\psi}$, $\hat{\boldsymbol{\Sigma}}$, $\hat{\boldsymbol{q}}$ and $\hat{\boldsymbol{H}}$ denote the response functions.

It is obvious that in the constitutive assumptions (2.6) we make use of the principle of equipresence[2].

The internal state tensor $\boldsymbol{A}$ is interpreted as the inelastic deformation gradient and it is assumed that

$$\hat{\boldsymbol{H}}(\boldsymbol{\Gamma}, \boldsymbol{g}; \boldsymbol{A}) = \gamma(\vartheta)\, \mathrm{K}(\boldsymbol{\Gamma})\, [\boldsymbol{T} - \boldsymbol{T}^0], \tag{2.7}$$

where $\gamma(\vartheta)$ is the viscosity coefficient of a material, $\mathrm{K}(\boldsymbol{\Gamma})$ is a fourth order tensor and $\boldsymbol{T}^0$ denotes the static stress tensor which depends on the stress tensor $\boldsymbol{T}$ and satisfies a certain additional condition

$$\mathfrak{F}(\boldsymbol{T}^0, \vartheta, \boldsymbol{A}) = 0. \tag{2.8}$$

The condition (2.8) is called the static yield condition. As a result of this condition we have the relation for $\boldsymbol{T}^0$ as follows

$$\boldsymbol{T}^0 = \hat{\boldsymbol{T}}^0(\boldsymbol{T}, \vartheta, \boldsymbol{A}). \tag{2.9}$$

---

[1] See COLEMAN and GURTIN [1967, 1] and VALANIS [1967, 4]. In the basic system of constitutive equations describing a material with internal state variables COLEMAN and GURTIN [1967, 1] and VALANIS [1967, 4] assumed the vector parameter $\alpha = (\alpha_1, \ldots, \alpha_n)$ defining the internal state. The introduction of the tensor internal parameter $\boldsymbol{A}$ is due to PERZYNA and WOJNO [1968, 1].

[2] Cf. COLEMAN and MIZEL [1963, 2], [1964, 1] and TRUESDELL and NOLL [1965, 1].

The eqs. $(2.6)_4$ and (2.7) indicate that the rate of the inelastic deformation gradient is proportional to the excess over the static yield condition[1].

The differential eq. $(2.6)_4$ with the function $\hat{\boldsymbol{H}}$ given by eq. (2.7) may be written as the Volterra integral equation[2]

$$\boldsymbol{A} = \boldsymbol{A}^0 + \int_0^t \gamma(\vartheta)\, \mathrm{K}(\boldsymbol{\Gamma})\, [\boldsymbol{T} - \boldsymbol{T}^0]\, d\xi, \tag{2.10}$$

where $\boldsymbol{A}^0 = \boldsymbol{A}(0)$ denotes the initial value for $\boldsymbol{A}$.

We shall say that the thermodynamic process in B described by the system of functions $\{\chi, \boldsymbol{T}, \psi, \boldsymbol{q}, \eta, \vartheta, \boldsymbol{A}\}$ is called an admissible process in B if it is compatible with the constitutive eqs. (2.6) at each point X of B and for all time t.

We shall require that for any time t for a thermodynamic process in B the thermodynamic postulate of positive production of entropy be satisfied. This is equivalent to the following inequality[3]

$$-\dot{\psi} + \boldsymbol{\Sigma} \cdot \dot{\boldsymbol{\Gamma}} - \frac{1}{\varrho\vartheta}\, \boldsymbol{q} \cdot \boldsymbol{g} \geqq 0, \tag{2.11}$$

which must be satisfied for every particle X in B.

As a consequences of the thermodynamic postulate (2.11) we have

$$\begin{aligned}[\hat{\boldsymbol{\Sigma}} - \partial_{\boldsymbol{\Gamma}} \hat{\psi}] \cdot \dot{\boldsymbol{\Gamma}} - \partial_{\boldsymbol{g}} \hat{\psi}(\boldsymbol{\Gamma}, \boldsymbol{g}; \boldsymbol{A}) \cdot \dot{\boldsymbol{g}} - \\ - \operatorname{tr}\{\partial_{\boldsymbol{A}} \hat{\psi} \dot{\boldsymbol{A}}\} - \frac{1}{\varrho\vartheta}\, \boldsymbol{q} \cdot \boldsymbol{g} \geqq 0.\end{aligned} \tag{2.12}$$

Choosing arbitrary values $\dot{\boldsymbol{\Gamma}}$ and $\dot{\boldsymbol{g}}$ it is possible to determine an admissible process in B. Hence, to satisfy the inequality (2.12) we must assume[4]

$$\partial_{\boldsymbol{g}} \psi = \boldsymbol{0}, \tag{2.13}$$

$$\hat{\boldsymbol{\Sigma}} = \partial_{\boldsymbol{\Gamma}} \hat{\psi}(\boldsymbol{\Gamma}; \boldsymbol{A}), \tag{2.14}$$

$$\boldsymbol{q} \cdot \boldsymbol{g} \leqq \varrho \vartheta^2 \sigma \tag{2.15}$$

where σ is called the internal dissipation of an elastic/viscoplastic material and is defined as follows

$$\begin{aligned}\sigma &= -\frac{1}{\vartheta} \operatorname{tr}\{\partial_{\boldsymbol{A}} \hat{\psi}(\boldsymbol{\Gamma}; \boldsymbol{A}) \dot{\boldsymbol{A}}\} = \\ &= -\frac{\gamma(\vartheta)}{\vartheta} \operatorname{tr}\{\partial_{\boldsymbol{A}} \hat{\psi}(\boldsymbol{\Gamma}; \boldsymbol{A})\, \mathrm{K}(\boldsymbol{\Gamma})\, [\boldsymbol{T} - \boldsymbol{T}^0]\}.\end{aligned} \tag{2.16}$$

---

[1] For footnote 1 see p. 208.

[2] For footnote 2 see p. 208.

[3] This inequality is called the Clausius-Duhem inequality (cf. COLEMAN and NOLL [1963, 1].

[4] Cf. COLEMAN and GURTIN [1967, 1] and VALANIS [1967, 4].

As a result of eqs. $(2.6)_4$ and (2.7), we have

$$T = T^0 + \frac{1}{\gamma(\vartheta)} \mathrm{K}^{-1}(\Gamma)\,[\dot{A}], \tag{2.17}$$

which may be treated as an expression for the dynamical stress tensor[1].

3. Reduced Constitutive Equations

The principle of material frame indifference[2], which states that an admissible process must remain admissible after a change of frame, imposes further restrictions on the response functions.

A change of frame is characterized by a time-dependent orthogonal tensor Q. The scalars ϑ, ψ and η are unaffected by a change of frame, but the vectors g and q and the tensors F and T transform as follows

$$g \to Qg, \quad q \to Qq, \quad F \to QF, \quad T \to QTQ^T. \tag{3.1}$$

In our present theory the internal state tensor A is interpreted as the inelastic deformation gradient and therefore transforms as follows

$$A \to QA. \tag{3.2}$$

After satisfying the principle of material frame indifference, we can write the system of constitutive equations for an elastic/viscoplastic material in the following reduced form

$$\begin{aligned} \psi &= \overline{\psi}(C, \vartheta; B), \\ \tilde{T} &= 2\varrho_R \partial_C \overline{\psi}(\mathbf{c}, \vartheta; B), \\ \eta &= -\partial_\vartheta \overline{\psi}(C, \vartheta; B), \\ q_R &= \overline{q}_R(C, \vartheta, g; B), \\ \dot{B} &= \gamma(\vartheta)\,\overline{\mathrm{K}}(C, \vartheta)\,[\tilde{T} - \tilde{T}^0], \end{aligned} \tag{3.3}$$

where $C = F^T F$ is the right Cauchy-Green deformation tensor, $\tilde{T} = (\det F)\, F^{-1} T (F^{-1})^T$ denotes the second Piola-Kirchhoff stress tensor, $B = A^T A$ the inelastic deformation tensor, $q_R = (\det F)\, F^{-1} q$ is the heat flux vector per unit surface in the reference configuration R, and ϱ_R the mass density in the reference configuration R.

---

[1] The eq. (2.17) shows that $\gamma(\vartheta) \to \infty$ yields $T = T^0$. In this limit case, the material is insensitive to the rate of deformation and behaves as an elastic-plastic material.

[2] Cf. NOLL [1958, 1] and TRUESDELL and NOLL [1965, 1].

Thus, an elastic/viscoplastic material is described in a thermodynamic process by the response functions $\overline{\psi}$, $\overline{q}_R$, $\overline{\mathrm{K}}$ and

$$\tilde{T}^0 = \overline{T}^0(\tilde{T}, \vartheta, B), \tag{3.4}$$

and by viscosity coefficient $\gamma(\vartheta)$.

4. Relaxation Process

Let us consider the thermodynamic process characterized by the constancy of the deformation tensor $C = C^*$ and temperature $\vartheta = \vartheta^*$. Such a process is called the isothermal relaxation process[1].

The isothermal relaxation process for an elastic/viscoplastic material is described by the equations as follows

$$\begin{aligned} \dot{B} &= \gamma(\vartheta^*)\,\overline{\mathrm{K}}(C^*, \vartheta^*)\,[\tilde{T} - \tilde{T}^0], \\ \tilde{T} &= 2\varrho_R\,\partial_C\overline{\psi}(C^*, \vartheta^*, B), \\ \tilde{T}^0 &= \overline{T}^0(\tilde{T}, \vartheta^*, B). \end{aligned} \tag{4.1}$$

The eqs. (4.1) can be reduced to one differential equation for the internal state tensor

$$\dot{B} = \overline{M}(C^*, \vartheta^*, B), \tag{4.2}$$

with the initial value $B(0) = B^0$.

The triple (C^*, ϑ^*, B^*) which satisfies the condition

$$\overline{M}(C^*, \vartheta^*, B^*) = 0 \quad \text{and} \quad g = 0 \tag{4.3}$$

is called the equilibrium state of a material at X.

For an elastic/viscoplastic material the equilibrium state may be reached if

$$\tilde{T} = \tilde{T}^0. \tag{4.4}$$

It has been shown that the asymptotically stable equilibrium state[2] (C^*, ϑ^*, B^*) may be reached for an elastic/viscoplastic material only in the isothermal relaxation process[3].

The creep phenomenon for an elastic/viscoplastic material is described by the system of two differential equations[4].

---

[1] Cf. Perzyna and Wojno [1968, 1].

[2] The asymptotically stable equilibrium state is understood according to the definition given by Coleman and Gurtin [1967, 1].

[3] Cf. Perzyna and Wojno [1968, 1].

[4] Cf. Perzyna and Wojno [1968, 2].

References

1958 [1] Noll, W.: A Mathematical Theory of the Mechanical Behavior of Continuous Media. Arch. Ratl. Mech. Anal. **2**, 117—226.

1963 [1] Coleman, B. D., and W. Noll: The Thermodynamics of Elastic Materials with Heat Conduction and Viscosity. Arch. Ratl. Mech. Anal. **13**, 167—178.

[2] Coleman, B. D., and V. Mizel: Thermodynamics and Departures from Fourier's Law of Heat Conduction. Arch. Ratl. Mech. Anal. **13**, 245—261.

[3] Perzyna, P.: The Constitutive Equations for Rate Sensitive Plastic Materials. Quart. Appl. Math. **20**, 321—332.

[4] Perzyna, P.: The Constitutive Equations for Workhardening and Rate Sensitive Plastic Materials. Proc. Vibr. Probl. **4**, 281—290.

1964 [1] Coleman, B. D., and V. Mizel: Existence of Caloric Equations for State in Thermodynamics. J. Chem. Phys. **40**, 1116—1125.

1965 [1] Truesdell, C., and W. Noll: The Non-Linear Field Theories of Mechanics. Encyclopedia of Physics, Vol. III/3. Berlin-Göttingen-Heidelberg: Springer.

1967 [1] Coleman, B. D., and M. E. Gurtin: Thermodynamics with Internal State Variables. J. Chem. Phys. **47**, 597—613.

[2] Olszak, W., and P. Perzyna: General Constitutive Equations for an Elastic/viscoplastic Material. Recent Progress in Applied Mechanics, p. 383—390. Stockholm: Almqvist and Wiksell.

[3] Perzyna, P.: On Thermodynamic Foundations of Viscoplasticity. Symposium on the Mechanical Behavior of Materials under Dynamic Loads, September, San Antonio.

[4] Valanis, K. C.: Unified Theory of Thermomechanical Behavior of Viscoelastic Materials. Symposium on the Mechanical Behavior of Materials under Dynamic Loads, September, San Antonio.

1968 [1] Perzyna, P., and W. Wojno: Thermodynamics of a Rate Sensitive Plastic Material. Arch. Mech. Stos. **20**, 499—511.

[2] Perzyna, P., and W. Wojno: Notes on an Elastic/viscoplastic Material, (In preparation).

Representation of Inelastic Mechanical Behavior by Means of State Variables

By

E. T. Onat

New Haven, Conn. (U.S.A.)

Summary

Previously introduced notion of the space of state and orientation is used to discuss the nature of differential equation representation of mechanical behavior in the presence of finite deformations. An initially isotropic material element loses some of its symmetry during the course of deformation so that, in general, only a small set of superimposed rigid-body rotations leave the state and orientation of the material invariant at a given time. This observation implies that the state and orientation of the material can, and probably must, be represented by a set of tensors of various rank. The general form of the law which governs the "growth" of these variables during the course of deformation is obtained. Applications of these general results to elasticity, viscoelasticity and plasticity are briefly discussed.

1. Introduction

Deformation of an inelastic solid is accompanied by relaxation processes which take place within the solid. The present and future behavior of such a solid is therefore affected by the events in the past so that histories of deformation, temperature and stress play an important role in the representation of thermomechanical behavior. Indeed most of the recent work on constitutive relations is concerned with the study of the functional relationships which exist between these histories. (See, for instance, Rivlin [1], Coleman [2]). The present paper is devoted to a study of differential equation representation of these functional relationships and constitutes a sequel to [3]. For the sake of simplicity, only isothermal deformations are considered in the present paper. Thermodynamical considerations could be added to the present work along the lines discussed in [3].

In Sections 2 and 3 we introduce the notion of the space of state and orientation from purely phenomenological considerations. The stress in an element then becomes a function of its state and orientation and the representation of mechanical behavior involves statements concerning the motion of the state and orientation point as a function of the applied deformation. In Section 3, we study structure of the state and orientation space. In particular we discuss the role of superimposed rigid-body rotations. If the material possesses symmetry at some stage of deformation, then certain rigid-body rotations applied to the element may not cause any motion of the corresponding state and orientation point. The specification of the motions induced by the rigid-body rotations in the space of state and orientation can be made in a simple way if the state space is assumed to be the direct sum of Euclidean spaces, the coordinates of each Euclidean space being components of a tensor of certain rank. We do not prove that the consideration of rigid-body rotations must lead to state and orientation variables which are tensors. However it is likely that this is indeed the case.

Section 4 considers the representation question in terms of tensorial state and orientation variables. It is seen that the "growth" law for these variables possesses rather strong properties of invariance. The paper closes with the consideration of some familiar materials from the point of view of the present paper.

A possible advantage of the present method of representation lies in the hope that the tensorial state and orientation variables can be related to the physics of the deformation process in a rather natural way. These variables can be interpreted as average quantities describing the present arrangement of particles within an element. An elaboration of this point of view can be found in the discussion of examples cited at the end of the paper.

2. Phenomenological Study of Mechanical Behavior

We assume that there exists a supply of identical test specimens and that these specimens are oriented in the same way with respect to a fixed coordinate frame. These specimens carry no stress at time $\tau = 0$ and have the temperature $\theta = \theta_0$. For simplicity we consider isothermal deformations so that the temperature of specimens will be kept at the value θ_0 during testing. We imagine that a typical test consists of the application of a homogeneous time-dependent deformation to the specimen and the observation of the resulting stresses. The applied deformation is measured in terms of the displacement gradients $D_{ij}(\tau)$ defined as

$$x_i = D_{ij}(\tau) X_j, \tag{1}$$

where X_j and x_i are initial and current coordinates of the material point X_j in a fixed rectangular Cartesian frame. We denote by $\sigma_{ij}(\tau)$ the stress components observed in the same coordinate frame. A single test involves two tensor histories

$$\boldsymbol{D}(\tau), \boldsymbol{\sigma}(\tau) \quad \text{on} \quad o \leq \tau < \infty,$$

where bold face letters constitute a shorthand notation for the corresponding components. We assume that $\boldsymbol{D}(\tau)$ is continuous and bounded in its components and we focus attention on materials for which the resulting $\boldsymbol{\sigma}(\tau)$ is also continuous.

From a mathematical point of view a given solid acts as an operator $\boldsymbol{F}$ which assigns to each continuous tensor valued function $\boldsymbol{D}(\tau)$, a function $\boldsymbol{\sigma}(\tau)$:

$$\boldsymbol{\sigma}(\tau) = \boldsymbol{F}(\boldsymbol{D}(\tau)). \tag{2}$$

The purpose of phenomenological studies is to discover the nature of $\boldsymbol{F}$ and to arrive at its mathematical representation.

The present paper is devoted to a further discussion of the differential equation representation of mechanical behavior or, equivalently, of the operator $\boldsymbol{F}$. This mode of representation is intimately connected with the notion of state and orientation developed in [3]. In the following section we give a brief summary of this notion.

3. State and Orientation

We consider a deformation history and focus the attention on times $\tau \geq t$. We introduce displacement gradients measured with respect to the configuration at time t. These are defined as

$$D^*_{ij}(\tau) = \frac{\partial x_i}{\partial y_j}, \quad \tau \geq t, \tag{3}$$

where x_i and y_i are the coordinates of a material point at times τ and t respectively. It is easily seen that

$$\boldsymbol{D}(\tau) = \boldsymbol{D}^*(\tau)\, \boldsymbol{D}(t), \tag{4}$$

where $\boldsymbol{D}^* \boldsymbol{D}$ denotes matrix multiplication of $\boldsymbol{D}^*$ and $\boldsymbol{D}$. Note that $\boldsymbol{\sigma}(\tau)$ for $\tau \geq t$ will depend both on $\boldsymbol{D}(\tau)$ on $[o, t]$ and on $\boldsymbol{D}^*(\tau)$ on $[t, \infty]$. This statement can be written as

$$\underset{[t,\infty)}{\boldsymbol{\sigma}} = \boldsymbol{F}(\underset{[o,t]}{\boldsymbol{D}};\ \underset{[t,\infty)}{\boldsymbol{D}^*}). \tag{5}$$

Now consider two sets of specimens. Subject the first set to displacement gradients $\boldsymbol{D}^{(1)}(\tau)$ on $[o, t]$ and the second test $\boldsymbol{D}^{(2)}(\tau)$ on $[o, t]$.

The specimens in two sets are said to be in the same state and orientation at time t if identical deformations with respect to the configurations at time t produce identical stresses in both sets for times $\tau \geq t$. More precisely and in the notation of (5) we say that $\boldsymbol{D}^{(1)}$ and $\boldsymbol{D}^{(2)}$ on $[o, t]$ produce specimens which are in the same state and orientation at time t if

$$F\,[\underset{[o,t]}{\boldsymbol{D}^{(1)}};\; \underset{[t,\infty)}{\boldsymbol{D}^{*}}] = F\,[\underset{[o,t]}{\boldsymbol{D}^{(2)}};\; \underset{[t,\infty)}{\boldsymbol{D}^{*}}] \tag{6}$$

for any $\boldsymbol{D}^{*}$. (Note that $\boldsymbol{D}^{*}(t) = \boldsymbol{I}$ by definition, where $\boldsymbol{I}$ is the identity matrix).

We now consider the set D^t of all displacement gradient histories on $[o, t]$. We decompose this set into subsets S such that histories belonging to S produce the same state and orientation. S now be considered as an element of a set Σ^t. We say that material elements or specimens subjected to displacement gradients belonging to S are in the state and orientation S.

If the number of elements in the set Σ^t are much less than the number of elements in D^t then there would be advantage in working with elements of Σ^t rather than the distinct $\boldsymbol{D}(\tau)$.

It was shown in [3] that the set Σ^t can be converted into a metric space, by introducing a physically meaningful notion of distance between its elements.

In the present paper we restrict the attention to non-aging materials[1]. In this event Σ^t becomes a subspace of a fixed state and orientation space Σ and time ceases to play an explicit role in further developments.

It follows from the definition of the state and orientation that the stress σ in an element is a function of its state and orientation S and moreover the state and orientation $S(t + \Delta t)$ of an element at time $(t + \Delta t)$ is a function of its state and orientation at time t and the history of $\boldsymbol{D}^{*}$ on the interval $[t, t + \Delta t]$. These statements can be written as

$$\sigma = \boldsymbol{f}(S), \tag{7}$$

$$S(t + \Delta t) = G(S(t);\; \underset{[t,t+\Delta t]}{\boldsymbol{D}^{*}}). \tag{8}$$

Equations (7) and (8) constitute the basis of our further considerations. To explore the implications of these equations we must introduce further hypotheses concerning the nature of the space Σ. We shall first of all assume that Σ is a linear space. A less stringent hypothesis would

[1] For a definition of non-aging materials see [3].

be to assume that Σ is locally homeomorphic to a linear space, or, in other words, Σ is a manifold. Most of what we say in the sequel would remain valid if we were to take this second hypothesis as a starting point. Since Σ is assumed to be a linear space, it is meaningful to speak of $S^{\cdot}$, the derivative of S. We note from (8) that $S^{\cdot}(t)$ may depend only on $S(t)$ and on the values of $\boldsymbol{D}^*$ in the neighborhood of t. Here we shall make the strong assumption that $S^{\cdot}(t)$ depends in addition to S only on $\frac{d}{d\tau}\boldsymbol{D}^*$ evaluated at time t. (We remember, of course, that $\boldsymbol{D}^*(t) = I$).
It is easily seen that

$$\dot{\boldsymbol{D}^*}(t) = \boldsymbol{V} + \boldsymbol{\Omega}. \tag{9}$$

where V and Ω are the rate of deformation and the rate of rotation tensors at the instant of interest[1].
Thus, under the above assumptions, (8) becomes

$$S^{\cdot} = g(S, \boldsymbol{V} + \boldsymbol{\Omega}). \tag{10}$$

In order to be more specific with respect to (7) and (10) one must introduce the notion of coordinates in the Σ-space. Before doing this, however, it is desirable to discuss certain requirements of invariance on (7) and (10) which emanate from a consideration of the role of superimposed rigid-body rotations in mechanics.

For this purpose we consider the displacement histories

$$\boldsymbol{D}(\tau) \quad \text{and} \quad \boldsymbol{D}^R(\tau) = \boldsymbol{R}(\tau)\,\boldsymbol{D}(\tau) \quad \text{on} \quad [o, t], \tag{11}$$

where $\boldsymbol{R}(\tau)$ is an arbitrary proper orthogonal transformation except that it satisfies the requirements

$$\boldsymbol{R}(t) = \boldsymbol{Q}, \tag{12}$$

where $\boldsymbol{Q}$ is a given proper orthogonal transformation.

$\boldsymbol{D}^R(\tau)$ is obtained by superimposing a time dependent rigid-body rotation on the deformation characterized by $\boldsymbol{D}(\tau)$. Suppose now that $\boldsymbol{D}(\tau)$ on $[o, t]$ gives rise to the state and orientation S together with the stress σ at time t:

$$\boldsymbol{D}(\tau) \to S \quad \text{and} \quad \sigma(=\boldsymbol{f}(S)). \tag{13}$$

The question we now ask is concerned with the state and orientations produced by $\boldsymbol{D}^R$, the displacement gradients derived from $\boldsymbol{D}$ by super-

---

[1] $$V_{ij} = \frac{1}{2}\left(\frac{\partial v_i}{\partial x_j} + \frac{\partial v_j}{\partial x_i}\right), \quad \Omega_{ij} = \frac{1}{2}\left(\frac{\partial v_i}{\partial x_j} - \frac{\partial v_j}{\partial x_i}\right),$$

where v_i the components of the velocity vector and x_i are the current coordinates. Note also that

$$\dot{\boldsymbol{D}^*}(t) = \boldsymbol{D}^{\cdot}(t)\,\boldsymbol{D}^{-1}(t).$$

imposed rigid-body rotations. The usual hypotheses concerning the influence of superimposed rigid-body rotations in mechanics yield the following answers which we introduce as assumptions:

(a) $\boldsymbol{D}^R$ defined by (11) and subject to (12) produce *a single* state and orientation, say S_1, at time t.

(b) S_1 depends only on S and $\boldsymbol{Q}$. We express this property by the equation

$$S_1 = P_{\boldsymbol{Q}} S, \tag{14}$$

where $P_{\boldsymbol{Q}}$ can be interpreted as a transformation in the Σ space. (14) has the meaning that a rotation $\boldsymbol{Q}$ applied to an element in S produces an element in the state and orientation $P_{\boldsymbol{Q}} S$. We, of course, have $P_{\boldsymbol{I}} S = S$.

(c) The transformation $P_{\boldsymbol{Q}}$ has the property

$$P_{\boldsymbol{QR}} = P_{\boldsymbol{Q}} P_{\boldsymbol{R}}. \tag{15}$$

(d) If S_1 is associated with the stress σ_1 then

$$\sigma_1 = \boldsymbol{Q} \sigma \boldsymbol{Q}_{\boldsymbol{T}}. \tag{16}$$

In other words

$$\boldsymbol{f}(P_{\boldsymbol{Q}} S) = \boldsymbol{Q}\, \boldsymbol{f}(S)\, \boldsymbol{Q}^T. \tag{17}$$

The above assumptions introduce the following invariance properties for the growth law (10):

If

$$S^{\cdot} = g(S, \boldsymbol{V} + \boldsymbol{\Omega}),$$

then

$$(P_{\boldsymbol{R}} S)^{\cdot} = g(S, \boldsymbol{V} + \boldsymbol{\Omega} + \boldsymbol{R}^{\cdot}), \quad \boldsymbol{R} = \boldsymbol{I} \tag{18}$$

and

$$(P_{\boldsymbol{Q}} S)^{\cdot} = g(P_{\boldsymbol{Q}} S, \quad \boldsymbol{Q}(\boldsymbol{V} + \boldsymbol{\Omega})\boldsymbol{Q}^T), \quad \boldsymbol{Q} = \text{const.} \tag{19}$$

In order to further explore the implications of the invariance properties (17), (18) and (19) we must make explicit statements on the nature of the transformations $P_{\boldsymbol{R}}$ in the Σ space. Let us elaborate on this point by considering *the case of inviscid gases.* For this class of materials the orientation of an element is of no importance and therefore

$$P_{\boldsymbol{Q}} S = S \quad \text{whatever} \quad \boldsymbol{Q} \text{ and } S. \tag{20}$$

When (20) is employed in (17), (18) and (19) the following representation of mechanical behavior results:

$$\sigma = \boldsymbol{I} f(S) \tag{21}$$

and

$$S^{\cdot} = g(S; \operatorname{tr} \boldsymbol{V}, \operatorname{tr} \boldsymbol{V}^2, \operatorname{tr} \boldsymbol{V}^3). \tag{22}$$

Another interesting limiting case is that of *strongly anisotropic solids*. For this class we have

$$P_Q S \neq S \quad \text{whenever} \quad \boldsymbol{Q} \neq \boldsymbol{I}.$$

for all points in Σ. The representation of thermomechanical behavior for this class was discussed in [3] and will not be repeated here.

Most materials of interest fall between the cases of gases and anisotropic solids. It is this intermediate case that we discuss in detail below.

4. Material Symmetry and the Space of State and Orientation

Let us consider a material which is isotropic in its initial state. Let S_0 denote the state and orientation of the material at $\tau = o$. We have by isotropy

$$P_Q S_0 = S_0, \tag{23}$$

whenever

$$\boldsymbol{Q} \in 0^+,$$

where 0^+ denotes the group of proper orthogonal transformations. We now deform the material so that at time t it is in the state and orientation S. We now look for P_Q which leave S invariant. These satisfy

$$P_Q S = S. \tag{24}$$

It can be shown by using (15) that $\boldsymbol{Q}$ associated with (24) constitute a subgroup of 0^+. Let us denote this subgroup by 0_S^+.

The "size" of this group depends on S. For certain S it may be as large as 0^+, but for certain other S it may contain just the identity transformation.

We could see this point more clearly by the following observation on 0_S^+. Let us consider the stress tensor σ associated with S. Consider the set of proper orthogonal transformations which leave σ invariant. These satisfy

$$\boldsymbol{Q} \sigma \boldsymbol{Q}^T = \sigma, \tag{25}$$

and constitute a subgroup of 0^+ which we denote by 0_σ^+. We see from the definition of state and orientation and (17) that 0_σ^+ contains 0_S^+:

$$0_S^+ \subseteq 0_\sigma^+. \tag{26}$$

When σ has distinct principal values then 0_σ^+ is generated solely by 180° rotations about the principal axes and therefore it is a finite group. In such a case 0_S^+ will also be a finite group in view of (25). Of course 0_S^+ can be smaller than 0_σ^+; in the case where no symmetry is left in an element in S, 0_S^+ will contain only the identity element.

As remarked before in order to make use of the previously mentioned invariance requirements one must specify P_Q for each S and for all $\boldsymbol{Q} \in 0^+$. This requires, in turn, the specification or representation of the group 0_S^+ associated with S, the knowledge of 0_S^+ enabling one to determine the distinct state and orientations produced from S by the transformations P_Q.

We shall presently see that all this can be done in a rather economical way with the help of tensors. We have just observed that a second rank tensor such as the stress tensor σ defines a subgroup of 0^+ via (25). The nature of the group 0_σ^+ depends on σ. If σ is of the form $p\boldsymbol{I}$ where p is a scalar than $0_\sigma^+ = 0^+$. If, as seen before, σ has distinct principal values then 0_σ^+ is a small finite group.

To take advantage of this pleasant aspect of tensors we introduce the following assumptions concerning the space of state and orientation Σ.

5. Representation of Mechanical Behavior

We assume that the linear space Σ is a direct sum of a number of Euclidean spaces, the coordinates of each Euclidean space being components of a tensor (under the rotations of the associated element) of a certain rank. Thus each state and orientation S is defined by a number of scalars $\boldsymbol{a}_0^{(i)}$ $(i = 1, \ldots, n_0)$, of vectors $\boldsymbol{a}_1^{(i)}$ $(1 = 1, \ldots, n_1)$, tensors of second rank $\boldsymbol{a}_2^{(i)}$ $(i = 1, \ldots, n_2)$, tensors of third rank, etc. For reasons of simplicity we shall, henceforward, omit the superscripts in the above quantities so that, say, $\boldsymbol{a}_2$ in the future will denote the set $\boldsymbol{a}_2^{(i)}$ $(i = 1, \ldots, n_2)$ and also its generic element. With this convention we have the following one to one correspondence

$$S \leftrightarrow (\boldsymbol{a}_0, \boldsymbol{a}_1, \ldots, \boldsymbol{a}_n). \tag{27}$$

The action of P_Q on S is now defined in the following manner

$$P_Q S \leftrightarrow (P_Q \boldsymbol{a}_0,\ P_Q \boldsymbol{a}_1, \ldots, P_Q \boldsymbol{a}_n), \tag{28}$$

where

$$\begin{aligned} P_Q \boldsymbol{a}_0 &= \boldsymbol{a}_0, \\ P_Q \boldsymbol{a}_1 &= Q_{ip}(\boldsymbol{a}_1)_p = \boldsymbol{Q}\boldsymbol{a}_1, \\ P_Q \boldsymbol{a}_2 &= Q_{ip}Q_{jq}(\boldsymbol{a}_2)_{pq} = \boldsymbol{Q}\boldsymbol{a}_2\boldsymbol{Q}^T. \end{aligned} \tag{29}$$

We observe that P_Q as defined by (28) and (29) satisfies the property (15). We also observe that the subgroup 0_S^+ is created by Q which satisfy the equations

$$a_1 = Q a_1, \quad a_2 = Q a_2 Q^T, \ldots. \tag{30}$$

We are now in a position to write the representation of mechanical behavior in a more explicit form.

In view of (27), (7) becomes

$$\sigma = f(a_0, a_1, \ldots, a_n) \tag{31}$$

where f has the following property of invariance because of (17) and (29):

$$Q^T f(a_0, Q a_1, Q a_2 Q^T, \ldots Q = f(a_0, a_1, \ldots, a_n). \tag{32}$$

The growth law (10) becomes

$$a_i^{\cdot} = g_i(a_0, a_1, \ldots, a_n; \quad V + \Omega), \quad i = 1, \ldots, n \tag{33}$$

The invariance requirement (18) now takes, in view of (33), the form

$$(P_R a_i)^{\cdot} = g_i\big(a_0, \ldots, a_n; \quad V + (\Omega + R^{\cdot})\big), \quad R = I. \tag{34}$$

Let us consider explicitly the left hand side of this equation for the case of, say, $i = 3$. We have, with $R = I$,

$$(R_{ip} R_{jq} R_{kr} (a_3)_{pqr})^{\cdot} = (a_3^{\cdot})_{ijk} + R_{ip}^{\cdot} (a_3)_{pjk} + R_{jq}^{\cdot} (a_3)_{iqk} + R_{kr}^{\cdot} (a_3)_{ijr}. \tag{35}$$

Note that $R^{\cdot}$ is antisymmetric since $R = I$. To arrive at the restriction on the form of (33) we choose, without loss of generality, $R^{\cdot} = -\Omega$ in (34). (34) becomes, when combined with (35)

$$(a_3^{\cdot})_{ijk} = g_3(a_0, \ldots, a_n; \quad V + O) + \Omega_{ip} (a_3)_{pjk} + \Omega_{jq} (a_3)_{iqk} + \\ + \Omega_{kr} (a_3)_{ijr}, \tag{36}$$

where O denotes the tensor with zero components. The right hand side of this equation is the new form g_3 which satisfies (34).

The above process can be repeated for $i = 0, 1, 2, \ldots$ to obtain the following results, where we replace $g_i(V + 0)$ by $f_i(V)$:

$$\begin{aligned} a_0^{\cdot} &= f_0(a_0, \ldots, a_n; V), \\ a_1^{\cdot} &= f_1(a_0, \ldots, a_n; V) + \Omega a_1, \\ a_2^{\cdot} &= f_2(a_0, \ldots, a_n; V) + \Omega a_2 + a_2 \Omega^T, \\ a_n^{\cdot} &= f_n(a_0, \ldots, a_n; V) + T_\Omega a_n, \end{aligned} \tag{37}$$

where $T_\Omega \boldsymbol{a}_n$ is a shorthand notation for the type of terms which contain Ω in (36).

The invariance requirement (19), on the other hand, implies that $\boldsymbol{f}_i$ must obey the following relation for any P_Q:

$$\boldsymbol{f}_i(P_Q\boldsymbol{a}_0, P_Q\boldsymbol{a}_1, \ldots, P_Q\boldsymbol{a}_n; P_Q\boldsymbol{V}) = P_Q\boldsymbol{f}_i(\boldsymbol{a}_0, \ldots, \boldsymbol{a}_n; \boldsymbol{V}),\ i = 0, 1, \ldots, n \tag{38}$$

Here $P_Q\boldsymbol{V}$ and $P_Q\boldsymbol{f}_i$ are defined as in (29), and express tensors which result from $\boldsymbol{V}$ (a second rank tensor) and $\boldsymbol{f}_i$ ($a\, i^{\text{th}}$ rank tensor) by the application of the orthogonal transformation Q.

(38) takes the following explicit form for $i = 0, 1, 2$.

$$\begin{aligned}
\boldsymbol{f}_0(\boldsymbol{a}_0, Q\boldsymbol{a}_1, Q\boldsymbol{a}_2Q^T, \ldots; Q\boldsymbol{V}Q^T) &= \boldsymbol{f}_0(\boldsymbol{a}_0, \boldsymbol{a}_1, \boldsymbol{a}_2, \ldots; \boldsymbol{V}),\\
\boldsymbol{f}_1(\boldsymbol{a}_0, Q\boldsymbol{a}_1, Q\boldsymbol{a}_2Q^T, \ldots; Q\boldsymbol{V}Q^T) &= Q\boldsymbol{f}_1(\boldsymbol{a}_0, \boldsymbol{a}_1, \boldsymbol{a}_2, \ldots; \boldsymbol{V}),\\
\boldsymbol{f}_2(\boldsymbol{a}_0, Q\boldsymbol{a}_1, Q\boldsymbol{a}_2Q^T, \ldots; Q\boldsymbol{V}Q^T) &= Q\boldsymbol{f}_2(\boldsymbol{a}_0, \boldsymbol{a}_1, \boldsymbol{a}_2, \ldots; \boldsymbol{V})Q^T.
\end{aligned} \tag{39}$$

The eqs. (31) and (37), together with invariance requirements (32) and (38) provide a framework for the description of mechanical behavior of a large class of materials in the presence of finite deformations. (31) expresses the fact that the stress at any instant is a function of the state and orientation variables $\boldsymbol{a}_0, \boldsymbol{a}_1, \ldots, \boldsymbol{a}_n$. (37) constitute the growth law for these variables. These equations indicate the direction which the state and orientation point must take in the Σ space as a function of the rate of deformation and rotation applied to the element. We note that the invariance requirements associated with rigid body rotations result in a particularly simple dependence of $\dot{\boldsymbol{a}}_i$ on Ω. These requirements also put restrictions[1] on the forms of $\boldsymbol{f}$ and $\boldsymbol{f}_i$.

The symmetry that the material may possess in its initial state produces certain restrictions on the values of $\boldsymbol{a}_i$ at the initial instant. If the material is isotropic at $\tau = o$, then, in view of (30), $\boldsymbol{a}_i$ must be isotropic tensors at $\tau = o$. This in turn implies that $\boldsymbol{a}_1(o) = 0$, $\boldsymbol{a}_2(o) = c\boldsymbol{I}$, etc. where c is a scalar. This comment has important consequences in the study of small deformations. We shall leave the consideration of this and related points to another paper.

The task of representing mechanical behavior of a given inelastic solid is however hardly begun by the findings of this paper. What we have here is only a framework which must be filled by experimental work of phenomenological nature and also by physical considerations. The choice of the number and the rank of the variables which define

---

[1] For a discussion of the implications of these restrictions see WINEMAN and PIPKIN [4].

the state and orientation of the material is the first step of the problem of representation. This step can benefit greatly from the knowledge of physical processes which take place within the element.

At this point it may be helpful to consider certain familiar materials from the point of view of the present paper.

6. Examples

We start with an elastic solid which is isotropic in its initial state. The well-known stress strain relations for this solid are [5].

$$\sigma = A\boldsymbol{I} + B\boldsymbol{c} + C\boldsymbol{c}^2 \tag{40}$$

where

$$\boldsymbol{c} = \boldsymbol{D}\boldsymbol{D}^T \tag{41}$$

and A, B and C are functions of the invariants $\operatorname{tr}\boldsymbol{c}$, $\operatorname{tr}\boldsymbol{c}^2$, $\operatorname{tr}\boldsymbol{c}^3$. We do not pause here to discuss the restrictions imposed upon A, B and C by the existence of a strain-energy function. To recast the above relations into the form considered in this paper we take the time derivative of (41) and obtain

$$\boldsymbol{c}^{\cdot} = \boldsymbol{V}\boldsymbol{c} + \boldsymbol{c}\boldsymbol{V} + \boldsymbol{\Omega}\boldsymbol{c} + \boldsymbol{c}\boldsymbol{\Omega}^T. \tag{42}$$

(40) and (42) show that for the present material, the state and orientation is characterized by the symmetric second rank tensor $\boldsymbol{c}$:

$$S \leftrightarrow \boldsymbol{c}.$$

We observe that (39) satisfies the requirement (32) and (42) has the form of the third equation in (37) and satisfies the third equation in (39).

It may be of interest to dwell briefly on the physical meaning of $\boldsymbol{c}$. As is well known, the deformation $\boldsymbol{D}$ converts a material element of cubical form and of certain orientation into a right prism. The tensor $\boldsymbol{c}$, by its principal directions and values defines the orientation and shape of this right prism. It can also be seen that $\boldsymbol{c}^{-1}$ is a measure of length change from the deformed configuration to the undeformed one. If one considers the material as composed of identical discrete particles then $n_i c_{ij}^{-1} n_j$ could be considered as a measure of the average distance between particles in the $\boldsymbol{n}$-direction. It is interesting to observe that c_{ij}^{-1} can be measured, in principle and by microscopic means, by studying the deformed element. One would expect that the state and orientation variables $\boldsymbol{a}_i$ discussed in the previous section would possess this property of measurability.

We next consider a *generalized Maxwell solid*. We assume that the stress in an element of this material is again controlled by the present arrangement of its particles and that a symmetric second rank tensor $\boldsymbol{c}$ is again a measure of this arrangement. We further assume that the state and orientation of the material is defined by $\boldsymbol{c}$ alone. We have therefore as a special case of (31) and (37)

$$\sigma = A\boldsymbol{I} + B\boldsymbol{c} + C\boldsymbol{c}^2, \tag{43}$$

where again by (32) A, B, and C are functions of the tr $\boldsymbol{c}$, tr $\boldsymbol{c}^2$, tr $\boldsymbol{c}^3$, and

$$\boldsymbol{c}^{\cdot} = \boldsymbol{f}(\boldsymbol{c}, \boldsymbol{V}) + \boldsymbol{\Omega}\boldsymbol{c} + \boldsymbol{c}\boldsymbol{\Omega}^T, \tag{44}$$

where $\boldsymbol{f}$ is a form invariant function of $\boldsymbol{c}$ and $\boldsymbol{V}$.

When $\boldsymbol{f} = \boldsymbol{V}\boldsymbol{c} + \boldsymbol{c}\boldsymbol{V}$ we have the case of elasticity where $\boldsymbol{c}$ follows the deformation applied to the element without "lag". In a Maxwell solid, the applied deformation may cause grain boundary slip, and may move dislocations without affecting $\boldsymbol{c}$ a great deal. Thus there would develop a lag between $\boldsymbol{c}$ and the total deformation. That the function $\boldsymbol{f}(\boldsymbol{c}, \boldsymbol{V})$ is different from $\boldsymbol{V}\boldsymbol{c} + \boldsymbol{c}\boldsymbol{V}$ is an indication of such a lag.

We close the paper by a few remarks on *elastic-plastic* solids. It is customary to assume that the stress is again controlled by a second rank symmetric tensor $\boldsymbol{c}$ so that (43) remains valid in the present case. However, the state of the material is no longer represented by $\boldsymbol{c}$ alone. One has to introduce a number of tensors which characterize the distribution of dislocations within the element. Let $\boldsymbol{d}$ be a shorthand notation for these tensors. We have

$$S \leftrightarrow \boldsymbol{c}, \boldsymbol{d}$$

and the growth law

$$\begin{aligned} \boldsymbol{c}^{\cdot} &= \boldsymbol{f}(\boldsymbol{c}, \boldsymbol{d}; \boldsymbol{V}) + \boldsymbol{\Omega}\boldsymbol{c} + \boldsymbol{c}\boldsymbol{\Omega}^T \\ \boldsymbol{d}^{\cdot} &= \boldsymbol{g}(\boldsymbol{c}, \boldsymbol{d}; \boldsymbol{V}) + T_{\Omega}\boldsymbol{d}, \end{aligned} \tag{45}$$

where T_{Ω} has the meaning in (37). As is well known, $\boldsymbol{f}$ and $\boldsymbol{g}$ are not any longer continuous functions of $\boldsymbol{V}$ but exhibit discontinuities to take account of "loading" and "unloading" of the element. In order to specify as to which one of these conditions would occur one normally introduces the yield criterion

$$F(\boldsymbol{c}, \boldsymbol{d}) \leq 0.$$

In a particular theory which deals with "isotropically strain-hardening" materials, $\boldsymbol{d}$ is taken to be just a scalar which presumably measures a scalar dislocation density. In some other theories $\boldsymbol{d}$ is assumed to be a tensor of second rank. In view of the recent explosive growth in the understanding of dislocations and their distributions (see, for instance [6]) one hopes that one will be able to introduce a physically meaningful and less ad hoc set of tensors $\boldsymbol{d}$, including third and possible higher rank tensors, to represent the state and orientation of an elastic-plastic material.

References

[1] Rivlin, R. S.: Nonlinear Viscoelastic Solids. SIAM Review **7** (3), 323–340 (1965).

[2] Coleman, B.: Thermodynamics of Materials with Memory. Arch. Ratl. Mech. Anal. **17**, 1–46 (1964).

[3] Onat, E. T.: The Notion of State and Its Implications in Thermodynamics of Inelastic Solids. Proceedings of the IUTAM Symposia, Vienna (1966), p. 292 – 314. Wien-New York: Springer. 1968.

[4] Wineman, A. S., and A. C. Pipkin: Material Symmetry Restrictions on Constitutive Equations. Arch. Ratl. Mech. Anal. **17**, 184–214 (1964).

[5] Rivlin, R. S.: Constitutive Equations for Classes of Deformations. Viscoelasticity: Phenomenological Aspects, p. 93–108. New York: Academic Press. 1960.

[6] Amelinckx and Delavignette: Direct Observation of Imperfections in Crystals. New York: Interscience. 1962.

Discussion

Odqvist: In a paper read before the Iutam Symposium on Irreversible Aspects of Continuum Mechanics, Vienna 1966, Prof. Y. N. Rabotnov introduced a series of state variables, all being scalars—as far as I now remember – and aimed at being sufficient but certainly not necessary criteria entering the constitutive equations. Is it correctly understood that your state variables in a sense will generalize his variables to comprise also tensors and are aiming at producing necessary criteria for the processes considered?

Onat: The main point of the paper is that the presence of symmetry in the material and the invariance requirements associated with the superimposed rigid-body rotations lead one to introduce state variables of tensorial character. However as we noted in the paper we do not offer a proof that under the above mentioned circumstances state variables must be tensors. Nevertheless tensorial state variables prove very convenient in considering the implications of invariance requirements.

Some Stochastic Problems of Thermoviscoelasticity[1]

By

H. Parkus and **J. L. Zeman**

Vienna (Austria)

Summary

Two problem groups are discussed. First, a straight bar is considered, with both ends fixed and exposed, from time $t = 0$ on, to a uniform temperature fluctuating in a random manner about mean value zero. The material of the bar is assumed to obey Norton's law of nonlinear viscoelasticity. Viscosity is supposed to be temperature-independent. The Fokker-Planck equation is set up for two different temperature processes, and is solved numerically. The problem of first-passage time is discussed.

Second, viscosity-dependence on temperature is introduced in the form of thermorheologically simple behavior. Reduced time is then a stochastic process. Some basic properties of this process are discussed. A Fokker-Planck equation for the joint conditional probability density of reduced time and temperature is set up for the two temperature processes. Moments for the first process are given. For the second process solution in terms of Hermite polynomials is indicated.

Introduction

Stochastic problems of thermoelasticity and thermoviscoelasticity are closely connected with the question of lifetime and failure of structures. Let a structure be loaded by given applied forces, constant or varying with time, and, in addition, let it be exposed to a temperature field varying with time in a random manner. Then a double effect will appear as a consequence of the temperature variation. First, random thermal stresses will be created in the structure, and will be superimposed on the stresses due to the loads. Second, those material properties which are temperature-dependent will vary in a random manner. The second effect is of particular importance in the case of viscoelastic mate-

[1] The research reported in this paper has been sponsored in part by the United States Government.

rials as a consequence of the high temperature-sensitivity of viscosity. In general, both effects will contribute to reduce the expected lifetime of the structure, or to increase its probability of failure.

Some special problems of this class have already been treated by the senior author (H. P.) and one of his coworkers. These include the probability of buckling of an elastic plate [1], the probability of buckling of a linear viscoelastic plate with temperature-independent viscosity [2], and the lifetime of a bar with temperature-dependent viscosity [3][1]. In the present paper the following two problem groups will be discussed: (*A*) Behavior of a bar of linear or nonlinear viscoelastic material with temperature-independent viscosity, under stress solely due to random temperature fluctuations. The investigation, which is also valid for a plate under uniform stress, will be based on the Fokker-Planck equation. (*B*) Influence of the temperature dependence of viscosity in a linear thermorheologically simple material. Again, the Fokker-Planck equation will be used as a basis. Hence, it will be assumed in both cases that temperature is a Markov process.

1. Viscoelastic Bar with Temperature-Independent Viscosity

We consider a straight bar with both ends fixed, Fig. 1, and exposed, from time $t = 0$ on, to a uniform temperature $T(t) = T_0 + \theta(t)$, where T_0 is a constant initial mean temperature and $\theta(t)$ represents the stochastic fluctuation. The material of the bar is assumed to obey Norton's law

$$\dot{\varepsilon} = \frac{\dot{\sigma}}{E} + c\sigma^k + \alpha\dot{\theta} \quad (k = 1, 3, 5, \ldots) \tag{1.1}$$

Fig. 1

with temperature-independent coefficient of viscosity c. In the present case $\dot{\varepsilon} = 0$. In addition, we assume the mean of $\theta(t)$ to be zero:

$$\langle\theta(t)\rangle = 0. \tag{1.2}$$

Two different temperature processes will be discussed.

a) *Temperature Corresponding to a Wiener Process*

In that case $\dot{\theta}$ corresponds to white noise, $\dot{\theta} = w(t)$, with correlation function given by

$$R_w(t_1, t_2) = 2D\delta(t_2 - t_1). \tag{1.3}$$

[1] The paper "Effect of random temperature distribution on creep in circular plates" by T. T. Soong and F. A. Cozzarelli, Int. J. Non-Linear Mech. **2**, 27 (1967) should also be mentioned here.

Putting $\sigma = y$ we have from eq. (1.1)

$$\dot{y} = -Ecy^k - Ecw. \tag{1.4}$$

Since $y(t)$ represents a Markov process we have a Fokker-Planck equation [4] for the corresponding conditional probability density $p(y; t \mid y_0; 0)$

$$\frac{\partial p}{\partial t} = DE^2 \alpha^2 \frac{\partial^2 p}{\partial y^2} + Ec \frac{\partial}{\partial y} (y^k p) \tag{1.5}$$

with initial condition

$$p(y; 0 \mid y_0; 0) = \delta(y - y_0). \tag{1.6}$$

The process $y(t)$ is nonstationary. It becomes stationary, however, after a sufficiently long time, $t \to \infty$. Putting $\partial p / \partial t = 0$ in eq. (1.5), we obtain for the stationary solution

$$p_{st} = \frac{(k+1)^{k/k+1}}{2\Gamma\left(\frac{1}{k+1}\right)} \frac{1}{a} \exp\left\{-\frac{1}{k+1}\left(\frac{y}{a}\right)^{k+1}\right\}, \qquad a^{k+1} = DE\alpha^2/c. \tag{1.7}$$

In addition, a simple recursion formula for the stationary moments may be found from eq. (1.5) by multiplying both sides by y^n, integrating between $-\infty$ and $+\infty$ and using integration by parts:

$$\left.\begin{array}{lll} m_{n+k-1} = (n-1)a^{k+1} m_{n-2} & & (n = 2, 4, 6, \ldots) \\ m_0 = 1, & m_n = 0 & (n = 1, 3, 5, \ldots). \end{array}\right\} \tag{1.8}$$

For the *linear* material, $k = 1$, the solution of the nonstationary eq. (1.5) is

$$p_1(y; t \mid y_0; 0) = \frac{1}{\sqrt{2\pi[1 - \exp(-2Ect)]}} \frac{1}{a} \exp\left\{-\frac{1}{2a^2} \frac{(y-\mu)^2}{1 - \exp(-2Ect)}\right\} \tag{1.9}$$

where

$$\mu = y_0 \exp(-Ect) \tag{1.10}$$

is the mean stress at time t.

For the corresponding moments the following differential equation is obtained from eq. (1.5):

$$\frac{dm_n}{d\tau} = -n m_n + a^2 n(n-1) m_{n-2}, \qquad \tau = Ect.$$

The solution of this infinite set of equations is, assuming zero initial stress, i.e., $y_0 = 0$,

$$\left.\begin{aligned} m_n(\tau) &= \frac{a^n}{\sqrt{2^n}}\, n! \sum_{r=0,2,4\ldots}^{n} \frac{(-1)^{r/2}}{\left(\frac{n-r}{2}\right)!\left(\frac{r}{2}\right)!}\, e^{-r\tau}, \qquad \text{if } n = 0, 2, 4, \ldots \\ m_n(\tau) &= 0 \qquad \text{if } n = 1, 3, 5 \ldots \end{aligned}\right\} \tag{1.11}$$

In particular, $m_2 = a^2(1 - e^{-2\tau})$.

The solution of eq. (1.5) for the *nonlinear* material, $k = 3, 5 \ldots$, can not be given in closed form. To facilitate numerical work we put

$$\left.\begin{aligned} p_k(y; t \mid y_0; 0) &= p_1(y; t \mid y_0; 0) + \overline{p}_k(y; t) \\ \overline{p}_k(y; 0) &= 0 \end{aligned}\right\} \tag{1.12}$$

and find, after substituting into eq. (1.5),

$$\begin{aligned} &\frac{\partial \overline{p}_k}{\partial t} - E c \frac{\partial}{\partial y}(y^k \overline{p}_k) - D E^2 \alpha^2 \frac{\partial^2 \overline{p}_k}{\partial y^2} \\ &= D E^2 \alpha^2 \frac{\partial^2 p_1}{\partial y^2} + E c \frac{\partial}{\partial y}(y^k p_1) - \frac{\partial p_1}{\partial t}. \end{aligned} \tag{1.13}$$

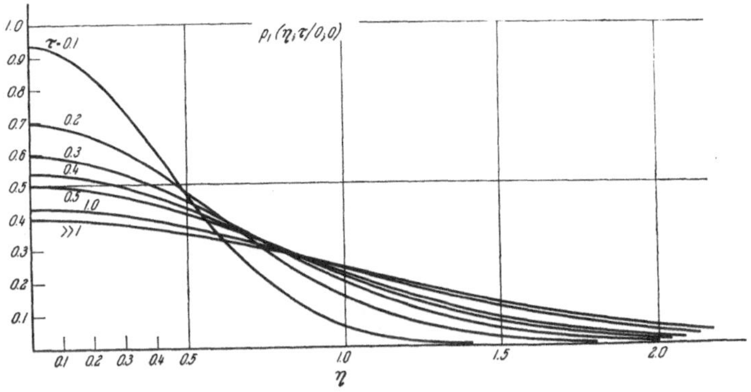

Fig. 2

This equation was solved for $k = 3$ and $k = 5$, using finite differences is y and t. Figs. 2 through 4 show the results in terms of the dimensionless quantities $\eta = y/a$, $\tau = t/b$, where $b = a^2/D E^2 \alpha^2$.

With p known the important problem of *first-passage probability*, closely connected with failure, may be solved.

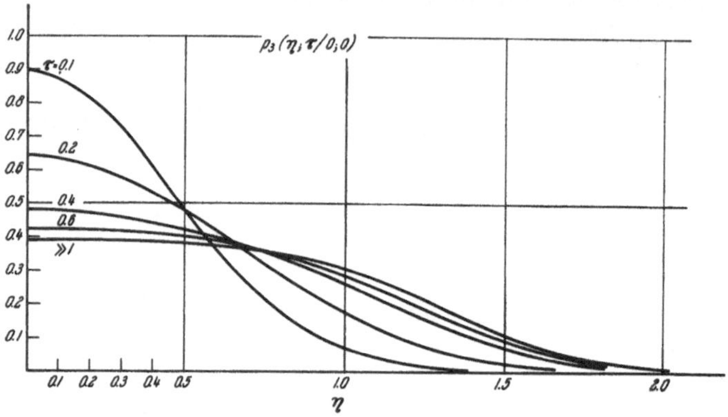

Fig. 3

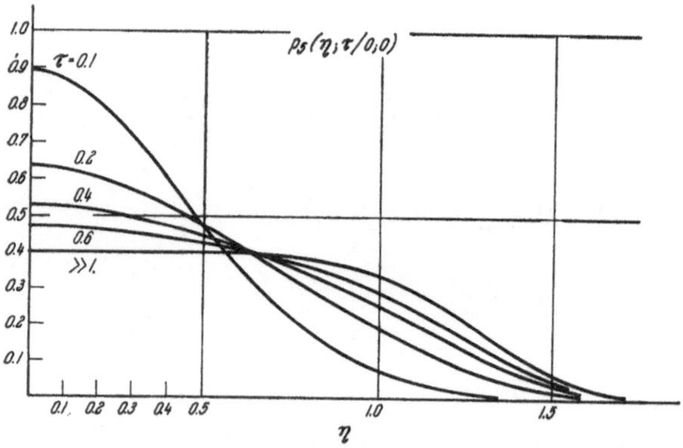

Fig. 4

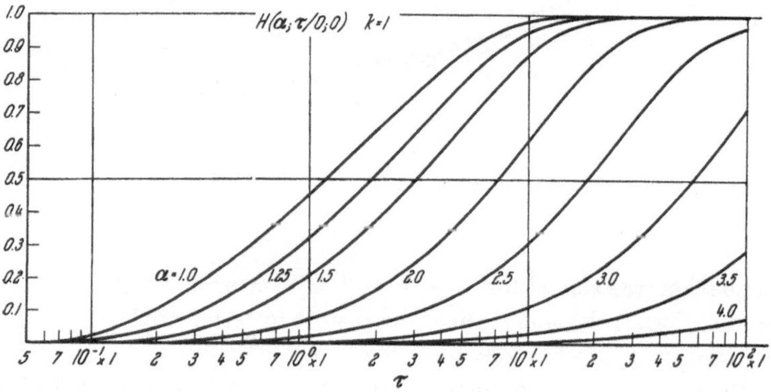

Fig. 5

Let $h(y; t \mid 0; 0)$ denote the density of the first-passage probability. Then

$$H(A; t \mid 0; 0) = \int_0^t h(A; \tau \mid 0; 0) d\tau \qquad (1.14)$$

represents the first-passage probability, i.e., the probability that the stress $y(t)$ will reach a given value A for the first time within the time interval $[0, t]$, after having started at $y_0 = 0$ at time $t = 0$.

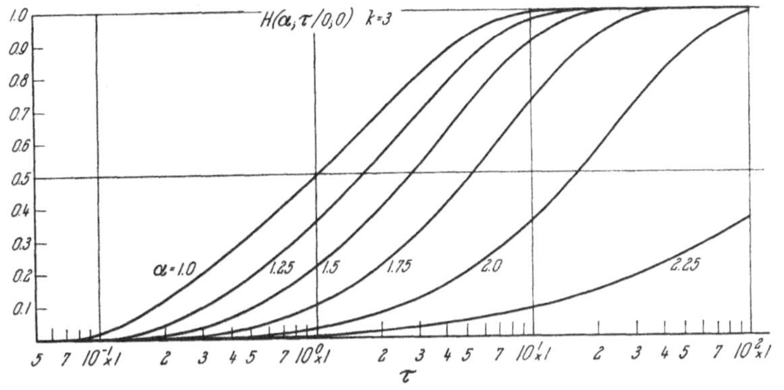

Fig. 6

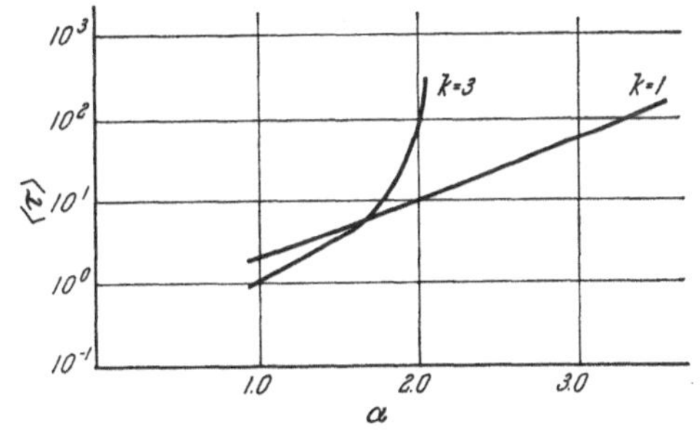

Fig. 7

For $h(y; t/0; 0)$ one immediately finds the integral equation

$$p(Y; t \mid 0; 0) = \int_0^t h(y; \tau \mid 0; 0) p(Y; t \mid y; \tau) d\tau, \qquad Y > y. \qquad (1.15)$$

This equation was solved numerically using finite differences and replacing the integral by a finite sum:

$$p(Y; n\Delta t \mid 0; 0) = \sum_{m=1}^{n-1} h(y; m\Delta t \mid 0; 0) p(Y; n\Delta t \mid y; m\Delta t) \Delta t. \qquad (1.16)$$

The results (in dimensionless quantities τ and $\alpha = A/a$) are represented in Figs. 5 and 6. Fig. 7 gives the expected time $\langle t \rangle$ for the first passage through value A:

$$< t > = \int_0^\infty \tau h(A; \tau \mid 0; 0)\, d\tau. \tag{1.17}$$

b) Temperalure Corresponding to Filtered White Noise

Since temperature in the preceding section is a Wiener process its probability density is

$$p_\theta(\theta; t \mid 0; 0) = \frac{1}{2\sqrt{\pi D t}} \exp\left\{-\frac{\theta^2}{4Dt}\right\}. \tag{1.18}$$

Thus its variance is equal to $2Dt$ and, hence, increases with time beyond all bounds. The same holds for all higher moments. This is not a very relastic situation. In order to improve on it we shall assume in this section that the temperature process is obtained by passing white noise through filters:

$$\ddot{\theta} + \gamma\dot{\theta} + \nu^2\theta = w(t) \tag{1.19}$$

where γ and ν^2 are positive constants. In addition, attention will be restricted to *linear* viscoelastic materials. Putting $\sigma = y_1$, $\theta = y_2$, $\dot{\theta} = y_3$ we have then from eq. (1.1) with $k = 1$, and eq. (1.19),

$$\left.\begin{aligned} \dot{y}_1 &= -Ecy_1 - E\alpha y_3 \\ \dot{y}_2 &= y_3 \\ \dot{y}_3 &= -\nu^2 y_2 - \gamma y_3 + w. \end{aligned}\right\} \tag{1.20}$$

The Fokker-Planck equation for the joint conditional probability density $p(y_1, y_2, y_3; t \mid 0, 0, 0; 0)$ corresponding to the Markov process (1.20) is

$$\begin{aligned} \frac{\partial p}{\partial t} = E\alpha y_3 \frac{\partial p}{\partial y_1} + Ec\frac{\partial}{\partial y_1}(y_1 p) - y_3\frac{\partial p}{\partial y_2} + \nu^2 y_2 \frac{\partial p}{\partial y_3} \\ + \gamma\frac{\partial}{\partial y_3}(y_3 p) + D\frac{\partial^2 p}{\partial y_3^2}. \end{aligned} \tag{1.21}$$

Putting $E = 0$ we obtain the Fokker-Planck equation for the temperature process (1.19) which is equivalent to that of a linear damped oscillator under white noise excitation. Its solution may be found in the literature (see, for instance, [5], [6]). For comparison with the Wiener process above we give here the variance $R_{22}(t)$ of the temperature

$$R_{22}(t) = \frac{D}{\nu^3}\frac{1}{\lambda\omega^2}\, e^{-\lambda\nu t}\,[4\nu^2 - \nu^2\lambda^2\cosh\omega t + \omega^2 e^{\lambda\nu t} + \omega\nu\lambda\sinh\omega t] \tag{1.22}$$

which, for $t \to \infty$, takes on the stationary value $D/\lambda\nu^3$. The following abbreviations have been used

$$\lambda = \frac{\gamma}{\nu}, \qquad \omega = \sqrt{\gamma^2 - 4\nu^2}. \tag{1.23}$$

The general solution of eq. (1.21), although being Gaussian, is too involved to be of much practical use.

The quantity of foremost interest here is the stress $\sigma = y_1$. With $y_1 = y_2 = y_3 = 0$ initially its mean value remains zero all the time while for its variance one finds

$$\begin{aligned} R_{11}(t) = \frac{DE^2\alpha^2}{\omega^2} \Big\{ & \nu_1^2 \left[\frac{1}{\mu_1}(e^{2\mu_1 t} - 1) - \frac{4}{\varkappa_1}(e^{\varkappa_1 t} - 1) - \frac{1}{Ec}(e^{-2Ect} - 1)\right] \\ & + \nu_2^2 \left[\frac{1}{\mu_2}(e^{2\mu_2 t} - 1) - \frac{4}{\varkappa_2}(e^{\varkappa_2 t} - 1) - \frac{1}{Ec}(e^{-2Ect} - 1)\right] \\ & + 2\nu_1\nu_2 \left[\frac{1}{Ec}(e^{-2Ect} - 1) + \frac{2}{\varkappa_1}(e^{\varkappa_1 t} - 1) + \frac{2}{\varkappa_2}(e^{\varkappa_2 t} - 1) \right. \\ & \left.\left. + \frac{2}{\gamma}(e^{-\gamma t} - 1)\right]\right\}, \end{aligned} \tag{1.24}$$

where

$$\left.\begin{aligned} \mu_1 &= -(\gamma - \omega)/2, & \mu_2 &= -(\gamma + \omega)/2, \\ \nu_1 &= \mu_1/(Ec + \mu_1), & \nu_2 &= \mu_2/(Ec + \mu_2), \\ \varkappa_1 &= \mu_1 - Ec, & \varkappa_2 &= \mu_2 - Ec. \end{aligned}\right\} \tag{1.25}$$

Figs. 8 through 10 show the time-dependence of $R_{11}(t)$, for various values of λ and Ec/γ. One notes first a steep increase from the initial value zero, then a short period of near constancy, and, finally, a much less pronounced increase towards the stationary value. The latter is also shown in Fig. 11. It can be seen that $R_{11}(\infty)$ increases with decreasing λ. A less severe filtering, $\gamma = 0$, would not be sufficient to keep temperature and stress variances bounded at all times.

Figs. 12 and 13 exhibit the first-passage probability for stress. However, the curves are not based on the exact relation (1.5) but were computed from an approximation formula given in [7]; see also [6]. The

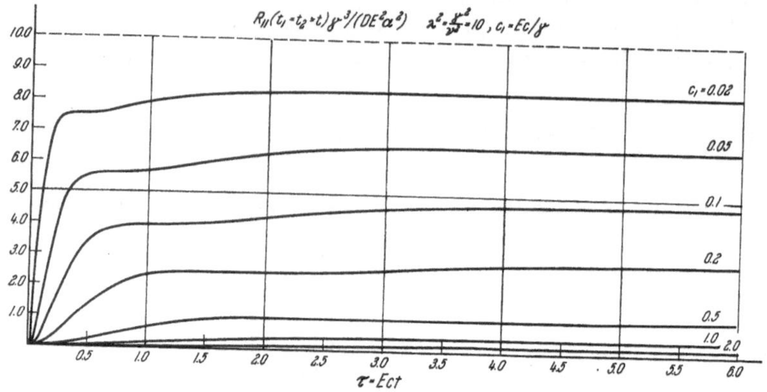

Fig. 8

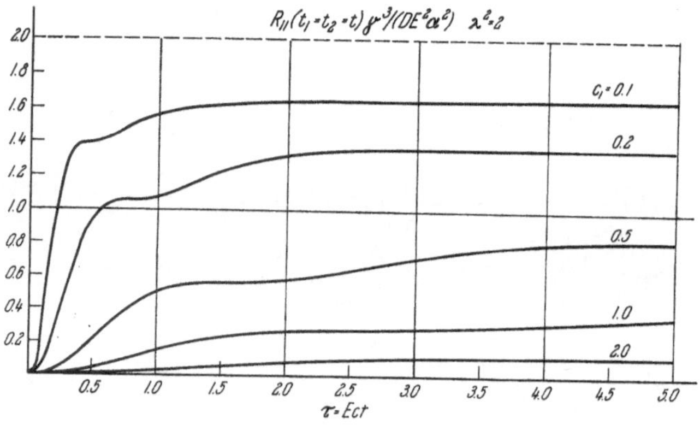

Fig. 9

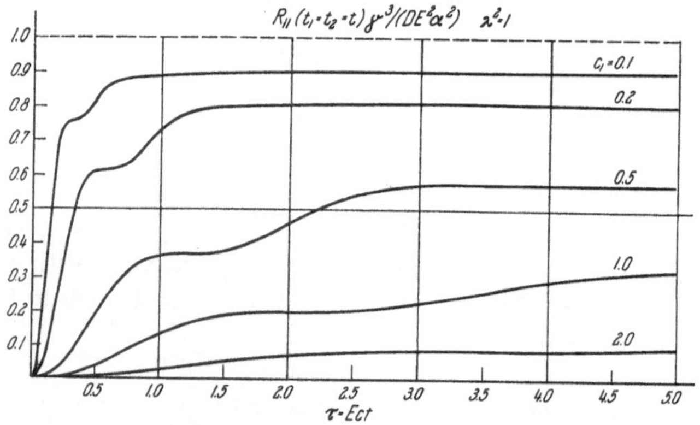

Fig. 10

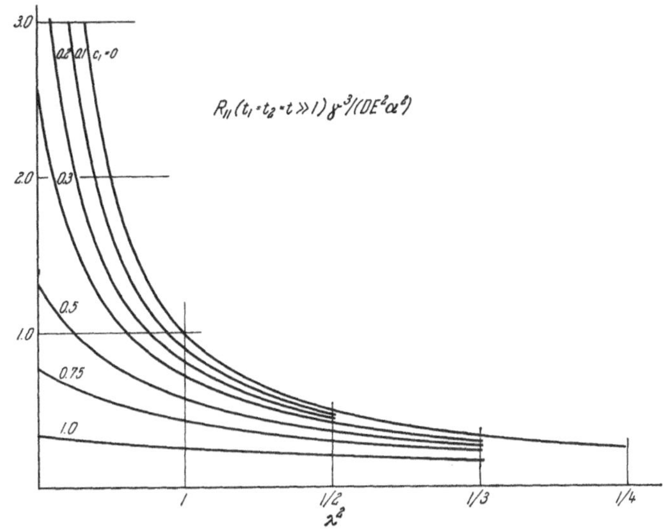

Fig. 11

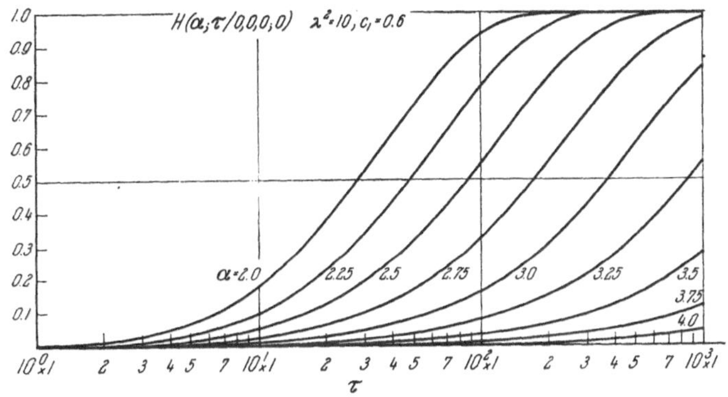

Fig. 12

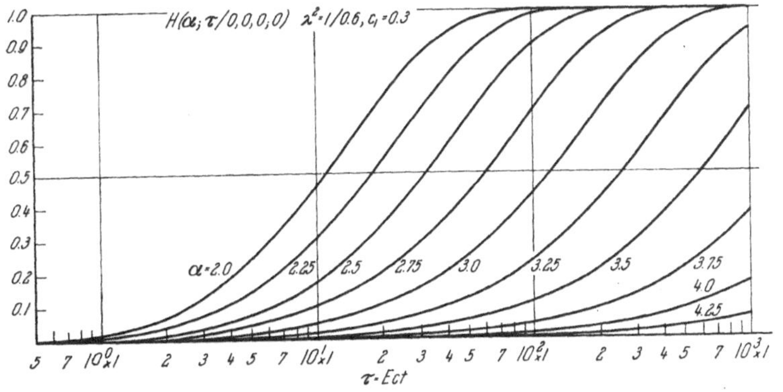

Fig. 13

variances $R_{12}(t)$ and $R_{22}(t)$ necessary, in addition to eq. (1.24), in the analysis were obtained numerically from eqs. (1.20) with the aid of the convolution integral representation.

Fig. 14 shows the expected time of first crossing.

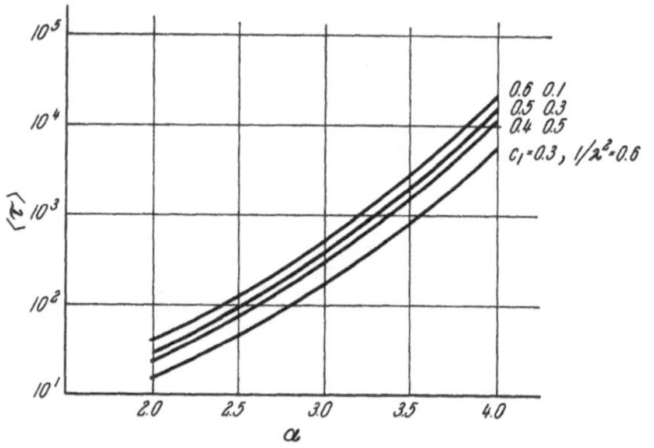

Fig. 14

2. Linear Viscoelastic Material with Temperature-Dependent Viscosity

As pointed out in the Introduction we turn now to the investigation of the influence of temperature-dependent viscosity with particular reference to thermorheologically simple materials. The constitutive equations of these materials are, in terms of the stress and strain deviators s_{ij} and e_{ij}, respectively,

$$s_{ij}(t) = \int_0^t A\,[\vartheta(t) - \vartheta(\tau)]\,\frac{\partial e_{ij}(\tau)}{\partial \tau}\,d\tau \tag{2.1}$$

$$s(t) = \int_0^t B[\vartheta(t) - \vartheta(\tau)]\,\frac{\partial}{\partial \tau}\,[e(\tau) - 3\alpha T(\tau)]\,d\tau \tag{2.2}$$

s and e are the first invariants of the stress and strain tensor, respectively, and $\vartheta(t)$ is a reduced time defined by

$$\vartheta(t) = \int_0^t \varphi[T(\tau)]\,d\tau, \tag{2.3}$$

where $\varphi(T)$ is a positive, monotonically increasing function characterizing the material under consideration. If, as in the preceding section, T_0 denotes the constant (absolute) reference temperature then $\varphi(T_0) = 1$.

For metals at high temperature, the temperature-dependence of the viscosity can be represented, cf. [8], by

$$\varphi(T) = \exp\left\{\frac{c}{T_0} \frac{T - T_0}{T}\right\}. \tag{2.4}$$

Care should be exercised, however, in applying the concept of thermo-rheologically simple materials to metals.

For sufficiently small temperature fluctuations $T - T_0 = \theta(t)$, eq. (2.4) may be approximated by

$$\varphi(\theta) = e^{\beta\theta} \tag{2.5}$$

with $\beta(T_0) = c/T_0^2$.

Since the temperature variation $\theta(t)$ is random reduced time $\vartheta(t)$ is a stochastic process. It may be seen from eqs. (2.1) and (2.2) that the study of the stochastic behavior of a viscoelastic structure of the thermo-rheologically simple type essentially reduces to the study of the stochastic behavior of the reduced time ϑ.

a) *Some Properties of ϑ for Stationary Temperature Fluctuation*

Let $\theta(t)$ be stationary with zero mean, $\langle\theta(t)\rangle = 0$. The function $\varphi(\theta)$ is then also stationary and $\vartheta(t)$ is, from eq. (2.3), a process with stationary increments, cf. [9], characterized, in the sense of correlation theory, by its structure function $D(t_1, t_2)$ rather than its correlation function:

$$D(t_1, t_2) = \langle[\vartheta(t_1) - \vartheta(t_2)]^2\rangle , \tag{2.6}$$

Using the relation

$$\langle\vartheta(t_1)\,\vartheta(t_2)\rangle = \int_0^{t_1} d\varepsilon \int_0^{t_2} R_\varphi(\lambda - \varepsilon)\, d\lambda,$$

where $R_\varphi(\omega)$ is the correlation function of the stationary process $\varphi[\theta(t)]$ one finds, after integration by parts,

$$D(t_1, t_2) = 2\left[(t_2 - t_1)\int_0^{t_2-t_1} R_\varphi(\omega)\, d\omega - \int_0^{t_1-t_2} \omega R_\varphi(\omega)\, d\omega\right] \tag{2.7}$$

which does not depend on the individual t_1 and t_2 but on their difference only.

If, in addition to being stationary, the process $\theta(t)$ is also ergodic then

$$\lim_{t\to\infty} \frac{1}{t} \int_0^t \varphi[\theta(\tau)]\, d\tau = \langle \varphi \rangle = \mu = \text{const}.$$

with probability one. Hence, we have the asymptotic relation

$$\lim_{t\to\infty} \frac{\vartheta(t)}{t} = \mu. \tag{2.8}$$

b) Temperature Corresponding to a Wiener Process

Putting $\vartheta = x_1$, $\varphi(\theta) = x_2$, and using approximation (2.5) we arrive at the two-dimensional Markov process

$$\left.\begin{aligned} \dot{x}_1 &= x_2 \\ \dot{x}_2 &= \beta x_2 w. \end{aligned}\right\} \tag{2.9}$$

The process is governed by the Fokker-Planck equation

$$\frac{\partial p}{\partial t} = -x_2 \frac{\partial p}{\partial x_1} + D\beta^2 \frac{\partial}{\partial x_2}\left(x_2 \frac{\partial (x_2 p)}{\partial x_2}\right) \tag{2.10}$$

for the joint conditional probability density $p(x_1, x_2; t \mid 0, 1; 0)$.

From eq. (2.10) one obtains the following equation for the moments

$$\frac{dm_{r,n}}{dt} = r\, m_{r-1,n+1} + D\beta^2 n^2 m_{r,n} \tag{2.11}$$

with initial condition

$$m_{r,n}(0) = \begin{cases} 1 & \text{for} \quad r = 0 \\ 0 & \text{for} \quad r > 0. \end{cases} \tag{2.12}$$

The first and second cross moments of ϑ are

$$m_{1,n} = \frac{1}{(2n+1) D\beta^2} \left[e^{D\beta^2(n+1)^2 t} - e^{D\beta^2 n^2 t}\right] \tag{2.13}$$

$$m_{2,n} = \frac{1}{4(n+1)(2n+1)(2n+3) D^2\beta^4} \big[(2n+1)\, e^{D\beta^2(n+2)^2 t} - 4(n+1)\, e^{D\beta^2(n+1)^2 t} + (3+2n)\, e^{D\beta^2 n^2 t}\big] \tag{2.14}$$

while for $\varphi(\theta)$ one has

$$m_{0,n} = e^{D\beta^2 n^2 t}. \tag{2.15}$$

As was to be expected all moments grow rapidly beyond all bounds with increasing time t. Since approximation (2.5) was used the solution is, therefore, valid for small t only.

c) *Temperature Corresponding to Filtered White Noise*

Letting $\vartheta = x_1, \theta = x_2, \dot{\theta} = x_3$ we have the three-dimensional Markov process

$$\left.\begin{aligned} \dot{x}_1 &= \varphi(x_2) \\ \dot{x}_2 &= x_3 \\ \dot{x}_3 &= -\nu^2 x_2 - \gamma x_3 + w \end{aligned}\right\} \tag{2.16}$$

with the corresponding Fokker-Planck equation

$$\frac{\partial p}{\partial t} = -\varphi(x_2)\frac{\partial p}{\partial x_1} - x_3\frac{\partial p}{\partial x_2} + \nu^2 x_2 \frac{\partial p}{\partial x_3} + \gamma \frac{\partial}{\partial x_3}(x_3 p) + D\frac{\partial^2 p}{\partial x_3^2}. \tag{2.17}$$

Upon introduction of new dimensionless variables

$$x_2 = \sqrt{\frac{D}{\nu^3}}\, x, \quad x_3 = \sqrt{\frac{D}{\nu}}\, y, \quad x_1 = \frac{z}{\nu}, \quad \tau = \nu t \tag{2.18}$$

this equation goes over into

$$\frac{\partial p}{\partial \tau} = -\varphi(x)\frac{\partial p}{\partial z} - y\frac{\partial p}{\partial x} + \frac{\partial}{\partial y}(x + \lambda y)\, p + \frac{\partial^2 p}{\partial y^2} \tag{2.19}$$

for $p(x, y, z; t \mid x_0, y_0, z_0; 0)$, with $\lambda = \gamma/\nu$.

In obtaining the solution of this equation the following procedure is used. First it is assumed that large variations of the temperature θ and, hence, of x are improbable. The function $\varphi(x)$ can then be replaced by the first few terms of its Taylor series expansion

$$\varphi(x) = 1 + c_1 x + c_2 x^2 + c_3 x^3 + \cdots \tag{2.20}$$

Second, a series expansion in terms of Hermite polynomials is used to represent the solution of eq. (2.19). Fortunately, what is needed in the applications of the theory is not the complete density p but rather the following two conditional densities

$$q(x, z; \tau \mid x_0, y_0, z_0; 0) = \int_{-\infty}^{+\infty} p(x, y, z; \tau \mid x_0, y_0, z_0; 0)\, dy \tag{2.21}$$

$$r(z; \tau \mid x_0, y_0, z_0; 0) = \int_{-\infty}^{+\infty} q(x, z; \tau \mid x_0, y_0, z_0; 0)\, dx. \tag{2.22}$$

The series expansion for r, for instance, will then read, with $x_0 = y_0 = z_0 = 0$,

$$r(z; \tau \mid 0, 0, 0; 0) = \delta(z - \tau) + \frac{1}{\sqrt{\pi}} e^{-z^2} \sum_{n=1}^{\infty} \frac{A_n(\tau)}{2^n n!} H_n(z). \qquad (2.23)$$

More details and applications will be published elsewhere.

References

[1] PARKUS, H.: Stability of Elastic Plates under Random Membrane Stress. Bull. Acad. Polon. Sc., série sc. techn., **14**, 125 (1966).

[2] BARGMANN, H.: Buckling of Viscoelastic Maxwell Plates under Random Temperature. Acta Mechanica **7**, 16 (1969).

[3] PARKUS, H.: On the Lifetime of Viscoelastic Structures in a Random Temperature Field. In: Recent Progress in Applied Mechanics (The Folke Odqvist Volume), p. 391. Stockholm and New York 1967.

[4] GRAY, A. H. JR., and T. K. CAUGHEY: A Controversy in Problems Involving Random Parametric Excitation. J. Math. Phys. **44**, 288 (1965).

[5] CHANDRASEKHAR, S.: Stochastic Problems in Physics and Astronomy. In: Selected Papers on Noise and Stochastic Processes (N. WAX, editor). New York: Dover Publications.

[6] PARKUS, H., and J. L. ZEMAN: A Stochastic Problem of Thermoelasticity. Proc. VIIIth Euromech Symposium. Warsaw 1967 (to appear).

[7] RICE, J. R., and F. P. BEER: First-Occurrence Time of High-Level Crossings in a Continuous Random Process. J. Acoust. Soc. America **39**, 323 (1966).

[8] ODQVIST, F. K. G., and J. HULT: Kriechfestigkeit metallischer Werkstoffe, p. 214. Berlin-Göttingen-Heidelberg: Springer. 1962.

[9] JAGLOM, A. M.: An Introduction to the Theory of Stationary Random Functions. Englewood Cliffs, N. J.: Prentice-Hall, Inc. 1962.

Yield Surfaces of Pure Aluminum at Elevated Temperatures[1]

By

Aris Phillips

New Haven, Conn. (U.S.A.)

Summary

Results of experiments are presented in which thin-walled tubes of pure aluminum are loaded in combined tension and torsion, in the plastic range, at room temperature and at elevated temperatures to 325 °F. Yield surfaces in stress-space are obtained at several temperatures for the virgin material and for the material prestrained in torsion to three different levels of prestraining. It is shown that no cross effect exists at all tested temperatures and levels of prestressing. It is also shown that the yield surfaces do not pass through the prestressing point. The paper ends with an analytical evaluation of the experimental results.

Introduction

This paper presents the results of some experiments in which thin-walled tubes of commercially pure aluminum in the annealed condition were loaded in combined tension and torsion in the plastic region at elevated temperatures. The purpose of these tests was to determine the yield surface of pure aluminum in the tension-torsion space as a function of temperature for several levels of torsional prestressing. Although several papers have been written in which yielding in simple tension or simple torsion as a function of temperature was considered, (e.g., [1], [2], [3]), to the author's knowledge the literature does not include a systematic experimental study of the dependency of the yield surface on the temperature. In this paper we attempt to provide such a systematic study for commercially pure aluminum at the temperature range

---

[1] This work was sponsored by the National Science Foundation of the United States government.

$70\,°F \leqq T \leqq 325\,°F$. The paper ends with a discussion of the implications of the experimental results for the theory of plasticity.

A complete description of the experimental set up and procedure will be given in [4] to which the reader is referred for details. Here we mention that the thin-walled tubes have dimensions of $8^1/_2''$ length, $2^1/_{16}''$ I. D., and 0.050″ wall thickness. They were annealed at 650 °F for one hour and furnace cooled. The load was applied by a special dead load machine described in [4]. The strains were measured by means of Bakelite type foil gages.

The definition of yielding was that of the proportional limit. In order to obtain this limit it was necessary to produce some plastic deformation but the amount of this deformation was limited to a maximum of 5×10^{-6} in/in each time a proportional limit was obtained. A special procedure described in [4] was developed for obtaining the proportional limit. By means of this procedure two independent observers would obtain the same results. Thus the criticism raised by MAIR and PUGH [5] on the use of the proportional limit for the definition of yielding was not valid in our experiments.

The yield curves were determined at the temperatures of 70 °F, 151 °F, 227 °F, and 305 °F. These curves were obtained at zero prestress and at three additional levels of prestress. The two first levels of prestress were accompanied by a very limited amount of prestrain, of the order of 10^{-4} in/in. At this level no creep strains were encountered.

The third level of prestress was accompanied by a prestrain of the order of 10^{-2} in/in. At this level some amount of creep strain was present while prestressing. Creep was absent, however, when probing the yield surface, since this probing involved motion of the stress point within the elastic region and piercing of the yield surface by only 5×10^{-6} in/in.

Initial Yield Surfaces

In Fig. 1 we see the yield curves of the annealed specimen S-3 at the four temperatures of 70°, 151°, 227°, 305 °F. These curves in the s_{11}, s_{12} plane were obtained under isothermal conditions at each indicated temperature. The small circles show the values of stress at which yielding occurred. The 70 °F curve was obtained first, then the temperature was raised to 151 °F and the corresponding curve was obtained; then the curve at 227° and finally that at 305° were determined.

The stress paths by means of which the probing of each yield curve was made are illustrated in Fig. 2. Beginning with a very small tensile stress at point 1, representing weight of equipment, loading in torsion proceeded to point 2 where yielding occurred; then the stress path

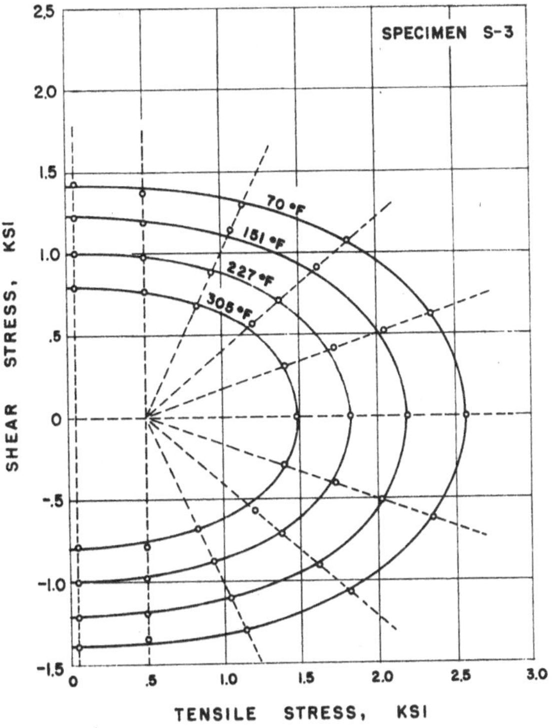

Fig. 1. Initial yield curves at elevated temperatures

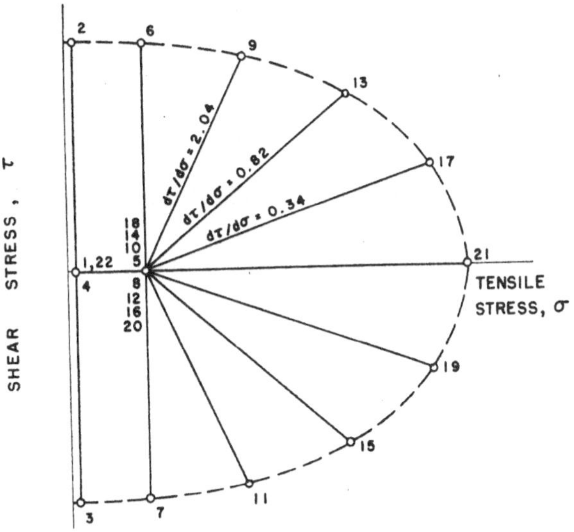

Fig. 2. Loading paths for initial yield curves

included unloading to point 1 and loading in reverse torsion to point 3 where yielding occurred again. The path follows the straight line segments connecting the points with increasing numbers until finally unloading ends at point 22 which coincides with point 1.

The curves shown on Fig. 1 are ellipses intermediate between the Mises and Tresca ellipses. They are very near to the Mises ellipse. The ratios max s_{11}/max s_{12} are as follows: at 70 °F: 1.80, at 151 °F: 1.80, at 227 °F: 1.83, at 305 °F: 1.89. These ellipses are the isothermals of a truncated elliptical cone in the s_{11}, s_{12}, T space. The basis of the cone is on the stress plane at 70 °F and the apex is on the temperature axis at $T = 620$ °F which is approximately the annealing temperature of this material. Since we have data only between 70° and 305 °F we can consider only a truncated cone between these two temperatures. The equation of this cone can be described by

$$\frac{s_{11}^2}{2580^2} + \frac{s_{12}^2}{1430^2} = \left(\frac{620 - T}{550}\right)^2, \quad 70^\circ \leqq T \leqq 305^\circ. \tag{1}$$

Yield Surfaces after Prestrain

Fig. 3 gives the yield curves for the same specimen S-3 as before, but after prestressing at $s_{12} = 2262$ psi, $s_{11} = 50$ psi, $T = 70$ °F. The amount of plastic shear prestrain is only 70×10^{-6} in/in.

After prestressing, the torsional stress was decreased until a value of shear stress was obtained for which yielding occured in reverse shear. Then the shear stress was increased until yielding occurred in the positive shear direction. It was found that this last yield stress was only 1830 psi which is substantially less than the prestress value. We conclude that the yield surface does not pass through the prestress point, a fact observed also by SIERAKOWSKI and PHILLIPS [6] and by JUSTUSSON and PHILLIPS [7].

The yield curves in Fig. 3 were obtained again for the temperatures of 70 °F, 151 °F, 227 °F, and 305 °F; they were determined under isothermal conditions at each indicated temperature. The 70° curve was obtained first and then the curves at each successively higher temperature were determined. The stress path used for obtaining each of these yield curves is shown in Fig. 4. The path follows the straight line segments connecting the points with increasing numbers from 1 to 23.

From Fig. 3 it is obvious that the yield surface is no longer a symmetric one about an axis parallel to the s_{11} axis. The yield curves are no longer ellipses. It is interesting that the isothermals are crowded in the reverse shear direction and that there is a decrease in the diameter of each isothermal in the direction of the shear stress.

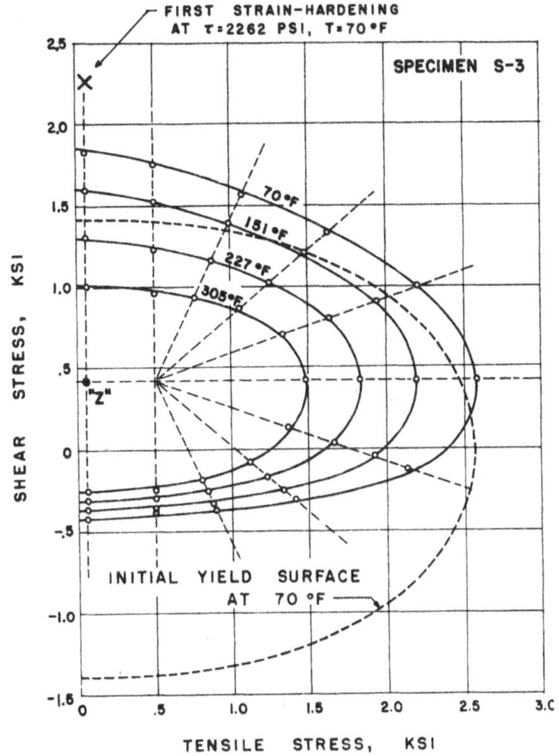

Fig. 3. Yield curves at elevated temperatures for first prestressing

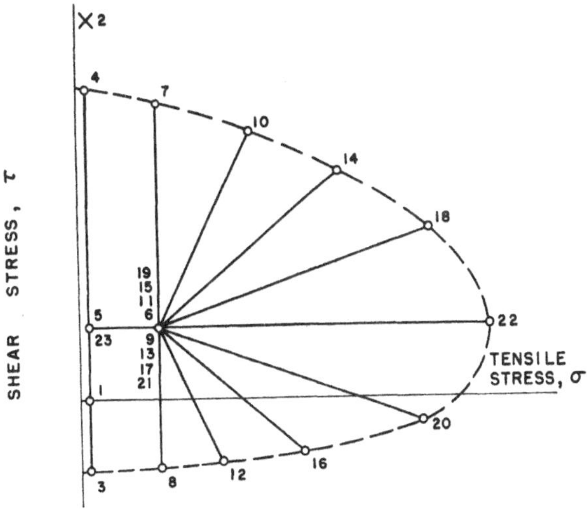

Fig. 4. Loading paths for yield curves at first prestressing

From the tests of IVEY [8], MAIR and PUGH [5] and NAGHDI, ESSENBURG and KOFF [9], it is seen that prestressing in torsion produces no cross effect in tension. This means that the maximum tensile stress reached by the yield curve after prestressing in torsion is the same as the maximum tensile stress reached by it before prestressing. Fig. 3 verifies

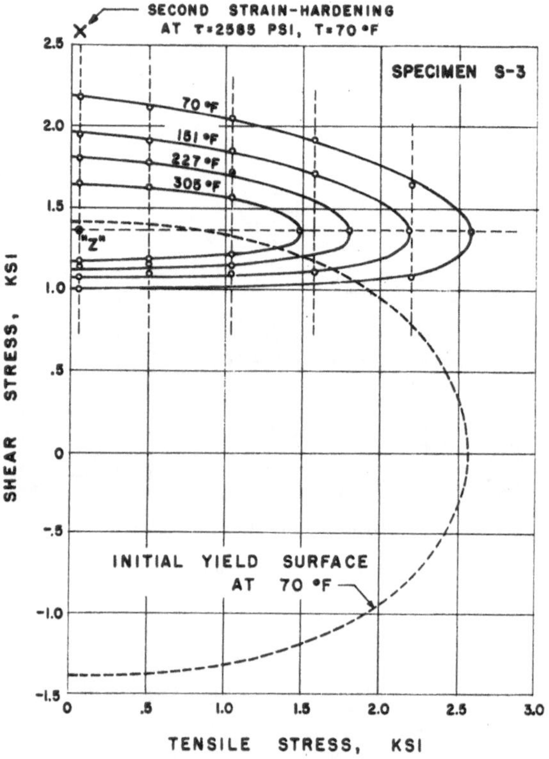

Fig. 5. Yield curves at elevated temperatures for second prestressing

the lack of cross effect for the 70 °F yield curve. Another observation is that the values of max s_{11} for each isothermal occur at the same value of the shear stress s_{12} independently of the temperature. In the next section it will be shown that this observation leads the theory to a conclusion of considerable interest.

Fig. 5 shows the yield curves for the same specimen S-3 but after a second prestressing in torsion to $s_{12} = 2585$ psi, $s_{11} = 50$ psi, $T = 70\,°$F. The amount of prestrain was again very small, only $\gamma''_{12} = 300 \times 10^{-6}$ in/in. The stress paths used for obtaining the four yield curves were as shown in Fig. 6. It is seen again that the prestress point is at considerable distance from the 70° curve. In addition, there is again lack of cross effect for the 70 °F curve, and the value of max s_{11} for each isothermal occurs again at the same value of s_{12} independently of the value of T.

An interesting result is that the 70 °F curve does not enclose the origin. Such a result was shown also by IVEY [8] and by JUSTUSSON and PHILLIPS [7]. However, this is the first time in which it is shown that the yield surface does not enclose the origin at such a small value of prestraining.

From Fig. 5 it is seen that the yield surface is even less symmetric than the previous one about an axis parallel to the s_{11} axis. The isothermals are again crowded in the reverse shear direction and there is an even greater decrease in the diameter of each isothermal in the direction of the shear stress.

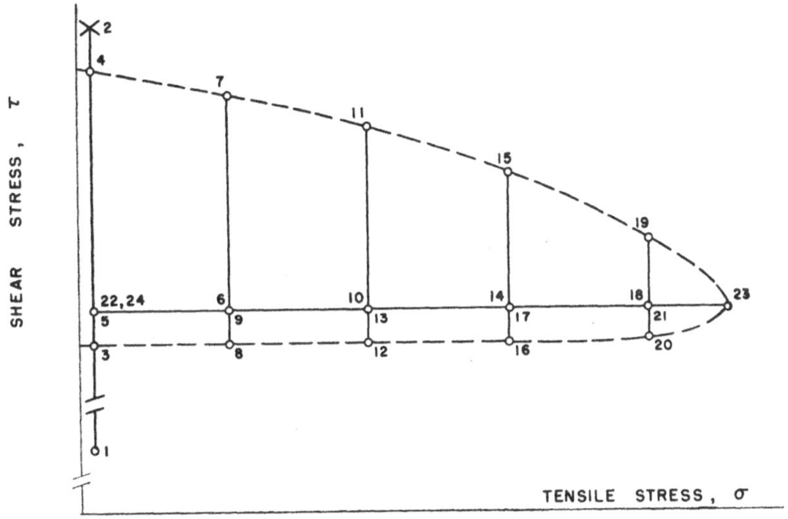

Fig. 6. Loading paths for yield curves at second and third prestressing

Fig. 7 shows the three sets of curves superimposed. We see that there is no cross effect at each tested temperature. It is seen that we have a shrinking in the size of the yield surface in the direction of prestressing as the prestressing increases, while in the s_{11} direction the size of the yield surface does not change. In a sense the lack of cross effect is an important element of stability of the yield surface as prestressing increases.

Fig. 8 shows the intersection of the three yield surfaces with the plane A_3A_3 shown in Fig. 7. It is seen that the intersections consist of straight lines. This is also true for the intersection of each surface with the plane perpendicular to the stress plane, passing through the max s_{11} values. This last conclusion is a result of the lack of cross effect.

Fig. 9 shows the yield curves for another specimen, S-4, which was prestressed in torsion to $s_{12} = 4200$ psi, $s_{11} = 50$ psi, $T = 70\,°F$. This amount of prestressing is approximately twice the maximum prestress-

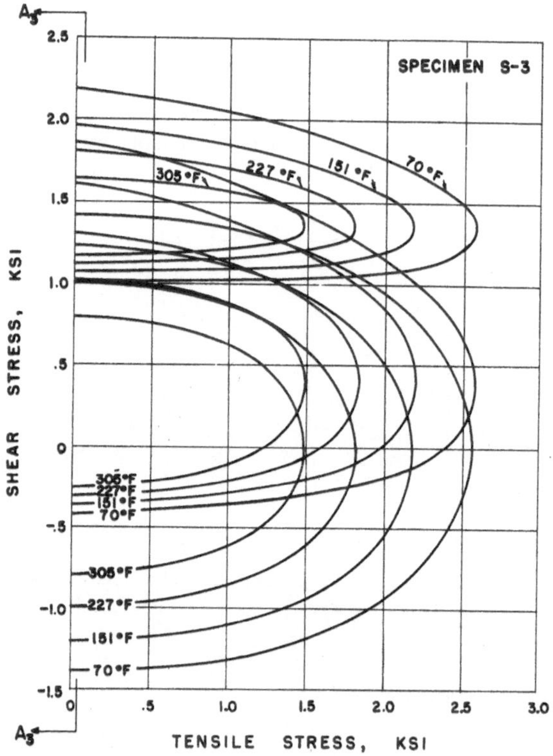

Fig. 7. Superposition of the three sets of yield curves indicating the absence of cross effect

ing of the previous specimen. While for the previous specimen the prestraining was small, now the prestraining is substantial, equal to approximately 1%. The procedure used for obtaining the yield curve is as shown in Fig. 6. The testing temperatures are now slightly different, as indicated in the figure.

We see again that the prestress point is at considerable distance from the 70 °F curve and a comparison with the yield curves of the virgin S-4 specimen (not shown in the figure) shows that there is again lack of cross

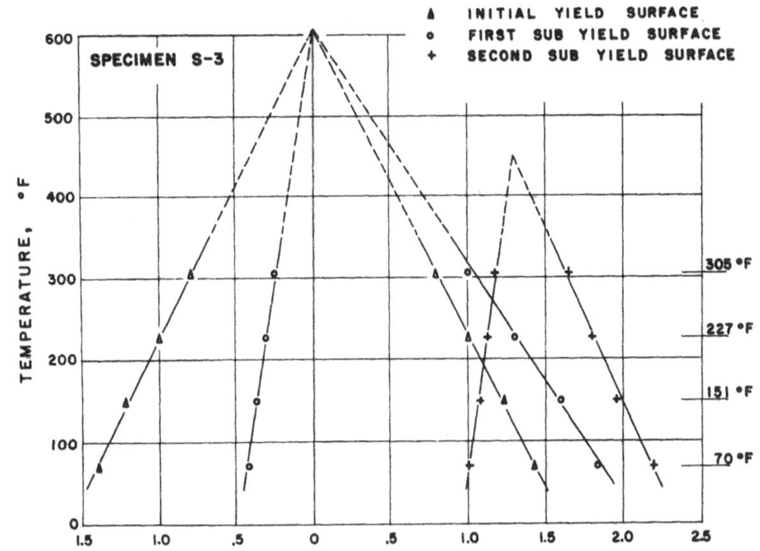

Fig. 8. Section A_3A_3

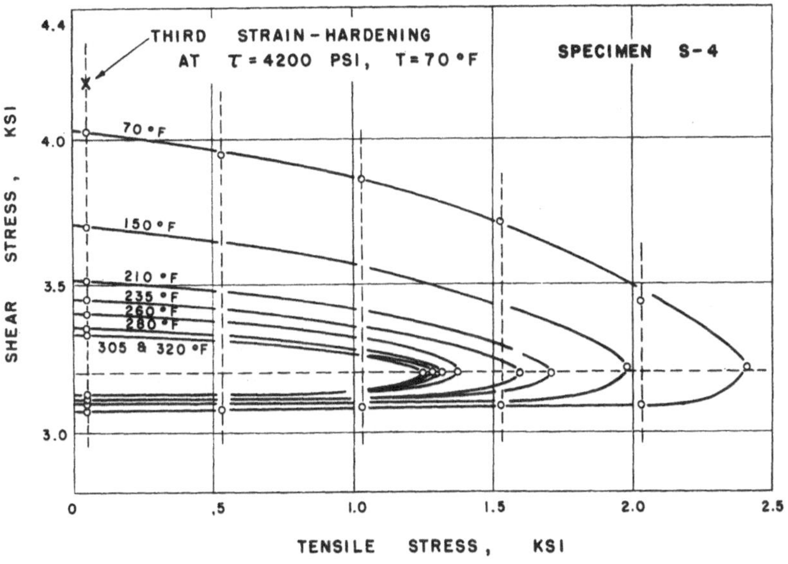

Fig. 9. Yield curves at elevated temperatures for third prestressing-3A

effect at all temperatures. The shrinking of the yield surface in the direction of prestressing as prestressing increases is more pronounced than before. In addition, the value of max s_{11} for each isothermal occurs again at the same value of s_{12} independently of the value of the temperature.

In Fig. 10 we observe the results of prestressing the same specimen S-4 a second time to the same values $s_{12} = 4200$ psi, $s_{11} = 50$ psi, $T = 70\,°F$ as before. This second prestressing produces a small shifting of the yield curves towards the prestress point. Such a shifting was to be expected according to previous results in Sierakowski and Phillips [6].

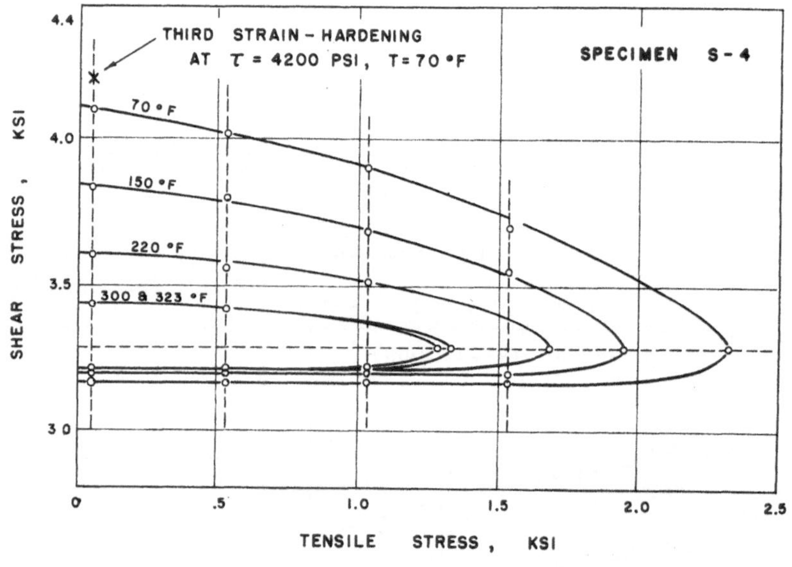

Fig. 10. Yield curves at elevated temperatures for third prestressing - 3B

Fig. 11 shows the intersection of the two yield surfaces in Fig. 9 and 10 with the plane A_3A_3 defined previously. It is seen that this intersection is no longer a set of straight lines but a set of two curves. The shifting of the yield surface because of reloading to the same prestress point is also shown.

Fig. 12 shows the intersection of the yield surfaces in Fig. 9 and 10 with the planes normal to the stress plane and which pass through the max s_{11} values. These intersections are straight lines very near to each other. Fig. 13 shows the stress-strain curve for specimen S-4, indicating the history of prestraining of this specimen.

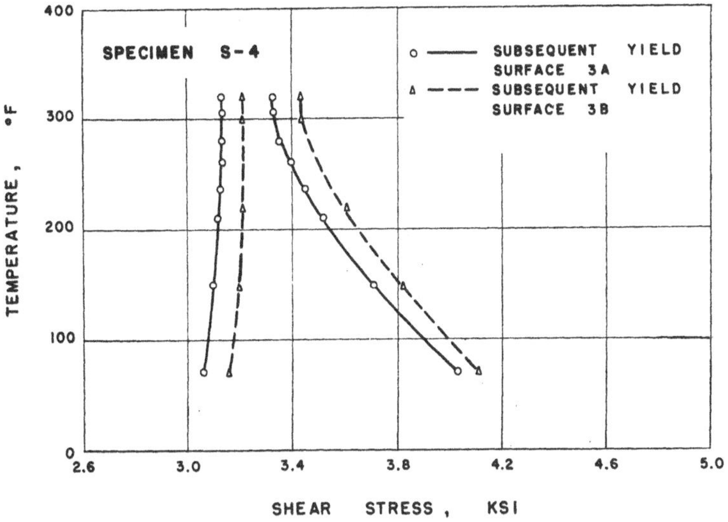

Fig. 11. Section A_3A_3 for third prestressing

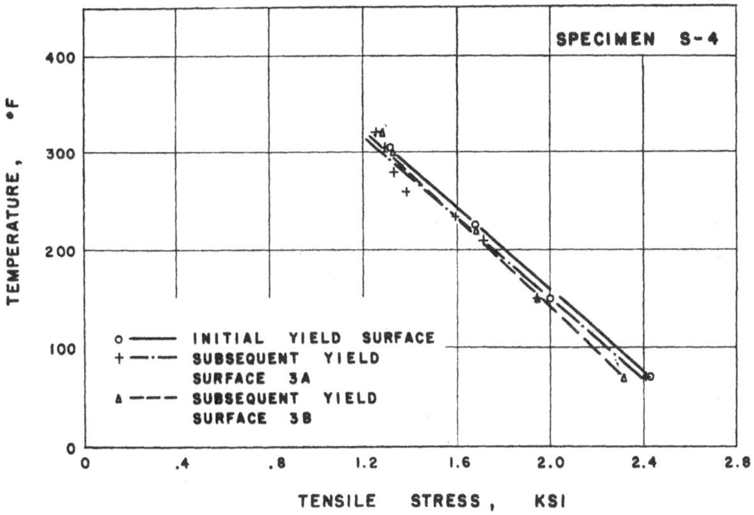

Fig. 12. Intersection of the yield surfaces in Fig. 9 and 10 with the planes normal to the stress plane and passing through the max s_{11} values

Theoretical Considerations

In this section we shall consider some implications of the previously presented experimental results for the theory of plasticity.

We adopt the GREEN-NAGHDI formulation of the theory of plasticity [10]. However, only infinitesimal strains will be considered. Then the

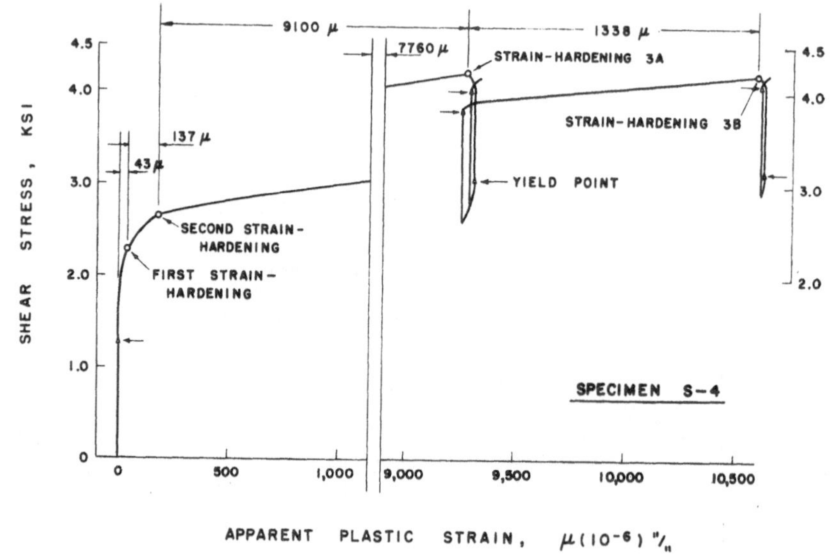

Fig. 13. Stress-strain curve for specimen S-4

total strain e_{KL} consists of an elastic part e'_{KL} and a plastic part e''_{KL} such that

$$e_{KL} = e'_{KL} + e''_{KL}. \tag{2}$$

If U is the internal energy per unit mass, T is the temperature, and S is the entropy per unit mass, we introduce the free energy function

$$A = U - TS \tag{3}$$

and with the help of the first and second laws of thermodynamics we obtain

$$\varrho_0 r - \varrho_0(\dot{A} + \dot{T}S + \dot{S}T) - Q_{K,K} + s_{KL}\dot{e}_{KL} = 0 \tag{4}$$

$$-\varrho_0(\dot{A} + S\dot{T}) + s_{KL}\dot{e}_{KL} - \frac{Q_k T_{,k}}{T} \geqq 0 \tag{5}$$

in which r is the heat supply function per unit mass, ϱ_0 is the mass density of the reference state, and Q_K is the heat flux.

We now introduce the yield function

$$f(s_{KL}, e''_{KL}, T) \tag{6}$$

and the strain-hardening function

$$\dot{\varkappa} = \dot{\varkappa}(s_{KL}, e''_{KL}, T, \dot{s}_{KL}, \dot{e}''_{KL}, \dot{T}) \tag{7}$$

from which for a given history of s_{KL}, e_{KL} and T we find a specific $\varkappa$. For given values of $\varkappa$ and e''_{KL}, say $\varkappa^{(1)}$ and $e''^{(1)}_{KL}$ equation

$$f(s_{KL}, e''^{(1)}_{KL}, T) = \varkappa^{(1)} \tag{8}$$

represents a surface in seven-dimensional space (stress-temperature space).

The actual state of stress and temperature at an element of the body is represented by a point P in the seven-dimensional space. We postulate that this state can be represented only by a point P *on or within* the yield surface. Suppose that the point P lies inside the yield surface. Any elemental change of stress and/or temperature will move P without disturbing the yield surface; the values of e''_{KL} and $\varkappa$ remain unchanged. On the other hand, suppose that P lies on the yield surface; then an elemental motion of P may move it on the yield surface or inwards without change in the yield surface and in the values of e''_{KL} and $\varkappa$, or may move it towards the outside of the yield surface in which case the yield surface must move with P, and will remain in contact with P. For a motion of the yield surface to occur it is necessary that additional plastic strains appear. Therefore, at the end of the motion of P with the yield surface, new values of e''_{KL} and $\varkappa$ exist, say $e''^{(2)}_{KL}$ and $\varkappa^{(2)}$ and the new equation of the yield surface will be

$$f(s_{KL}, e''^{(2)}_{KL}, T) = \varkappa^{(2)} \tag{9}$$

Loading, neutral loading, and unloading are defined by

$$\frac{\partial f}{\partial s_{KL}} \dot{s}_{KL} + \frac{\partial f}{\partial T} \dot{T} > 0 \tag{10a}$$

$$\frac{\partial f}{\partial s_{KL}} \dot{s}_{KL} + \frac{\partial f}{\partial T} \dot{T} = 0 \tag{10b}$$

$$\frac{\partial f}{\partial s_{KL}} \dot{s}_{KL} + \frac{\partial f}{\partial T} \dot{T} < 0, \tag{10c}$$

respectively. When we have loading, additional plastic strains are produced and are given by

$$\dot{e}''_{KL} = g_{KL}(s_{MN}, \dot{s}_{MN}, e''_{MN}, T, \dot{T}). \tag{11}$$

When we have unloading or neutral loading:

$$\dot{e}''_{KL} = 0 \tag{12}$$

that is, there is no change in the plastic strains. In addition, if P lies within the yield surface then an elemental motion of P will leave the yield surface intact and will produce zero plastic strains.

We now assume that

$$A = A(e'_{KL}, e''_{KL}, T) \tag{13}$$

$$S = S(e'_{KL}, e''_{KL}, T) \tag{14}$$

$$Q_K = Q_K(T, T_{,M}, e'_{MN}, e''_{MN}). \tag{15}$$

From the second law of thermodynamics we obtain

$$S = -\frac{\partial A}{\partial T} \tag{16}$$

$$s_{KL} = \varrho_0 \frac{\partial A}{\partial e'_{KL}}. \tag{17}$$

With the help of these two eqs. (4) and (5) become

$$\varrho_0 r + \left(s_{KL} - \varrho_0 \frac{\partial A}{\partial e''_{KL}}\right) \dot{e}''_{KL} - \varrho_0 \dot{S} T - Q_{K,K} = 0 \tag{18}$$

$$\left(s_{KL} - \varrho_0 \frac{\partial A}{\partial e''_{KL}}\right) \dot{e}''_{KL} - \frac{Q_K T_{,K}}{T} \geqq 0. \tag{19}$$

For an homogenously distributed temperature, $T_{,K} = 0$, and eq. (19) gives

$$\left(s_{KL} - \varrho_0 \frac{\partial A}{\partial e''_{KL}}\right) \dot{e}''_{KL} \geqq 0. \tag{20}$$

We see that the quantity

$$\varrho_0 \frac{\partial A}{\partial e''_{KL}}$$

must have the dimensions of stress and we shall write

$$\varrho_0 \frac{\partial A}{\partial e''_{KL}} = \overset{\circ}{s}_{KL} \tag{21}$$

which we shall call "thermodynamic reference stress" [11].

We shall now assume as in [10] that A has the form

$$A = A'(T, e'_{KL}) + A''(e''_{KL}). \tag{22}$$

Then

$$S = -\frac{\partial A'}{\partial T} \tag{23}$$

$$s_{KL} = \varrho_0 \frac{\partial A'}{\partial e''_{KL}}. \tag{24}$$

It follows that S, s_{KL} and e'_{KL} are independent of e''_{KL}. We also find that (20) becomes

$$\left(s_{KL} - \varrho_0 \frac{\partial A''}{\partial e''_{KL}}\right) \dot{e}''_{KL} \geqq 0 \tag{25}$$

and it is seen that

$$\overset{\circ}{s}_{KL} = \varrho_0 \frac{\partial A''}{\partial e''_{KL}} \tag{26}$$

depends on e''_{KL} only.

Green and Naghdi [10] assume that the function (11) is linear in the rates of stress:

$$\dot{e}''_{KL} = \alpha_{KLMN}\, \dot{s}_{MN} \tag{27}$$

where the coefficients α_{KLMN} may be functions of s_{MN}, e''_{MN}, and T. This assumption, together with the assumption that for neutral loading $\dot{e}''_{KL}$ must vanish, furnish the conclusion that $\dot{e}''_{KL}$ must have a direction independent of s_{KL}, [12].

At this stage we shall introduce the physical observation that at a given state of stress and temperature, the rate of stress $\dot{s}_{KL}$ cannot cause a plastic strain rate $\dot{e}''_{KL}$ which will be opposite in direction to that of $\dot{s}_{KL}$. Hence, $\dot{e}''_{KL}$ must be directed toward the exterior of the yield surface since its direction is fixed and independent of $\dot{s}_{KL}$. If this is not the case one can always find a loading stress rate $\dot{s}_{KL}$ which is opposite to $\dot{e}''_{KL}$.

Consider now for the moment the possibility that $\overset{\circ}{s}_{KL}$ lies outside the yield surface. Let s_{KL} be the stress state which lies on the yield surface and is closest to $\overset{\circ}{s}_{KL}$. Then inequality (25) implies that any non-zero $\dot{e}''_{KL}$ generated by loading from s_{KL} is directed towards the interior of the yield surface. Since such an eventuality has been ruled out, *we conclude that $\overset{\circ}{s}_{KL}$ must lie within or on the yield surface.*

Inequality (25) can be considered to be a generalisation of the irreversibility condition

$$s_{KL}\dot{e}''_{KL} > 0, \quad \text{if} \quad \dot{e}''_{KL} \neq 0 \tag{28}$$

introduced by Hodge and Prager [13] and by Prager [12], to which it reduces when $\overset{\circ}{s}_{KL}$ coincides with the origin. From (28) it follows that since the plastic strain rate vector is directed to the exterior of the yield surface, the yield surface should always enclose the origin. The present experiments show however, that even a very small amount of plastic strain will make the yield surface move to such an extent that the origin will be left outside. Thus, inequality (28) must be replaced by inequality (25) which follows from thermodynamics.

We have seen that $\overset{\circ}{s}_{KL}$ must be within or on the yield surface. On the other hand $\overset{\circ}{s}_{KL}$ within the framework of the theory developed here, is independent of temperature. Since with increasing temperature the yield curve shrinks in size it follows that $\overset{\circ}{s}_{KL}$ must be within or on the innermost yield curve.

It should be expected that at recrystallization temperature the yield curve should have been reduced to a point, and therefore, that $\overset{\circ}{s}_{KL}$ should lie at this point. On the other hand, at and beyond the recrystallization temperature the theory needs modification since there e''_{KL} and A'' change with time without additional loading.

Although no experimental results are available at present for the yield curves beyond 325 °F there is a way of estimating the position of $\overset{\circ}{s}_{KL}$. We remember that at each prestress value the maximum tensile stress at different temperatures occurs at the same shear stress value. This is true for both small and large values of plastic strain. If we assume that this observation holds also for temperatures above 325 °F we conclude that $\overset{\circ}{s}_{KL}$ is the value of s_{12} for which $s_{11} = \max$.

Knowledge of the position of $\overset{\circ}{s}_{KL}$ for each value of e''_{KL} enables us to find an expression of A''. Indeed, from (26) we obtain

$$\frac{\partial A''}{\partial e''_{KL}} = f(e''_{KL}) \tag{29}$$

where

$$f(e''_{KL}) = \varrho_0 \overset{\circ}{s}_{KL}. \tag{30}$$

By considering a limited number of stress paths eq. (29) will give a system of partial differential equations from which by integration the expression of A'' can be obtained.

In [10] it is proposed that for the infinitesimal theory

$$A'' = \frac{1}{2} A''_{KLMN} e''_{KL} e''_{MN}. \tag{31}$$

Differentiation gives

$$\frac{\partial A''}{\partial e''_{KL}} = A''_{KLMN} e''_{MN} = f(e''_{KL}) \tag{32}$$

a linear expression for $f(e''_{KL})$. Our experimental results indicate that such a linear expression does not exist. Indeed, Fig. 14 gives for specimen S-4 the actual $\overset{\circ}{s}_{KL}, \gamma''_{KL}$ curve which is not a linear one.

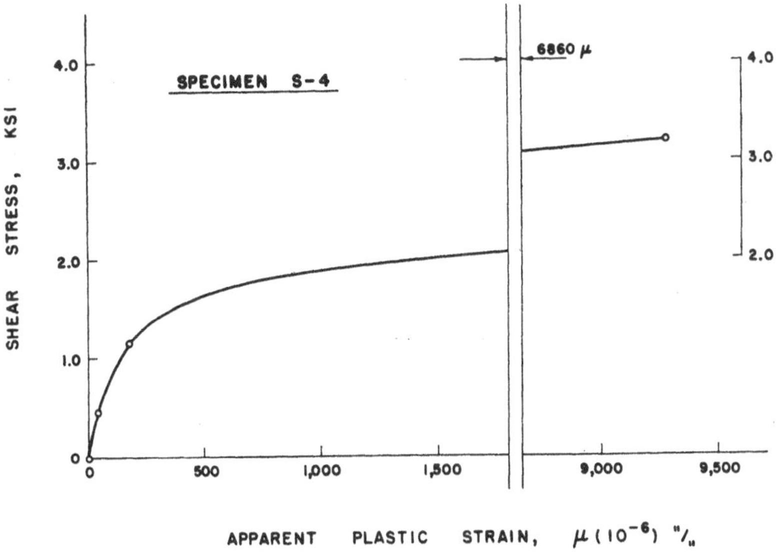

Fig. 14. $\overset{\circ}{s}_{KL}, \gamma''_{KL}$ curve for specimen S-4

The observation that the prestress point lies outside the yield surface is of considerable interest since it requires a modification of the classical theory of plasticity. Such a modification has been proposed by Phillips and Sierakowski [14], and further examined by Justusson and Phillips [15]. A mathematical theory of plasticity incorporating this modification has been developed by Eisenberg and Phillips [16].

References

[1] Smith, G. V.: Stress-Strain-Time-Temperature Relations in Metallic Materials. In: ASTM Special Technical Publication No. 325. Amer. Soc. Testing Materials, pp. 35–59 (1962).

[2] Lubahn, J. D.: Deformation Phenomena. In: Mechanical Behavior of Materials at Elevated Temperatures (J. E. Dorn, Editor), pp. 319–392. New York: McGraw Hill Book Co. 1961.

[3] Pugh, H. L. D., S. S. Chang, and B. E. Hopkins: Tensile Properties of a High Purity Iron from −196 °C to 200 °C at Two Rates of Strain. Phil. Mag. **9**, 753 to 768 (1963).

[4] Phillips, A., C. S. Liu, and J. W. Justusson: An Experimental Investigation of Yield Surfaces at Elevated Temperatures. (To be published).

[5] MAIR, W. U., and H. L. D. PUGH: Effect of Prestrain on Yield Surfaces in Copper. J. Mech. Eng. Science **6**, 150–163 (1964).
[6] SIERAKOWSKI, R. L., and A. PHILLIPS: The Effect of Repeated Loading on the Yield Surface. Acta Mechanica **6**, 217–231 (1968).
[7] JUSTUSSON, J. W., and A. PHILLIPS: An Experimental Investigation on Yield Surfaces. (To be published).
[8] IVEY, H. J.: Plastic Stress-Strain Relations and Yield Surfaces for Aluminum Alloys. J. Mech. Eng. Science **3**, 15–31 (1961).
[9] NAGHDI, P. M., F. ESSENBURG, and W. KOFF: An Experimental Study of Initial and Subsequent Yield Surfaces in Plasticity. J. Appl. Mech. **25**, 120 to 209 (1958).
[10] GREEN, A. E., and P. M. NAGHDI: A General Theory of an Elastic-Plastic Continuum. Arch. Rat. Mech. Anal. **18**, 251–281 (1965).
[11] PHILLIPS, A., and M. A. EISENBERG: Observations on Certain Inequality Conditions in Plasticity. Int. J. Non-Linear Mechanics **1**, 247–256 (1966).
[12] PRAGER, W.: Recent Developments in the Mathematical Theory of Plasticity. J. Appl. Phys. **20**, 235–241 (1949).
[13] HODGE, P., and W. PRAGER: A Variational Principle for Plastic Materials with Strain Hardening. J. Math. Phys. **27**, 1–10 (1948).
[14] PHILLIPS, A., and R. L. SIERAKOWSKI: On the Concept of the Yield Surface. Acta Mechanica **1**, 29–35 (1965).
[15] JUSTUSSON, J. W., and A. PHILLIPS: Stability and Convexity in Plasticity. Acta Mechanica **2**, 251–267 (1966).
[16] EISENBERG, M. A., and A. PHILLIPS: A Theory of Plasticity with Non-Coincident Yield and Loading Surfaces. (To be published).

Acknowledgement

The author wishes to express his appreciation to his colleague, Professor J. W. JUSTUSSON, and to his graduate students, Mr. C. S. LIU and Mr. J. L. TANG for their invaluable contribution in the execution of the experiments described in this paper. He also wishes to thank the National Science Foundation for the financial support which made this research possible.

Discussion

ODQVIST: Can you state, for example, in terms of the observed yield limit in torsion, the amount of prestressing that is necessary in order to have to generalize DRUCKER's postulate in the way you indicated?

PHILLIPS: Instead of a generalization of DRUCKER's postulate, I would prefer to say a specialization of DRUCKER's postulate and a generalization of PRAGER's irreversibility condition. Inequality (25) is both the above specialization and generalization and is derived from thermodynamics. On the other hand, neither DRUCKER's postulate nor PRAGER's irreversibility condition are derived from thermodynamics. When the yield surface no longer includes the origin of the stress space, PRAGER's irreversibility condition becomes invalid and must be replaced by inequality (25). In our experiments this happens at a prestress level between 2262 psi and 2585 psi whereas the proportional limit on torsion of the virgin tube is 1430 psi. The corresponding amount of prestrain lies between 70×10^{-6} in/in and 300×10^{-6} in/in.

On a Thermodynamic Constitutive Theory and Its Application to Various Nonlinear Materials

By

R. A. Schapery*

Lafayette, Ind. (U.S.A.)

Summary

The author's thermodynamic constitutive theory for nonlinear viscoelastic behavior is extended to account for rate-independent plastic flow and for nonlinear creep with strong stress-dependence. It is then shown that the resulting stress-strain equations are consistent with mechanical behavior reported for several different materials under small and large strains; although principal concern here is with metals and polymeric solids, preliminary indications are that a variety of other materials, such as soils and biological tissue, can be characterized using the same basic equations. In this theory, history effects, with uniaxial or multiaxial loading, are taken into account by means of single integrals which are very similar to the Boltzmann type in linear viscoelastic theory.

1. Introduction

Many different stress-strain equations have been developed for characterizing the complex mechanical behavior of nonlinear, inelastic solids. They include, for example, the experimentally-based equations of rate-independent plasticity [1] and nonlinear creep [2], as well as mathematically-based integral representations, such as those given by Green and Rivlin [3] and Coleman and Noll [4].

The mathematical theories are probably general enough to account for almost all types of observed nonlinearities. However, they can be quite impractical, even when thermodynamic restrictions are taken into consideration. In particular, Onat [5] has demonstrated that a prohibitively large number of terms would be required if the multiple integral expansion in [3] is used to characterize inelastic behavior of metals. Furthermore, approximations based on the fading memory assumption [4] are not appropriate for metal applications since they do not normally possess this property.

* Now at Texas A & M University, College Station, Texas.

The present situation appears to be such that very specialized constitutive equations must be used in order to be able to experimentally determine property functions and conduct a structural analysis in a practical manner. This specialization is with respect to both the particular material and the loading and temperature histories.

However, when one reviews existing information on actual inelastic behavior, it is found that many engineering materials possess certain common nonlinear characteristics over wide ranges of loading and temperature. This becomes especially clear when their behavior is examined in a thermodynamic framework.

In this paper we combine this information with irreversible thermodynamic principles to extend a previously developed constitutive theory. It describes nonlinear response under constant or varying loads and temperature for isotropic and anisotropic materials.

Although attention is restricted here to primarily polymer solids and metals, it is believed the proposed equations have more general applicability in view of their thermodynamic basis.

In Section 2 we introduce definitions needed in the thermodynamic development and discuss the physical significance of state variables.

The thermodynamic theory developed earlier by the writer [7, 8] is generalized in Section 3 in order to incorporate the type of strong nonlinearities observed for metals and plastics; in the author's earlier papers, the linear thermodynamic concepts used by BIOT [9] were extended to nonlinear, nonisothermal behavior, and then used to derive viscoelastic constitutive equations involving nonlinearities of the type observed over wide temperature and strain ranges for soft polymeric materials. Guided by observed nonlinear response, specific forms are assumed for free energy and entropy production. They lead to thermodynamic constitutive equations, which, when expressed in terms of certain reduced variables, are identical to those governing linear viscoelastic behavior.

Solution of these equations yields explicit stress-strain relations with temperature, which are given in Section 4. In order to not restrict ourselves at this point to a particular measure of small or large strain, they are expressed in terms of generalized coordinates and forces.

For some materials, experimental evidence indicates that one can assume all nonlinearity arises from a single entropy production coefficient. The stress-strain equations then become identical with those of *linear* thermoviscoelasticity, except the intrinsic time-scale for material response is distorted by temperature and stress and/or strain invariants. In Section 5, the validity of these simplified equations for plastics and metals is examined. We show, for example, that the classical, three-dimensional stress-strain equations of rate-independent, incremental plasticity and nonlinear creep of metals are included in this description. When the timescale is unaffected by mechanical invariants and ageing is excluded, one obtains in generalized notation the well-known equations for thermo-rheologically simple media under transient temperature [10, 11].

Finally, in Section 6 we allow for nonlinearity in the free energy and shown that a resulting stress-strain equation agrees with existing data on soft, amorphous polymer solids under various loading conditions. Because the stress-strain equation is so similar to the Boltzmann form of linear viscoelasticity, even with finite strains, it is easily applied.

The effect of straining on temperature is not studied here, but the governing energy equation has been derived in [8].

2. Definitions. Significance of Generalized Coordinates

A material element (specimen) with fixed mass is taken to represent the thermodynamic system. The *maximum* number of variables required to define its state is assumed to be N generalized coordinates q_i and absolute temperature T; the element's temperature is assumed to be spacewise constant. (For a discussion of the significance of state variables from an experimentalist's point of view see [5]). A generalized force Q_i, conjugate to the variable q_i, is defined by the condition that $Q_i \delta q_i$ (i not summed) is an incremental amount of work δW done on the system during a very small change δq_i.

The reference state (R) will be defined as a stable or neutrally stable thermodynamic equilibrium state for which all forces Q_i and coordinates q_i are zero, and the temperature is T_R. The element is assumed to have unit volume in this state. Neutral stability is allowed in order to accomodate materials that undergo steady-flow when subjected to constant forces.

The generalized coordinates are divided into the two groups of hidden and observed. Hidden coordinates are defined by the condition that their conjugate forces are always zero; the remaining coordinates are the observed variables.

The familiar summation convention will be used wherein a repeated index is summed out over it entire range, unless specified otherwise. We shall use three sets of indices in the thermodynamic development, each having a different range. Specifically, i and j range over all generalized coordinates:

$$i \text{ and } j = 1, 2, \ldots, N;$$

m and n range over observed coordinates only:

$$m \text{ and } n = 1, 2, \ldots, k;$$

and r and s range over hidden coordinates only:

$$r \text{ and } s = k + 1, k + 2, \ldots, N.$$

The material element is assumed to be statistically homogeneous and large compared to molecular dimensions and structural inhomogeneities (such as due to grains, fibers, or filler particles). For the purpose of developing stress-strain equations we can then assume its deformation is homogeneous and expressed in terms of a set of appropriately defined (large or small) strains. In the general three-dimensional case one could use the six components of the Lagrangian strain tensor [12], as in [8]; however, it may often be desirable to use other or fewer measures of deformation, depending on the particular application.

The number of observed generalized coordinates is equal to the number of independent strains required to define the class of deformations of interest; this number can not, of course, exceed six. For generality, we consider the observed coordinates to be functions of the strains, and not necessarily equal to them.

In the subsequent development of stress-strain equations, it will be seen that the set of hidden coordinates must be sufficiently large so that, together with observed coordinates and temperature, changes in the free energy are completely determined for all processes of interest.

Since the hidden coordinates do not appear explicitly in the final stress-strain equations, a precise physical interpretation is not actually needed. Nevertheless, they may be thought of as defining such things as lattice distortions and location of interstitial atoms in a metal, or configurations of the chainlike molecules in amorphous polymers. If we consider creep of a polycrystalline metal, for example, the observed distortion reflects the combined action of grain boundary sliding, internal slipping of grains, twinning, etc. In this case, the free energy of the element is not directly dependent on the strains (except for dilatation) but changes only as a result of changes in hidden coordinates. A similar situation exists with fluids in that the free energy is independent of shear strain.

In order to describe the response of a material over a limited range of temperature and/or time, only a certain portion of the total set of hidden coordinates will be actually needed. For example, creep of an amorphous polymer above the glass-transition temperature arises mainly from movement of relatively long chain segments. Only enough hidden coordinates are needed to describe the modes of motion of these segments. On the other hand, below this temperature, long-range motion of the chains is negligible, and the higher modes associated with internal motion of individual segments are dominant [13]. As could be expected, nonlinear response is considerably different, depending on whether or not the polymer is in its glassy or rubbery state.

3. Thermodynamic Constitutive Theory

We begin by stating an equation for entropy production dS'/dt derived earlier [7]

$$\frac{dS'}{dt} = \frac{1}{T}\left(Q_i - \frac{\partial F}{\partial q_i}\right)\frac{dq_i}{dt} \geq 0 \tag{1}$$

where $F = F(q_i, T)$ is the Helmholtz free energy. The so-called Clausius-Duhem inequality, $dS'/dt \geq 0$, must be satisfied for all processes followed by the element. An assumption made in the development of this equation is that thermodynamic state variables can be defined although the system is not in thermodynamic equilibrium; no approximation has been made with respect to nonlinear response.

This assumption is reasonable in the light of existing molecular models, but at this time it can be checked only indirectly for solids through its implications for stress-strain behavior. However, as a point of encouragement, it can be shown that eq. (1) is valid for chemical reactions of gases as long as the law of mass action holds [14]. This law leads generally to a highly nonlinear irreversible process.

We should add that kinetic energy of the system's mass center was not included in the derivation of (1). Had it been, it would also be necessary to account for kinetic energy when relating generalized forces to mechanical stresses by means of the definition of Q_i given above. It is a simple matter to show that the kinetic energy then disappears from the resulting stress-strain equations. Therefore, the results to be given here are not restricted to quasi-static problems, but are potentially applicable to wave propagation phenomena.

The coefficients

$$X_i \equiv Q_i - \frac{\partial F}{\partial q_i} \tag{2}$$

are, in general, nonlinear functions of all quantities which can affect irreversibility of the process. The state variables T and q_i may enter, as well as any other variables which are needed to completely define entropy production, including their time rates of change; even the time, t, may be needed to reflect ageing. Designating all of these additional variables by the single symbol λ, we write

$$X_i = X_i(q_j, \dot{q}_j, \ddot{q}_j, \ldots;\ T, \dot{T}, \ddot{T}, \ldots;\ \lambda, \dot{\lambda}, \ddot{\lambda}, \ldots) \tag{3}$$

where dots denote differentiation with respect to time.

It is important to note that eq. (1) can be satisfied for all possible processes in a neighborhood of an arbitrary equilibrium state if and only if[1]

$$X_i = b_{ij}^e \frac{dq_j}{dt} \tag{4}$$

where b_{ij}^e is a positive semidefinite matrix; viz.

$$b_{ij}^e \frac{dq_i}{dt} \frac{dq_j}{dt} \geq 0 \quad \text{when} \quad \frac{dq_i}{dt} \frac{dq_i}{dt} > 0. \tag{5}$$

The matrix b_{ij}^e in (4) depends only on the instantaneous values of q_i and T, and is symmetric according to Onsager's principle [14].

Without loss of generality, form (4) can be used even when the system is not close to equilibrium, but now the matrix, b'_{ij}, say, may depend on all quantities in (3) and is not necessarily symmetric. Thus, writing

$$X_i = b'_{ij} \frac{dq_j}{dt} \tag{6}$$

we obtain from (2) the following equations of motion:

$$\frac{\partial F}{\partial q_i} + b'_{ij} \frac{dq_j}{dt} = Q_i. \tag{7}$$

This is the same set as derived earlier [7] except the matrix b_{ij}^e was used.

Recall that $Q_i \equiv 0$ when $k + 1 \leq i \leq N$; therefore, if F and b'_{ij} were known functions, these $N - k$ equations in (7) could, in principle, be used to solve for the hidden coordinates and thereby eliminate them from the remaining k equations for which $i \equiv m$. These latter equations could then be interpreted as stress-strain equations relating Q_m, q_m and T, assuming the parameters λ were either known or equal to Q_m. This procedure would be analogous to that followed in [7] for linear behavior.

However, eq. (7) is actually of no practical use unless considerable simplification is introduced such that the hidden coordinates can be explicitly eliminated. As the first simplification, we assume all coefficients b'_{ij} are affected equally by the variables indicated in (3); viz.

$$b'_{ij} = a_D b_{ij} \tag{8}$$

where b_{ij} is a constant matrix corresponding to linear response in the neighborhood of the reference equilibrium state (R); viz., $b_{ij}^e \equiv b_{ij}$

[1] By definition, this neighborhood is where all rates-of-change shown in (3) are so small that only first order terms have to be retained. Linear terms, such as $\ddot{q}_i$, must be excluded from (4) in order to satisfy inequality (1) for arbitrary, slow processes.

when $q_i = 0$ and $T = T_R$. The function a_D is non-negative,

$$a_D \geq 0 \tag{9}$$

in view of (1). As a partial motivation for (8), we observe that in the particular case for which a_D depends only on temperature, $a_D \equiv a_T(T)$, the stress-strain equations of thermorheologically simple materials are obtained [7].

Now, consider the free energy function, F. In the elastic range of behavior of metals and glassy polymers these materials are often practically linear; this means all terms higher than second order can be neglected in F. With inelastic behavior, changes in F are still due to limited local molecular or atomic rearrangements. It seems reasonable, therefore, that a second-order approximation will prove to be adequate in many cases. An indication that it is valid, even in the presence of large plastic flow in metals, is that the majority of the work done during this flow normally goes into heat generation rather than crystal lattice distortion [15].

If the free energy is expanded at the reference state with respect to both temperature change, $\theta \equiv T - T_R$, and coordinates, and all terms higher than second order are dropped, it can be put in the following form [7],

$$F = -S_R\theta + \frac{1}{2}\, a_{ij} q_i q_j - \beta_i \theta q_i - (C_q/2T_R)\,\theta^2 \tag{10}$$

with the constant coefficients,

$$a_{ij} = a_{ji} = (\partial^2 F/\partial q_i\, \partial q_j)_R$$

$$\beta_i = -(\partial^2 F/\partial q_i \partial T)_R$$

S_R = entropy at the reference state

C_q = specific heat for constant coordinates.

The matrix a_{ij} is positive semidefinite [7]. Also, without loss of generality the free energy is assumed to vanish at the reference state.

Substitution of expansion (10) and coefficients (8) into (7) results in

$$a_{ij} q_j + b_{ij} \frac{dq_j}{d\psi} = Q_i + \beta_i \theta \tag{11}$$

where

$$d\psi \equiv dt/a_D$$

which, when integrated, yields the "reduced time":

$$\psi = \int_0^t dt'/a_D. \tag{12}$$

Equation (11) becomes identical to that given in [7] for a linear, thermo-rheologically simple viscoelastic material when a_D is assumed to depend only on temperature.

Observe that free energy (10) is *not* necessarily a second-order approximation with respect to strain since coordinates q_m are, at this stage, still arbitrary functions of mechanical strains. However, a more general representation may still be needed to characterize materials which undergo large changes in their free energy, such as rubbery and crystalline polymers.

Some simplification can be introduced if the material is assumed to be in (approximate) thermodynamic equilibrium under sufficiently slow strain rates[1]. In particular, the hidden coordinates can then be assumed to vanish at any strained equilibrium state for which $T = T_R$, without loss in generality, as shown in [8]. Thus, if the strain rates are not too high, we can expand the free energy with respect to hidden coordinates and neglect terms higher than second order, even though applied strains are large. The expansion is given in [8] as:

$$F = F_e - S_e\theta + \frac{1}{2}\, a'_{rs} q_r q_s - \beta'_r \theta q_r - (C'_q/2T_R)\theta^2 \tag{13}$$

where F_e, S_e, and C'_q are equilibrium values of the free energy, entropy, and specific heat, respectively, at a finite state of strain and $T = T_R$. These quantities, as well as coefficients $a'_{rs} = a'_{sr}$ and β'_r, depend on strains (i.e. observed coordinates q_m).

When assumptions analogous to (8) are introduced, viz.,

$$\begin{aligned} a'_{rs} &= a_F a_{rs} : a_F = a_F(q_m) \\ \beta'_i &= a_s \beta_i : a_s = a_s(q_m) \end{aligned} \tag{14}$$

equations of motion similar to (11) are obtained[2],

$$a_{ij} q_j + b_{ij} \frac{d q_j}{d\tilde{\psi}} = \tilde{Q}_i + \beta_i \tilde{\theta} \tag{15}$$

with a modified generalized force,

$$\tilde{Q}_i \equiv \frac{1}{a_F}\left(Q_i + a_F a_{in} q_n - \frac{\partial F_e}{\partial q_i}\right) : \tilde{Q}_r \equiv 0, \quad a_{mr} \equiv 0 \tag{16a}$$

[1] This assumption was not made in deriving (11) so as to not exclude plastic flow.

[2] By combining eqs. (15) and (16), it is found that $a_{mr} \equiv 0$ means the Q_m depend on strain history if and only if coefficients b_{ms} do not vanish. According to Curie's principle [1], $b_{ms} \neq 0$ implies the hidden coordinates in (15) are components of second order tensors; the same conclusion does not follow from (11) since a_{mr} may not vanish in (11). When hidden coordinates are not vector components, the constitutive equations derived here can be used in stress analysis of bodies with temperature gradients [8].

a modified temperature change,

$$\tilde{\theta} \equiv a_s \theta / a_F \tag{16b}$$

and a modified reduced-time

$$\tilde{\psi} \equiv \int_0^t \frac{a_F}{a_D} dt'. \tag{16c}$$

It is to be noted that $a_F \geq 0$ as long as the strained equilibrium state is not unstable [8]. However, the sign of a_S is not restricted by stability considerations.

Equations (15) and (16) were derived in [8], but a more general interpretation is given to a_D here; namely, it includes dependence on parameters λ and time derivatives, in addition to the generalized coordinates and temperature. Consistent with the above discussion, in which hidden coordinates are assumed to be small, their effect on the entropy production coefficient, a_D, will be neglected in all subsequent work.

The similarity between eqs. (11) and (15) and the equations of linear viscoelasticity in [7] leads immediately to explicit stress-strain equations with temperature, which will be discussed in subsequent Sections. A number of assumptions have been introduced up to this point, and we have tried to partially justify them. However, any explicit check on the assumptions must come through application on the resulting stress-strain equations.

4. Stress-strain Equations

In this Section, three-dimensional stress-strain relations will be given without explicitly restricting strain magnitude to small values.

For simplicity, we may consider the material element to be a rectangular parallelepiped whose edges are parallel to an orthogonal set of reference Cartesian axes, in which (x_1, x_2, x_3) are the coordinates of material points before deformation. Also, the usual double index notation for Cartesian tensors will be used in that the six observed generalized coordinates are now denoted by $q_{ab} = q_{ba}$. They are assumed to be second-order tensor functions of the Lagrangian strain tensor γ_{ab} [16]; these functions depend on the properties of the particular material of interest. When the strains are so small that these functions are linear, we can assume the generalized coordinates and strains are the same without loss in generality of the stress-strain equations; viz.:

$$q_{ab} \equiv \gamma_{ab} \quad \text{when} \quad |\gamma_{ab}| \ll 1. \tag{17}$$

If a material is isotropic in its unstrained state, the most general relations between q_{ab} and γ_{ab} are [17]:

$$q_{ab} = \alpha_1 \delta_{ab} + \alpha_2 \gamma_{ab} + \alpha_3 \gamma_{ac} \gamma_{cb} \tag{18}$$

where δ_{ab} is the Kronecker delta and $\alpha_1, \alpha_2, \alpha_3$ are scalar functions of the three strain invariants; for sufficiently small strains $\alpha_1 = 0$ and $\alpha_2 = 1$.

The relation between generalized forces $Q_{ab} = Q_{ba}$ and the true Cartesian stress tensor σ_{ab} referred to a fixed reference system y_a, is found from the condition that $Q_{ab} \delta q_{ab}$ must be an incremental amount of work done on the element for arbitrary δq_{ab}. The result for arbitrary material symmetry is [8]:

$$\sigma_{ab} = \frac{\partial y_a}{\partial x_c} \frac{\partial y_b}{\partial x_d} \tau^{cd} \tag{19}$$

where

$$\tau^{cd} \equiv \frac{\varrho}{\varrho_0} \frac{Q_{ab}}{2} \left(\frac{\partial q_{ab}}{\partial \gamma_{cd}} + \frac{\partial q_{ab}}{\partial \gamma_{dc}} \right) \tag{20}$$

and ϱ/ϱ_0 is the ratio of density after deformation to that before deformation. By definition, the q_{ab} in (20) are interpreted as functions of nine independent quantities γ_{ab} when performing the indicated differentiation. Also, (y_1, y_2, y_3) are the instantaneous coordinates of a material point which was located at (x_1, x_2, x_3) before deformation. Observe that when strains and rotations are sufficiently small (19) and (20) yield $\sigma_{ab} = Q_{ab}$.

It remains now to relate the generalized forces to generalized coordinates q_m through eq. (11) or (15). Consider set (11) first, which is identical to that of linear viscoelasticity apart from the reduced time, ψ.

The desired relations have already been derived in [7] by using $Q_r \equiv 0$ to eliminate the hidden coordinates. Results for both anisotropic and isotropic materials were given, and for reference purposes we shall record here only the solutions for initially isotropic media (i.e. isotropic in the reference state R).

Using notation analogous to that for elastic materials, and writing them in the familiar deviatoric and dilatational forms we have[1]:

$$Q'_{ab} = 2 \int_0^t G(\psi - \psi') \frac{d q'_{ab}}{d\tau} \, d\tau \tag{21 a}$$

$$Q_{aa} = 3 \int_0^t K(\psi - \psi') \frac{d q_{aa}}{d\tau} \, d\tau - 3 \int_0^t \beta(\psi - \psi') \frac{d\theta}{d\tau} \, d\tau \tag{21 b}$$

---

[1] All of the stress-strain equations in [7] were expressed in operational notation, rather than the integral form. If eqs. (21) and (22) are Laplace transformed with

and the inverse relations:

$$q'_{ab} = \frac{1}{2}\int_0^t J(\psi - \psi')\,\frac{dQ'_{ab}}{d\tau}\,d\tau \tag{22a}$$

$$q_{aa} = \frac{1}{3}\int_0^t B(\psi - \psi')\,\frac{dQ_{aa}}{d\tau}\,d\tau + 3\int_0^t \alpha(\psi - \psi')\,\frac{d\theta}{d\tau}\,d\tau \tag{22b}$$

where, by definition,

$$Q'_{ab} \equiv Q_{ab} - \frac{1}{3}\,Q_{cc}\,\delta_{ab} \tag{23a}$$

$$q'_{ab} \equiv q_{ab} - \frac{1}{3}\,q_{cc}\,\delta_{ab} \tag{23b}$$

and

$$\psi \equiv \int_0^\tau dt'/a_D\,, \quad \psi' \equiv \int_0^t dt'/a_D \tag{24}$$

When strains and rotations are sufficiently small, the quantities Q'_{ab} and q'_{ab} become the components of the familiar deviatoric stress and strain tensors.

The material property functions $G(\psi)$, $K(\psi)$, $\beta(\psi)$, etc., are those of linear viscoelasticity, except the time-scale may be a function of stress, strain, etc., as well as temperature. Thus, $G(\psi)$ and $K(\psi)$ are the familiar shear and bulk relaxation moduli, respectively, which can be evaluated experimentally in the linear response range. Similarly, $J(\psi)$ and $B(\psi)$ are the linear viscoelastic shear and bulk creep functions.

It is often acceptable to assume a material is elastic in its bulk response. For this case eqs. (21b) and (22b) reduce to the algebraic relation

$$Q_{aa} = 3K(q_{aa} - 3\alpha\theta), \tag{25}$$

where K and α are the constant bulk modulus and linear coefficient of thermal expansion, respectively.

The form of the material property functions in (21) and (22) is determined by the symmetric and positive-semidefinite properties of the con-

---

respect to ψ, after making the substitution $d\tau = a_D d\psi'$, they reduce to those given in [7]. Also, in applying these relations to nonuniformly strained bodies, the time derivative symbol d/dt is defined to be the rate of change of a quantity associated with a particular, infinitesimal material element. This material derivative is written as a partial derivative $\partial/\partial t$ when x_a and t are the independent variables; this definition is valid with large or small strains.

stants a_{ij} and b_{ij} in (11); viz.:

$$G(\psi) = \sum_s G^{(s)} e^{-\psi/\varrho_s} + G_e \tag{26a}$$

$$J(\psi) = \sum_s J^{(s)} (1 - e^{-\psi/\tau_s}) + J_0 + J_1 \psi \tag{26b}$$

$$\beta(\psi) = \sum_s \beta^{(s)} e^{-\psi/\varrho_s} + \beta \tag{26c}$$

$$\alpha(\psi) = \sum_s \alpha^{(s)} (1 - e^{-\psi/\tau_s}) + \alpha \tag{26d}$$

$K(\psi)$ and $B(\psi)$ have the same time dependence as $G(\psi)$ and $J(\psi)$, respectively, except the term proportional to ψ does not appear in $B(\psi)$ since it would predict "steady dilatational flow".

The "relaxation times", ϱ_s, "retardation times", τ_s, as well as all coefficients on the right-hand side of (26), are constant. Moreover, $\varrho_s > 0$, $\tau_s > 0$, and all coefficients are non-negative. For most materials there are a large number of time constants, ϱ_s and τ_s, which are closely-spaced. As a result the summation $\sum_s$ is often replaced by an integral.

It is to be noted also that an infinite value in the relaxation moduli at $t = 0$ (Dirac delta function) is thermodynamically admissible, but has been omitted here since it is not needed to represent actual behavior.

Since eqs. (21) and (22) are inverses of one another, the material properties are interrelated and certain internal conditions must be satisfied. For example, the steady-flow coefficient J_1 is zero if the long-time elastic modulus G_e is not zero.

The above equations are for a material whose only source of nonlinearity is in the entropy production coefficient a_D when generalized coordinates and forces are used. In order to account for nonlinearity in the free energy as well, the only thing we must do is replace Q_{ab}, θ, and ψ in the above equations by $\tilde{Q}_{ab}$, $\tilde{\theta}$, and $\tilde{\psi}$ as defined in (16). Explicit results for an initially isotropic material were derived in [8] and they will not be repeated here. However, a one-dimensional form will be given later in Section 6.

Recall in the development of eqs. (11) and (15) that somewhat different assumptions were used in arriving at each set. The latter equations have a more general free energy function; however, eq. (11) does not contain the explicit restriction that hidden coordinates vanish at all equilibrium states, as was assumed for (15). Had this restriction been used in developing (11), its only effect would be to make all coefficients a_{mr} vanish. The constants in (26) depend on a_{mr}, as can be seen in [7], but are not necessarily zero when a_{mr} vanishes.

Therefore, eqs. (21) and (22), with property functions (26), may be formally obtained as special cases from the general solutions in [8] by setting $a_F = a_S = 1$ and $F_e = a_{mn} q_m q_n / 2$. Consequently, we can think of eq. (15) as defining a *single constitutive equation* relating stress tensor, strain tensor, and temperature, from which all of the component stress-strain equations given here and in [8] can be derived.

As a further observation, we find that when eq. (21) is substituted into (20), incremental anisotropy can develop as a result of deformation. This is true even when the linear strain-displacement relations, $2\gamma_{ab} = (\partial u_a / \partial x_b + \partial u_b / \partial x_a)$, are applicable since geometric linearization does *not* necessitate material linearization $q_{ab} = \gamma_{ab}$.

It is also of interest to observe that the exponential and linear reduced time-dependence in the above relaxation and creep functions is a consequence of Onsager's principle $b_{ij} = b_{ji}$ and symmetry of a_{ij}. If this principle had not been applied we could still use the convolution forms (21) and (22) since the underlying equations are linear and time-invariant. However, as a check on the time-dependence, we note that molecular theories for isotropic polymers yield this same behavior [18], and thereby enable the isotropic relaxation and creep functions given here to be expressed in terms of molecular parameters.

This characteristic time dependence of the relaxation and creep functions implies that they can be represented by spring and dashpot models consisting of Maxwell elements in parallel and Voigt elements in series; see, for example [18, 19] for linear behavior and [8] for the nonlinear model corresponding to eq. (15). Models based on eq. (11) are identical with those of linear viscoelasticity except all dashpots have viscosities proportional to a_D.

Finally, it is to be noted that a_D is the only material property which depends on temperature in the above equations. This stems from the fact that they are based on a free energy expansion of the second order with respect to temperature change θ. With many materials this approximation is acceptable for engineering purposes since free energy is a weak function of temperature. However, eqs. (21) and (22) may be readily extended to reflect strong temperature effects in free energy (10) if they enter through a common factor in a_{ij} and through another one in β_i (see footnote 18 in [7]).

Temperature sensitivity of free energy can be expected with crystalline polymers, for example, when the crystals soften and/or the degree of crystallinity changes appreciably with temperature. When all elements of a_{ij} have a common temperature dependence and $a_D = a_D(T)$, mechanical response data at different constant temperatures can be shifted horizontally and vertically on a double-logarithmic plot to form a single "master curve"; such behavior has been reported for polyethylene [20].

We now turn to application of the stress-strain equations given here. In order to simplify the discussion we shall omit the thermal coefficients, β and α, in the following Sections.

5. Application to Some Hard Materials

It will be demonstrated that the stress-strain equations given in Section 4 are consistent with observed time-dependent and time-independent inelastic behavior of metals and of some polymeric materials below their glass transition temperature.

Let us now assume strains are so small that $Q_{ab} \equiv \sigma_{ab}$ and $q_{ab} \equiv \gamma_{ab}$. Therefore instead of γ_{ab}, the usual symbol, ε_{ab}, for small strains will be used.

All nonlinearity is now contained in the entropy production coefficient a_D. We have not yet specified its dependence on strains and the parameters λ, since this must be determined for each material. However, an important feature of this formulation is that often simple forms of a_D can be used to fit actual material response. This scalar can depend on stresses and strains, as determined by the particular material; for isotropic materials, only the three invariants of the stress and strain tensors (and their time derivatives) enter along with other non-mechanical parameters.

An additional parameter that is usually required is temperature. However, others may be needed, depending on the application. For example, the time scale for viscoelastic deformation of polymers is affected by swelling agents, such as water vapor [21]; expansion coefficients, analogous to those for thermal expansion, may also be needed in the stress-strain equations [21].

Uniaxial Creep and Relaxation of Polymers

When a single axial stress $\sigma = \sigma(t)$ is applied to a bar at $t = 0$, the strain response $\varepsilon = \varepsilon(t)$ is given by

$$\varepsilon = \int_0^t D(\psi - \psi') \frac{d\sigma}{d\tau}\, d\tau \tag{27}$$

if all nonlinearity is contained in reduced-time (24). Here $D(\psi)$ is the uniaxial, linear viscoelastic creep compliance, which is equal to the strain when a unit stress is applied at $t = 0$. The inverse of eq. (27) is

$$\sigma = \int_0^t E(\psi - \psi') \frac{d\varepsilon}{d\tau}\, d\tau \tag{28}$$

where $E(\psi)$ is the uniaxial relaxation modulus, defined as the stress response to a unit strain input.

When a stress σ_0 is applied at $t = 0$ and then held constant (creep test), eq. (27) yields, upon use of a Dirac delta function for $d\sigma/dt$:

$$\varepsilon = D(\psi)\,\sigma_0. \tag{29}$$

Similarly, a constant strain input ε_0 (relaxation test) reduces eq. (28) to

$$\sigma = E(\psi)\,\varepsilon_0. \tag{30}$$

In deriving (29) and (30) we have assumed a_D is independent of stress and strain rates, although it may depend on stress and/or strain. If a_D depends on stress alone, then in (29),

$$\psi = t/a_D(\sigma_0). \tag{31}$$

Similarly, if only strain dependence appears in a_D, reduced time ψ in (30) becomes

$$\psi = t/a_D(\varepsilon_0). \tag{32}$$

There is a considerable amount of creep and relaxation data on crystalline polymers and on amorphous polymers below their glass-transition temperature which can be represented by (29) and (30), respectively, using corresponding reduced times (31) and (32) [18, 22—26]. When the time-temperature superposition principle is applicable [18], we can account for different constant or variable temperatures by letting a_D depend on temperature as well as stress and/or strain.

As an example, creep data reported by FINDLEY and KHOSLA [24] on some thermoplastics can be represented very well by (29) and (31) with

$$D(\psi) = D_0 + m\,\psi^n \tag{33a}$$

where

$$a_D = \left[\frac{\sigma/\sigma_m}{\sinh \sigma/\sigma_m}\right]^{1/n} \tag{33b}$$

and D_0, m, n, and σ_m are constants; in all cases $0 < n < 1$. (A hyperbolic sine dependence was shown for the initial strain instead of $D_0\sigma_0$, but it can be replaced by the linear form with negligible error in most cases). The power-law dependence indicated in (33a) results when a continuous distribution of retardation times is used in $B(\psi)$ and $J(\psi)$, (26b).

The influence of stress-level on creep strain, as defined by (29) and (33), has been explained on the basis of Eyring's theory of rate processes and a highly idealized molecular picture [13, p. 250]; this molecular model leads to an apparent activation energy that is linearly dependent on stress.

In order to determine whether or not the more general creep compliance (29) with $a_D = a_D(\sigma)$ is valid for a given material, one can plot the ratio $\varepsilon(t)/\sigma_0$ at various stress levels against log t. Clearly, if $\varepsilon(o)/\sigma_0$ is constant and the separate plots can be shifted horizontally to form a single "master curve" of $D(t/a_D)$, (29) is a valid representation for creep. In addition, the amount of shifting yields the function log a_D; because of this significance, a_D is often called a "shift factor".

This behavior is not sufficient to verify (27) for time-dependent stresses. Additional stress histories, for example multiple-step loading and unloading, should be applied. Such a check on the application of stress-strain eq. (27) with properties (33) and data in [24] was made on polystyrene with good success [21].

It is interesting to notice that when nonlinear behavior of an anisotropic material is due primarily to the scalar a_D, a relatively simple multiaxial theory results. In this case, existing information from linear theory can be used to establish the particular form of the stress-strain equations, based on the material's geometric symmetry, and to relate gross properties to constituent properties in the case of composites. As shown in [26], some highly-nonlinear creep and recovery data on a fiber-reinforced phenolic resin can be described using this simple type of nonlinearity with a stress-dependent shift factor a_D.

In contrast with the creep data mentioned above, strain-dependence of a_D, rather than stress-dependence, has been reported in tensile relaxation studies of some crystalline polymers [22, 25]. The molecular origin of this dependence is not clear, but the free-volume theory has been offered as one explanation for materials which dilate appreciably [18, p. 352].

Unfortunately, existing molecular theories provide very limited information on nonlinear behavior, so that one must rely heavily on experiments. Experimental studies which are more critical than creep and relaxation tests may indicate that both stress and strain dependence in a_D are needed to accurately characterize a given material; as will be shown later, the strain-hardening creep theory of metals is characterized by using such a shift-factor. Moreover, if the present theory is valid for fatigue conditions, an additional parameter reflecting the number of cycles, for example, may be needed.

So far in this Section, we have examined only the case in which all nonlinearity arises from entropy production. However, nonlinearity due to large free energy changes exists in many materials. Stress-strain equations which account for this effect are discussed later in Section 6.

Inelastic Behavior of Metals

The thermodynamic theory contains rate-independent plastic flow and nonlinear creep theory of metals under uniaxial or multiaxial states of stress. First, we shall demonstrate rate-independent plastic flow behavior for a uniaxial stress state.

Still assuming all nonlinearity is contained in the non-negative scalar coefficient a_D, suppose that it is a function of temperature, T, stress, σ, strain, ε, and strain rate, $\dot{\varepsilon}$, and that only two terms need to be retained in an inverse expansion with respect to strain rate; viz.

$$\frac{1}{a_D} \simeq f(T, \sigma, \varepsilon) + g(T, \sigma, \varepsilon) \frac{d\varepsilon}{dt} \geq 0. \tag{34}$$

The first term, f, leads to nonlinear viscoelasticity, as shown above, while the second term will produce rate-independent plasticity. Whether or not only one term is dominant will, of course, depend on the temperature and on how far the system is from equilibrium.

We mentioned earlier that all dashpots in a mechanical model representation have viscosities proportional to a_D; thus, neglect of f, i.e.

$$\frac{1}{a_D} \simeq g \frac{d\varepsilon}{dt} \geq 0 \tag{35}$$

implies the viscosities are inversely proportional to strain rate. Further, the physical condition that the material response follow the initial elastic modulus during unloading from a plastic state is simply accomplished by setting $g = 0$, which stops all dashpot flow.

Observe that reduced time is now

$$\psi = \int_0^t dt'/a_D = \int_0^\varepsilon g\, d\varepsilon' = \psi(\varepsilon) \tag{36}$$

so that the uniaxial stress-strain equation in modulus form becomes

$$\sigma = \int_0^t E(\psi - \psi') \frac{d\varepsilon}{d\tau} d\tau = \int_0^\varepsilon E(\psi - \psi')\, d\varepsilon' \tag{37}$$

which is a nonlinear, rate-independent stress-strain equation; note that if stress rate, rather than strain rate, had been used in (35), we would have arrived at the same form; namely $1/a_D \simeq h(\sigma, \varepsilon)\, d\sigma/dt = h(\sigma, \varepsilon) \cdot (d\sigma/d\varepsilon)\, (d\varepsilon/dt) = g\, d\varepsilon/dt$.

As an example, let $E(\psi)$ be given by the familiar modified powerlaw form corresponding to a continuous distribution of relaxation times [28]:

$$E(\psi) = E_0[1 + \psi/\tau_0]^{-n} \tag{38}$$

where E_0 is the initial (elastic) modulus and τ_0 and n are positive constants; the long time modulus is assumed to be zero in order to allow for unlimited plastic flow when the stress is sufficiently high. Also, assume

$$\begin{aligned} d\varepsilon/dt > 0: \quad t > 0 \\ g = 0: \quad \varepsilon < \varepsilon_y \\ g = g_c = \text{positive constant}: \quad \varepsilon > \varepsilon_y \end{aligned} \tag{39}$$

where ε_y is the yield strain upon first loading of the specimen.

Substitution of (38) and (39) into eq. (37) gives

$$\frac{\sigma}{\sigma_y} = \begin{cases} \dfrac{\varepsilon}{\varepsilon_y} : \varepsilon \leq \varepsilon_y \\ \left\{1 + \left(\dfrac{\varepsilon}{\varepsilon_y} - 1\right)\varrho\right\}^{-n} + \dfrac{1}{\varrho(n-1)}\left\{1 - \left[1 + \left(\dfrac{\varepsilon}{\varepsilon_y} - 1\right)\varrho\right]^{1-n}\right\} : n \neq 1, \varepsilon \geq \varepsilon_y \\ \left\{1 + \left(\dfrac{\varepsilon}{\varepsilon_y} - 1\right)\varrho\right\}^{-1} + \dfrac{1}{\varrho}\ln\left[1 + \left(\dfrac{\varepsilon}{\varepsilon_y} - 1\right)\varrho\right] : n = 1, \varepsilon \geq \varepsilon_y \end{cases} \tag{40}$$

where $\varrho \equiv g_c \varepsilon_y/\tau_0 > 0$. Equation (40) is plotted in Fig. 1 and is seen to predict stress-strain behavior which is typical for metals. The line for $n = 40$ (narrow distribution of relaxation times) and $\varrho = 1/30$ (large time constant τ_0) as well as the curve for $n = 1$ (broad distribution of relaxation times) and $\varrho = 3$ (small time constant τ_0) are similar to mild steel curves at moderate temperature; the behavior of the model when the dashpots suddenly start flowing thus characterizes the actual situation at the yield point when there is a sudden reduction in solute atom resistance to dislocation motion.

If the strain rate is constant and we include a constant value for f in a_D, eq. (34), which vanishes below the yield strain ε_y, the curves in Fig. 1 are still applicable, but ϱ is a function of strain rate. Thus, this formulation can be used to describe rate-dependent plastic flow, in which the relative effect of strain rate depends on the ratio f/g.

When $f \equiv 0$ and the strain direction is reversed at any point along the curves in Fig. 1, the material will unload along a line having a slope equal to the initial elastic modulus if we set $g = 0$; if g takes the same constant value in (39) upon loading again to the last stress point reached on the original stress-strain curve, the same curve will be followed as long as $\dot{\varepsilon} > 0$.

It is important to observe that a Bauschinger effect as well as all subsequent behavior for loading into a plastic state will depend on the strain history-dependence assigned to g. In itself, thermodynamics does not restrict g to dependence on just instantaneous values of σ, T, and ε, as in (34), but requires only that $a_D \geq 0$; plastic flow during unloading from a plastic state is not actually excluded by this non-negative condition, since the sign of g could be assumed to change when $\dot{\varepsilon}$ changes sign. Separate physical conditions are needed to establish the form of this function which determines the nature of the plastic flow.

Indeed, if only rate-independent plastic flow is to be characterized, there is no actual need to use a model with many springs and dashpots, e.g. eq. (38). No apparent generality is lost by using a Maxwell model if

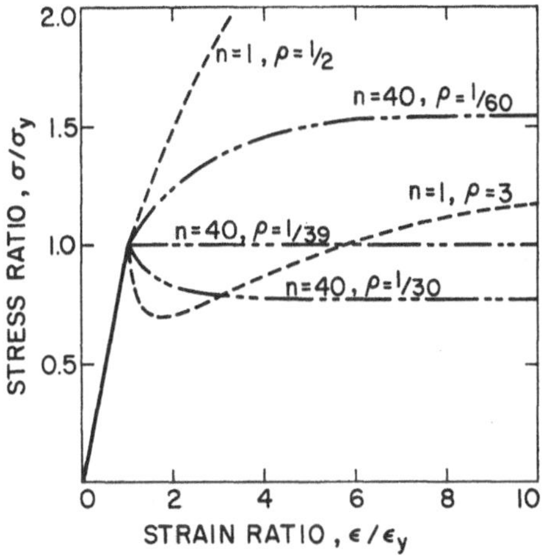

Fig. 1. Predicted stress-strain curves with plastic flow

g is considered to be completely arbitrary, except for the condition $a_D \geq 0$. Of course the model defined by (38) and (39) does enable characterization for limited strain histories in terms of a few constants.

However, the primary advantage gained in the present formulation is that elastic, viscous, and plastic mechanisms are all combined very simply and naturally for uniaxial and multiaxial loading, and it is done in a manner that is completely consistent with thermodynamic principles.

As a final point concerning the mechanical model interpretation of this one-dimensional example, we observe that eqs. (35) and (39) imply all dashpots have viscosities which are inversely proportional to strain rate above the yield point. This, in effect, reduces them to slip elements with strain-dependent friction coefficients, and all start slipping at the same time. In contrast, Iwan [29] has demonstrated the usefulness of models having elements which start slipping at different times, depending on their friction coefficients. In order to obtain Iwan's parallel-series model, for example, from thermodynamic eq. (7), it would be necessary to assume b'_{rs} and a_{rs} can be simultaneously diagonalized, even though all elements b'_{rs} do not have a common nonlinear factor.

Finally, we want to show that the classical three-dimensional theories of rate-independent plasticity and creep are contained in the present theory, and suggest a possible generalization. It is convenient to use eq. (22a) to accomplish this, in which generalized variables are replaced by mechanical stresses and strains. Bulk response is assumed elastic, as given by eq. (25).

Now assume a Maxwell representation for shear,

$$J(\psi) = J_0 + J_1 \psi \tag{41}$$

where J_0 and J_1 are positive constants. Differentiating (22a) with respect to ψ and then using $d\psi = dt/a_D$ gives

$$\frac{d\varepsilon'_{ab}}{dt} = \frac{J_0}{2} \frac{d\sigma'_{ab}}{dt} + \frac{J_1}{2a_D} \sigma'_{ab}. \tag{42}$$

Assume further that

$$\frac{1}{a_D} \simeq f + g \frac{d\bar{\gamma}_p}{dt} = f + g \frac{d\bar{\gamma}_p}{d\bar{\tau}} \frac{d\bar{\tau}}{dt} \tag{43}$$

where f and g are invariants, and $d\bar{\gamma}_p$ and $\bar{\tau}$ are defined to be the plastic components of the incremental octahedral shear strain and the octahedral shear stress, respectively; during loading on the yield surface $g \neq 0$, but $g \equiv 0$ during unloading or when the state of stress is within the yield surface.

Substitution of a_D into (42) yields

$$\frac{d\varepsilon'_{ab}}{dt} = \frac{J_0}{2} \frac{d\sigma'_{ab}}{dt} + \frac{J_1}{2} f \sigma'_{ab} + \frac{J_1}{2} g \frac{d\bar{\gamma}_p}{d\bar{\tau}} \frac{d\bar{\tau}}{dt} \sigma'_{ab}. \tag{44}$$

It is seen that the right side of eq. (44) now consists of the sum of elastic, creep, and plastic deviator strain rates, respectively. The latter two correspond to the von Mises condition for flow [2, p. 81; 30, p. 39]. When $f = f(T, \bar{\tau}, \bar{\gamma}_c)$, (where $\bar{\gamma}_c$ is the octahedral creep strain) the so-called strain-hardening theory of primary creep is included, while $f = f(T, \bar{\tau})$ defines secondary creep; age-hardening is accounted for if f depends explicitly on time. When rate-independent plastic flow exists $g = 1/J_1\bar{\tau}$. The tangent modulus $d\bar{\tau}/d\bar{\gamma}_p$ is measured in the absence of creep, and is commonly approximated by a power law in $\bar{\tau}$; similarly, f usually can be expressed as a power law [2].

The Maxwell representation (41) was used only to show that the present general theory contains the classical equations for inelastic behavior of metal. However, in some cases it may be more desirable and realistic to allow for a general shear creep compliance (or relaxation modulus) and not introduce the artificial decomposition of strain into plastic and creep components; a different rate in (43) would then be more appropriate, such as total octahedral shear strain rate or octahedral shear stress rate. For example, it may be possible to express the decreasing creep rate occuring during primary creep in terms of $J(\psi)$, without including strain-dependence in f. Strain response to arbitrary loading would then be calculated by means of eq. (22a).

In this regard, we have recently characterized creep of a hard polymer (polyvinylchloride) under combined tension and torsion in this manner [26]; nonlinearity was found to depend on only $\bar{\tau}$.

6. Application to Rubbery Polymers

Polymer solids above their glass transition temperature may exhibit very high strains before breaking. It is often found necessary to account for nonlinearity in the free energy, as well as in the entropy production, in order to adequately characterize their mechanical behavior. We shall briefly examine the validity of constitutive eq. (15) for initially isotropic materials in a state of uniaxial tension, although reference to multiaxial data will be made also.

The solution to this equation was given in [8], and for uniaxial loading, neglecting thermal expansion, it becomes:

$$\sigma = \frac{dF_e}{d\varepsilon} + a_F \frac{dq}{d\varepsilon} \int_0^t [E(\tilde{\psi} - \tilde{\psi}') - E_e] \frac{dq}{d\tau} \, d\tau \tag{45}$$

where

σ = axial force/initial area (engineering stress)

ε = extension/initial length (engineering strain)

$F_e = F_e(\varepsilon)$ = elastic strain energy

$E(\tilde{\psi})$ = linear viscoelastic relaxation modulus

$E_e = \mathrm{E}(\infty)$

$\tilde{\psi} = \int_0^t dt'/\bar{a} : \bar{a} \equiv a_D/a_F$

In contrast with glassy and crystalline polymers, the shift-factor $\bar{a}$ is generally quite insensitive to stress (or strain) in unfilled amorphous polymers above their glass-transition temperature; however, it is highly temperature-dependent as it reflects the influence of free volume on molecular mobility.

For comparison purposes it will be desirable to rewrite eq. (45) as

$$\sigma = f_e + a_F \frac{dq}{d\varepsilon} \int_0^t E(\tilde{\psi} - \tilde{\psi}') \frac{dq}{d\tau} d\tau \tag{46}$$

where

$$f_e \equiv dF_e/d\varepsilon - \frac{1}{2} E_e a_F dq^2/d\varepsilon.$$

This can be inverted to obtain the compliance form,

$$q = \int_0^t D(\tilde{\psi} - \tilde{\psi}') \frac{d}{d\tau} \left[\frac{\sigma - f_e}{a_F dq/d\varepsilon} \right] d\tau \tag{47}$$

where $D(\tilde{\psi})$ is the linear viscoelastic creep compliance. Let us now examine (46) and (47) when certain stress and strain histories are applied.

With constant strain (relaxation test) and assuming $\bar{a} = \bar{a}(\varepsilon, T)$,

$$\sigma = f_e + \frac{1}{2} a_F \frac{dq^2}{d\varepsilon} E(\tilde{\psi}); \quad \tilde{\psi} = t/\bar{a}. \tag{48}$$

When a constant stress is applied (creep test), (47) does not simplify. However, if the term in square brackets, $D(\tilde{\psi})$, and q have small curvature with respect to log ψ, the quasi-elastic approximation [11] may be used:

$$q \simeq D(\tilde{\psi}) \left[\frac{\sigma - f_e}{a_F dq/d\varepsilon} \right]. \tag{49}$$

Further, the incremental storage E_ε' and loss E_ε'' moduli for a specimen in a state of initial steady stress and strain and then subjected to small axial vibrations of frequency ω are

$$E_\varepsilon' = \frac{d^2 F_e}{d\varepsilon^2} - E_e a_F \left(\frac{dq}{d\varepsilon}\right)^2 + a_F \left(\frac{dq}{d\sigma}\right)^2 E'(\omega\bar{a}) \tag{50a}$$

$$E_\varepsilon'' = a_F \left(\frac{dq}{d\varepsilon}\right)^2 E''(\omega\bar{a}) \tag{50b}$$

where E' and E'' are the linear viscoelastic moduli.

If we set $f_e = 0$ and $\bar{a} = a_T(T)$, eqs. (48) and (49) take the familiar product form in time and strain observed in large strain, uniaxial tests of unfilled amorphous elastomers (e.g. [31, 32]). When it is further assumed that $q = \varepsilon$ in (46), a product form results for constant strain rate, which has also been widely reported [32, 33]. Consistent with these assumptions, the remaining nonlinear function a_F was found in [32] to be identical for creep, stress relaxation, and constant strain rate over a considerable range of time and temperature.

Equation (45), with $q = \ln(1 + \varepsilon)$, has been used successfully to predict isothermal uniaxial stress response of polyisobutylene under multiple step-straining [26].

Although the above relations (45)—(50) are stated for uniaxial loading, identical forms follow from the theory whenever there is a single independent applied strain, such as simple shearing or equal biaxial straining. In addition, if a material is initially isotropic and the bulk relaxation modulus is relatively constant in the *linear* range of behavior [or is proportional to $E(t)$ or $G(t)$] only one relaxation (or creep) function appears in the stress-strain relations; this is true regardless of the number of independent applied strains or stresses. Nonlinear data reported on multiaxial and uniaxial tests of unfilled and filled polymers in [34—37] are consistent with these observations; in some cases, especially with filled materials, $\bar{a}$ depends on strain and/or stress [36], but the limited isothermal data in [34] and [35] can be fitted using $\bar{a} = 1$. It is noteworthy that small-strain nonlinearity of some filled polymers under constant, combined tensile and shear strains [34] was expressed in terms of a single strain invariant, which can be shown to be the octahedral shear strain when dilatation is negligible; this result is analogous to that reported for creep of a plastic under constant, combined tensile and shear stresses [26].

Turning now to vibration tests, we observe that eq. (50), and a corresponding one for multiaxial loading [38], have a particular strain- and frequency-dependence that can be used to check applicability of the theory. For example, apart from the first two frequency-independent terms in E'_ε, the two moduli have the same product form with common strain-dependent functions.

Uniaxial vibration data having the form of eq. (50) were reported by Mason [39] for natural rubber. Also, shear vibrations superposed on lateral compression of a filled polymer [38] produced incremental moduli consistent with a multiaxial version of (50), in which $\bar{a}$ was found to decrease with initial strain. Also in [38], vibration data on another filled material under initial shear strain were shown, but agreement for the loss modulus was not good; it was later learned that the material was not fully cured when the tests were initiated.

While the above comparisons of constitutive eq. (45) with existing data are far from exhaustive, they do indicate that polymers often exhibit a simple type of nonlinearity which is contained in the present thermodynamic theory.

Clearly, the basic simplifying element present in this theory is that nonlinear viscoelastic processes are controlled by the same relaxation (or creep) functions as exist in the linear range under uniaxial or multiaxial loading. This behavior also forms the basis for the BUECHE-HALPIN theory of strength of gum elastomers [40]. Similar simplicity in relaxation behavior has been reported by KUBÁT [41], in which the data were normalized with respect to initial stress.

Concerning the validity of (45) for metals, a simplified version with $\bar{a} = \bar{a}(\varepsilon)$, was proposed by KELLY [42] for application to wave propagation in metals as an extension of the Sokolovsky-Malvern theory of strain-rate dependent plasticity. However, in view of our earlier discussion, the stress-strain equations given in Sections 4 and 5, with all nonlinearity in the entropy production coefficient a_D, and possibly in the relation between observed generalized coordinates and strains, may be more appropriate for metals.

7. Concluding Remarks

By introducing certain physically and experimentally motivated simplifications at an early stage in a thermodynamic theory, it has been shown that three-dimensional, nonlinear stress-strain equations can be derived which are very similar to the Boltzmann form in linear viscoelasticity theory, and are consistent with a considerable amount of data on polymers and metals. Furthermore, because of this similarity, the energy equation connecting mechanical and thermal variables is closely related to that in the linear theory [7, 8]. It certainly would be desirable to have sound molecular bases for this theory, but, unfortunately, molecular information is very limited; of particular interest would be a basis for the common nonlinear scalar coefficient, a_D, in entropy production.

Whenever these equations can be used, nonlinear viscoelastic characterization procedures are simplified since the only time-dependent creep or relaxation functions needed are those which exist in the linear response range. Development of structural analysis computer programs which are to account simultaneously for realistic nonlinear elastic, viscous, and plastic phenomena is also greatly aided because all of the effects are taken into account quite simply by the proposed equations.

Although we have discussed only metal and polymer applications in this paper, there is experimental evidence that indicates other materials can be characterized by means of the same equations. For example, strain rate response of soils to constant stress often follows eq. (33) [43, 44]. Furthermore, a special form of eq. (45) was reported in [6] for a biological material.

We observe that the thermodynamic theory given here is based on the use of strains, rather than stresses, as state variables. Consequently, all nonlinear terms in the stress-strain equations due to free energy are expressed as functions of strains instead of stresses. However, we have shown recently that stresses can normally be used as independent state variables, and have developed a theory based on a stress-dependent Gibbs free energy [45].

The design of experimental programs to evaluate nonlinear material properties has not been discussed here, but is considered in another paper [26]. It is shown, for example, that two-step relaxation tests can be used, together with a graphical procedure, to evaluate all properties in (45). Similarly, creep and recovery data provide sufficient information for evaluating properties in an analogous equation based on the stress formulation in [45].

Acknowledgement

This work was sponsored by the Air Force Materials Laboratory, Research and Technology Division, Air Force Systems Command, United States Air Force under Contract Number F 33615-67-C-1412.

The author is grateful to the project scientist, Dr. J. C. Halpin, for helpful discussions on the behavior of polymers.

References

[1] Fung, Y. C.: Fundamentals of Solid Mechanics. Englewood Cliffs, N. J.: Prentice-Hall. 1965.

[2] Hult, J. A. H.: Creep in Engineering Structures. Waltham, Mass.: Blaisdell. 1966.

[3] Green, A. E., and R. S. Rivlin: The Mechanics of Non-linear Materials with Memory — Part I. Arch. Ratl Mech. Anal. **1**, 1—21 (1957).

[4] Coleman, B. D., and W. Noll: Foundations of Linear Viscoelasticity. Rev. Mod. Phys. **33**, 239—249 (1961).

[5] Onat, E. T.: Description of Mechanical Behavior of Inelastic Solids. Proc. 5th U. S. National Congress of Applied Mechanics, ASME, p. 421—434 (1966).

[6] Fung, Y. C.: Biomechanics — Its Scope, History, and Some Problems of Continuum Mechanics in Physiology. Appl. Mech. Rev. **21**, 1—20 (1968).

[7] Schapery, R. A.: Application of Thermodynamics to Thermomechanical, Fracture, and Birefringent Phenomena in Viscoelastic Media. J. Appl. Phys. **35**, 1451—1465 (1964).

[8] Schapery, R. A.: A Theory of Non-linear Thermoviscoelasticity Based on Irreversible Thermodynamics. Proc. 5th U. S. National Congress of Applied Mechanics, ASME, p. 511—530 (1966).

[9] Biot, M. A.: Theory of Stress-Strain Relations in Anisotropic Viscoelasticity and Relaxation Phenomena. J. Appl. Phys. **25**, 1385—1391 (1954).

[10] MORLAND, L. W., and E. H. LEE: Stress Analysis for Linear Viscoelastic Materials with Temperature Variation. Trans. Soc. Rheology 4, 233–263 (1960).
[11] SCHAPERY, R. A.: Stress Analysis of Viscoelastic Composite Materials. J. Composite Materials 1, 228–267 (1967).
[12] SOKOLNIKOFF, I. S.: Mathematical Theory of Elasticity. New York: McGraw-Hill. 1956.
[13] BUECHE, F.: Physical Properties of Polymers. New York: Interscience. 1962.
[14] DEGROOT, S. R., and P. MAZUR: Non-Equilibrium Thermodynamics, p. 206. Amsterdam: North-Holland Publ. Co. 1962.
[15] TITCHENER, A. L., and M. B. BEVER: Progress in Metal Physics 7, Ed. by B. CHALMERS and R. KING. New York: Pergamon. 1958.
[16] GREEN, A. E., and W. ZERNA: Theoretical Elasticity, 2nd Ed. London: Oxford. 1968.
[17] ERINGEN, A. C.: Nonlinear Theory of Continuous Media. New York: McGraw-Hill. 1962.
[18] FERRY, J. D.: Viscoelastic Properties of Polymers. New York: Wiley. 1961.
[19] BIOT, M. A.: Linear Thermodynamics and the Mechanics of Solids. Proc. 3rd U. S. National Congress of Applied Mechanics, ASME. p. 1–18 (1958).
[20] HUSEBY, T. W. and S. MATSUOKA: Mechanical Properties of Solid and Liquid Polymers. Mater. Sci. Eng. 1, 321–341 (1967).
[21] HALPIN, J. C.: Composite Materials Workshop, p. 87. Ed. by S. W. TSAI, J. C. HALPIN, and N. J. PAGANO. Stamford, Conn.: Technomic. 1968.
[22] PASSAGLIA, E., and J. R. KNOX: Engineering Design for Plastics, Chapt. 3. Ed. by ERIC BAER. New York: Reinhold. 1964.
[23] THORKILDSEN, R. L.: Engineering Design for Plastics, Chapt. 5. Ed. by ERIC BAER. New York: Reinhold. 1964.
[24] FINDLEY, W. N., and G. KHOSLA: Application of the Superposition Principle and Theories of Mechanical Equation of State, Strain, and Time Hardening to Creep of Plastics under Changing Loads. J. Appl. Phys. 26, 821–832 (1955).
[25] PASSAGLIA, E., and H. P. KOPPEHELE: The Strain Dependence of Stress Relaxation in Cellulose Monofilaments. J. Polymer Sci. 33, 281–289 (1958).
[26] SCHAPERY, R. A.: On the Characterization of Nonlinear Viscoelastic Materials. Polymer Eng. Sci. 9, 295–310 (1969).
[27] SCHAPERY, R. A.: Unpublished Research.
[28] WILLIAMS, M. L.: Structural Analysis of Viscoelastic Materials. AIAA J. 2, 785–808 (1964).
[29] IWAN, W. D.: On a Class of Models for the Yielding Behavior of Continuous and Composite Systems. J. Appl. Mech. 34, 612–617 (1967).
[30] HILL, R.: The Mathematical Theory of Plasticity. London: Oxford. 1950.
[31] STAVERMAN, A. J., and F. SCHWARZL: Die Physik der Hochpolymeren IV, p. 139. Ed. by H. A. STUART. Berlin-Göttingen-Heidelberg: Springer. 1956.
[32] HALPIN, J. C.: Nonlinear Rubberlike Viscoelasticity — A Molecular Approach. J. Appl. Phys. 36, 2975–2982 (1965).
[33] SMITH, T. L.: Deformation and Failure of Plastics and Elastomers. Polymer Eng. Sci. 5, 270 (1965).
[34] BERGEN, J. T., D. C. MESSERSMITH, and R. S. RIVLIN: Stress Relaxation for Biaxial Deformation of Filled High Polymers. J. Appl. Polymer Sci. 3, 153–167 (1960).
[35] VALANIS, K. C., and R. F. LANDEL: Large Multi-axial Deformation Behavior of a Filled Rubber. Trans. Soc. Rheology 11, 243–256 (1967).

[36] Schapery, R. A.: Approximate Methods for Thermoviscoelastic Characterization and Analysis of Nonlinear Solid Rocket Grains. AIAA J., to be published (1970).
[37] Dickie, R. A., and T. L. Smith: Deformation and Rupture of Elastomers in Equal Biaxial and Simple Tension. Air Force Materials Laboratory, Wright-Patterson Air Force Base, Ohio, Tech. Rept. AFML-TR-68-112 (May 1968).
[38] Schapery, R. A.: An Engineering Theory of Nonlinear Viscoelasticity with Applications. Int. J. Solids and Structures **2**, 407—425 (1966).
[39] Mason, P.: The Viscoelastic Behavior of Rubber in Extension. J. Appl. Polymer Sci. **1**, 63—69 (1959).
[40] Bueche, F., and J. C. Halpin: Molecular Theory for the Tensile Strength of Gum Elastomers. J. Appl. Phys. **35**, 36—41 (1964).
[41] Kubát, J.: A Similarity in the Stress Relaxation Behavior of High Polymers and Metals. A Summary of Several Papers. Royal Institute of Technology, Stockholm (1965).
[42] Kelly, J. M.: Generalizations of Some Elastic-Viscoplastic Stress-Strain Relations. Trans. Soc. Rheology **11**, 55—76 (1967).
[43] Mitchell, J. K., R. G. Campanella, and A. Singh: Soil Creep as a Rate Process. J. Soil Mech. Foundations Div. A. S. C. E. **94**, 231—253 (1968).
[44] Singh, A., and J. K. Mitchell: General Stress-Strain-Time Functions for Soils. J. Soil Mech. Foundations Div. A. S. C. E. **94**, 21—46 (1968).
[45] Schapery, R. A.: Further Development of a Thermodynamic Constitutive Theory: Stress Formulation. Purdue U. Rept. AA & ES 69-2 (1969).

Creep Rupture at Strain Concentrations
Limitations of the Fracture Mechanics Approach

By

A. I. Smith, D. Murray and **R. H. King**

East Kilbride (U. K.)

Summary

The difficulties of extending fracture mechanics concepts to the cracking and fracture of metal components under creep conditions are considered, and the additional complications—both mechanical and metallurgical—which apply to materials under prolonged exposure to elevated temperatures, discussed. The role of notch-rupture testing and features of creep crack initiation and propagation are described, and suggestions made of the form of 'fracture-toughness' parameter which might be applicable to fracture at elevated temperatures.

1. Introduction

The difficulty of assessing the capacity of metallic structures and components to undergo plastic or creep deformation during service at elevated temperatures is analogous to the problem of measuring fracture toughness at ambient and sub-ambient temperatures. The materials used must have the maximum combination of strength with sufficient capacity for flow, without fracture, to relieve any stress concentration. Local overstressing may occur in a variety of circumstances, which include (individually or collectively):

Geometrical factors

Inhomogeneities in the material

Strain-induced effects

Thermally induced effects (including cyclic events).

A material should either have considerably enhanced creep strength at shorter times, i.e. the slope of its 'stress-parameter' curve will be quite steep, thus preventing excessive plastic deformation or, if deformation should occur, it should do so without severe 'strain-damage' effects which would lead to either microscopic or macroscopic creep cracking.

In the past, several properties have been considered when attempting to assess the likelihood of a material failing in a brittle manner, or with low ductility in high-temperature service. Components should be made of materials that have some or all of the following 'ductility' properties:

1. Good elongation and reduction of area values at rupture in both short-time and long-time creep rupture tests, and without 'ductility troughs'.

2. 'Notch strengthening', in comparative notched and plain bar stress rupture testing; again with particular emphasis on longer time properties.

3. Good thermal fatigue or high-strain fatigue properties at elevated temperatures, as revealed in, e.g. the 'Forrest reversed bending test'.

4. Characteristic of slowly accelerating creep during the phase of tertiary creep in simple creep-strain tests, particularly in long-time tests.

It is usually possible to classify materials into 'good' or 'bad' using the traditional measures of creep ductility and creep strength, but more severe design problems arise when a compromise has to be made in the selection of the most suitable material if those available cannot be classified quantitatively from 'best' to 'worst' using the above factors. The missing information is a quantitative factor which will provide a narrower band for final selection.

There is a distinction between the conventional study of fracture mechanics and the effects associated with time and temperature dependent forms of fracture toughness. This paper does not discuss the use of conventional fracture toughness testing at or below ambient temperature, e.g. to evaluate samples of material after exposure to creep-strain damage, but considers the application of those concepts, methods, and experience already evolved for ambient temperatures that are relevant to the problems of time/temperature dependent fracture modes in components (or testpieces) operating in or close to the creep range of temperature for the material.

2. Fracture Mechanics at Elevated Temperatures

In the quantifying of fracture toughness at ambient or elevated temperatures there are two distinct aspects which control failure in service or in a laboratory test:

(a) the initiation of a propagatable crack,

(b) propagation of that crack to rupture.

In fracture mechanics as developed in the past decade [1], there has been intensive study of the circumstances accompanying the progressive loading of testpieces that contain one of a number of forms of defect which constitute, or will lead to the development of, a propagatable crack. The extensive literature [2] reports broad studies of both theoretical stress/strain linear-elastic and elastic-plastic analyses of the regions surrounding such defects. The present stage of development is that a wide range of testpiece forms or component shapes and modes of loading can be made to yield the critical stresses and strains for the events observed experimentally. From these analyses the 'critical size' and form of defects may be deduced for various testpieces or components. The work covers the behaviour of both 'brittle' and 'ductile' materials in this temperature range. There are occasional references to the desirability of undertaking extended-time tests at reduced loads, to establish the possible occurrence of slow propagation of cracks.

In fracture mechanics the concentration has been primarily on 'brittle fracture' failures, with increasing attention latterly to the aspects of the phenomena observed in 'ductile' classes of material, e.g. low-alloy steels. The concepts of fracture mechanics have also been applied to the slower propagation of fatigue cracks and to stress-corrosion cracks, and a few investigations have been carried out for non-metallic visco-elastic materials, to evaluate the effects of time-dependent deformation on a region containing a crack-initiating defect [3, 4]. In 'brittle fracture', crack propagation tends to be considered on a 'go' or 'no-go' basis, and theoretical work has concentrated on the description of the circumstances of this transition. 'Griffiths' type formulae have been modified to include the effects of the energy absorbed by the processes of local plastic deformation [5]. The theoretical studies have ranged from the dislocation, micro-scale of processes, to the macro-scale of behaviour in actual components. Before examing the possible extension [6] of this approach to elevated temperature conditions, it is important to identify the part which temperature plays in influencing the mechanisms by which material sustains damage.

At elevated temperatures plastic-strain damage is no longer determined solely by the stress but is highly time-dependent, and is accumulated gradually under a constant stress. Metallurgical instability, inherent in material microstructures, is strongly influenced by time at elevated temperature, and this may lead to 'thermal' damage. The ability of material to withstand local plastic deformation without cracking, may be severely restricted by the precipitation, (perhaps strain-induced), of brittle intermetallic phases not normally present in lower temperature applications.

At lower temperatures, strain-hardening may arrest local deformation and cause a redistribution of stress following more general plastic deformation. At elevated temperatures, however, this is governed by the temperature dependence of the mechanisms of work-hardening and thermal softening. If the local plastically yielded region can work-harden sufficiently, the strain concentration will be dispersed over the surrounding material without serious loss of ductility.

Under steady creep conditions, where the 'flow-inhibiting' processes of primary creep are in balance with the thermally-activated 'flow-inducing' processes, the material is usually still coherent. The agglomeration of vacancies into voids and grain boundary cracks, as shown in Fig. 1, occurs relatively late in the secondary stage, and increases damage during tertiary creep (see Fig. 2). Thus, creep-strain damage in a test-piece or component can result in very different stress/strain distributions from those considered in elastic-plastic analyses applicable at lower temperatures.

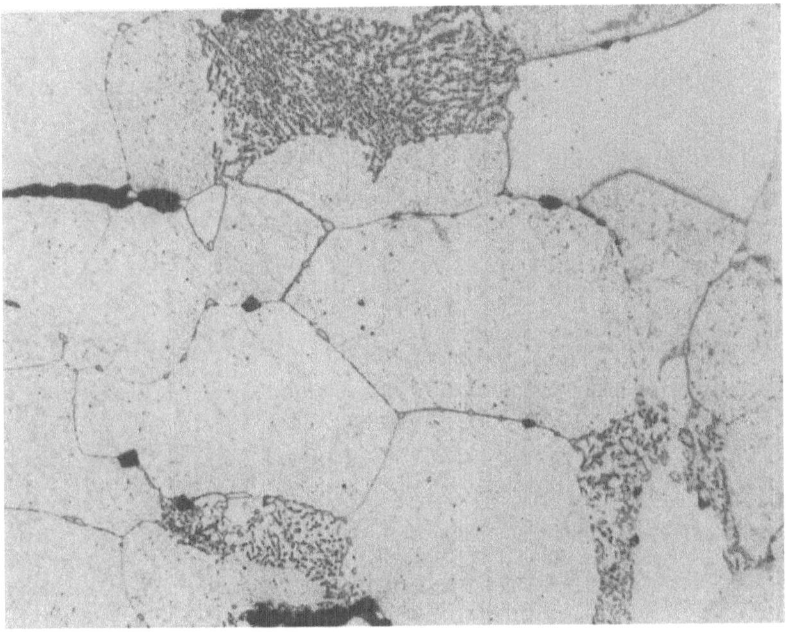

Fig. 1. Examples of grain boundary cracks; formed under creep straining conditions on boundaries normal to the direction of stress [9]

Factors which may give rise to local concentrations of creep-strain damage include, on the macro-scale, discontinuities in geometry, variations in the material at joints and welds, and the existence of thermal gradients. On the micro-scale, structural defects, inhomogeneity and anisotropy are significant, causing an accumulation of damage at elevated temperatures that differs only in extent and not in nature.

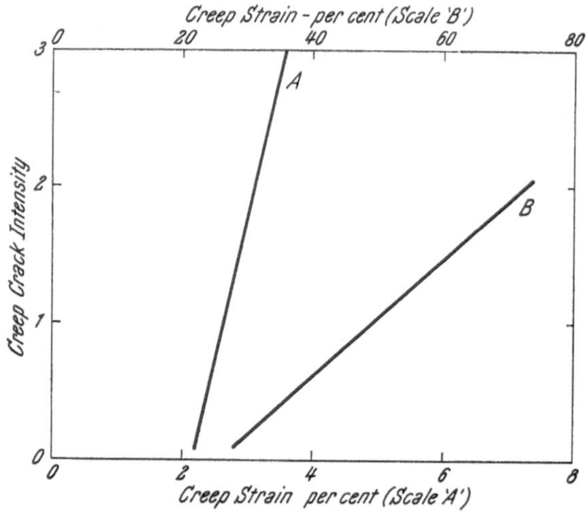

Fig. 2. The amount or intensity of creep cracking occurring in a brittle (*A*) and a ductile (*B*) material during tertiary creep [9]

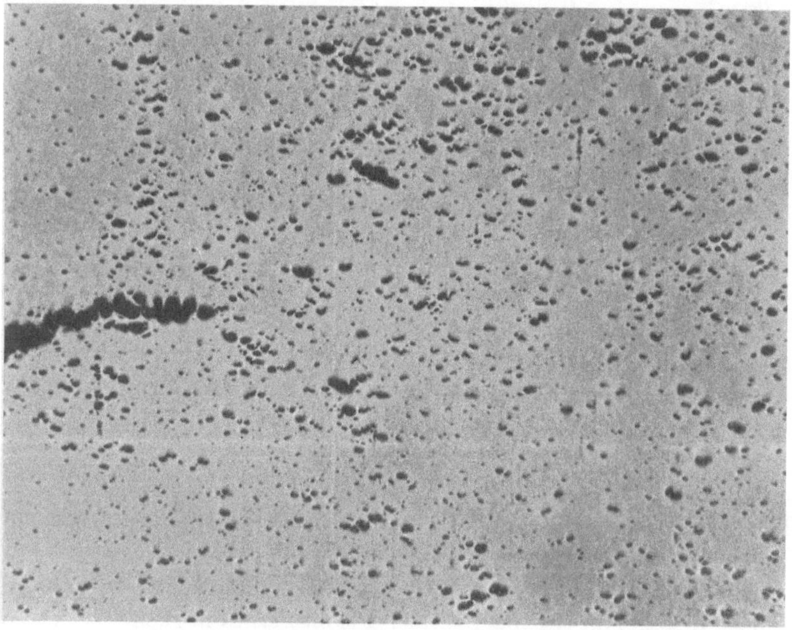

Fig. 3. An early stage in the propagation of a creep-rupture crack in a brittle material [9]

Local restraint and plastic deformation, caused by local variations in elastic modulus, yield point, and coefficient of thermal expansion for the material within a single component or between materials of connected components [7], will occur, even where the rate of change of temperature or stress is sufficiently slow for the operating conditions to be considered steady and uniform throughout the component. In addition, operating conditions involving cyclic variations of temperature or stress make the problem more severe, as the rate of strain damage will be accelerated by the cumulative and interactive mechanisms. During the earlier stages of this deformation, creep stress/strain analysis can predict the state of a component as accurately either as the 'creep laws' for a material allow, or as a computer can handle the numerical analysis. From the stage of the appearance of a propagating creep crack, (e.g. Fig. 3), it becomes important to know also whether the material will allow this to propagate easily under circumstances of over-stressing or over-heating compared with the design conditions for the component. It is at this point that a parameter is required to describe the fracture toughness under creep strain damaged conditions, which reflects the circumstances apertaining to elevated temperature service conditions.

3. The Role of Notch-rupture Testing

It is important to discuss this engineering test, in current use, in order to understand the elevated-temperature fracture problem. Past investigations [8] of the notch-rupture test have either provided useful 'descriptions of creep-fracture' as part of general data evaluation or have attempted to reveal the importance of 'initiation' or 'propagation' in determining the useful life of the tested material. In a notch-rupture testpiece (Fig. 4) the stress concentration is expected to locate the initiation of the rupturing crack (e.g. see Fig. 4, p. 965 of Ref. [10]) and the multi-axial stressing conditions to drive the crack to propagate (e.g. see Fig. 5, p. 966 of Ref. [10]) under creep conditions. If initiation or propagation were uniquely dominant, or both were equally important for all materials under all circumstances, then a parameter might be derived from a notch-rupture test programme that would provide an adequate criterion for defining materials as acceptable or unacceptable for long-time elevated-temperature service. The results from a typical material assessment programme, which should include both plain-bar and notch-rupture tests, cannot separate these two important aspects of fracture under creep conditions. At best, such results will provide a division into the categories 'notch-strengthened' or 'notch-weakened'. This classification is dependent upon testpiece geometry and size [8] but it is notable that 'notch-weakening' can be often correlated with poor ductility measured by elongation at rupture of plain-bar testpieces, as in, e.g. Fig. 2, p. 34 of Ref. [11]. The failure of the notch-rupture test to provide a unique description of fracture is associated with the several possible sequences of microscopic events (of different magnitudes of effect) for brittle and for ductile creep-resistant materials.

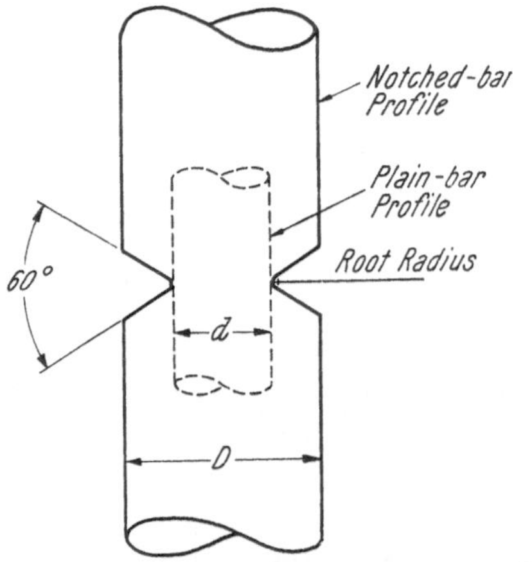

Fig. 4. Schematic drawing of the relative sizes of the testpiece sections of a notched and a plain stress-rupture testpiece

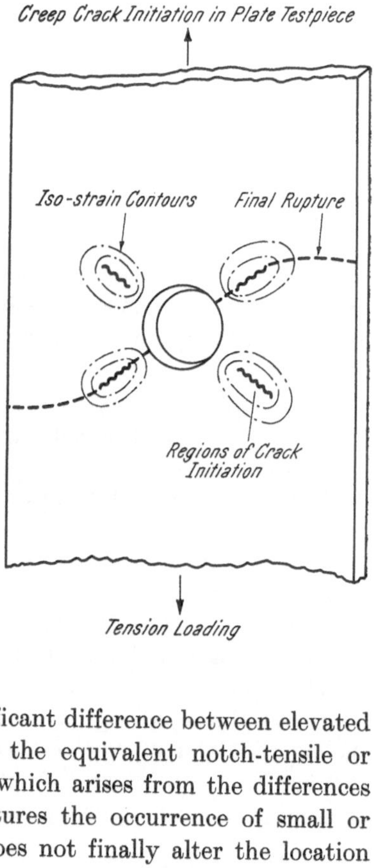

Fig. 5. A qualitative description of the development of creep strain and cracking around a hole in a tension sheet test [13]

It is relevant to illustrate the significant difference between elevated temperature notch-rupture tests and the equivalent notch-tensile or bending tests at lower temperatures, which arises from the differences in failure criteria. At lower temperatures the occurrence of small or larger zones of plastic deformation does not finally alter the location of crack initiation at the root of the notch and its direction of initiation, these being still controlled by the redistributed stress field in a fracture-toughness specimen as illustrated in the sections in Fig. 4, p. 251 of Ref. [12]. Conversely, in a testpiece strained under creep conditions, the site of the propagating creep crack will be strongly influenced by the extent and distribution of the time-dependent plastic strain when this approaches the level of creep-strain damage at which 'creep cracking', (usually localized 'intergranular cracking') effectively weakens material in these regions. Fig. 5 shows qualitatively that the local redistributions of stress in such a complex elastic-plastic time-dependent field could then cause the creep cracks to propagate in a direction determined only by the local strain distributions, rather than by the general stress distribution in the testpiece. In this example of a sheet testpiece with a central circular hole, lobular creep strain zones are formed, similar to a general yield effect at low temperature, but in the creep situation the propagated cracks are initiated in these zones. This indicates the importance of the proper choice of testpiece form which locates the initiation of a creep crack in a region where crack initiation and propagation may be monitored.

4. Fracture Toughness Testing of Creep-strained Material

Material which has been subject to creep strain may have undergone changes which make it more vulnerable to rupture under the following circumstances.

1. At, or near, ambient temperatures, fracture toughness of the material may be impaired and this should therefore be established, so that these parts of the plant service cycle are shown to be safe; (c.f. the considerable amount of post-creep test work on impact values). This can now be obtained by determination of stress intensity K_c or K_{1c} using current types of fracture-toughness testing, and special attention paid to samples of material simulating, or actually from, regions of the plant which may be especially susceptible to creep-strain damage, e.g. at welds and regions of unexpectedly high service temperatures. This type of investigation may be restricted to those situations in which sufficient sample material is available for test specimen manufacture.

2. At temperatures at or near the lower end of the 'creep range' of temperatures, and where creep occurs only at relatively high stresses and the strain rates are low and strongly stress-dependent, material properties and behaviour are less well defined. Creep may also occur after the incidence of delayed yield effects and there may be a strong tendency for only time-dependent strain-hardening flow to occur. Much plant is designed for service near the lower end of the creep temperature range and may operate for considerable periods in the temperature region under discussion when working at less than maximum service conditions.

3. At temperatures wholly in the creep range, in which the material characteristics are strongly time/temperature/stress/strain dependent, significant constitutional changes may occur during periods of extended service. In this temperature range most processes of fracture are highly time-dependent.

Evaluation of the elevated temperature 'fracture toughness' necessitates the investigation of at least three factors for samples of material which have been withdrawn from service at various stages of life (i.e. which have various degrees of creep-strain damage), or which have been given a 'service-simulating' treatment.

1. The magnitude of overstressing (mechanically or thermally-induced) which will initiate a crack that is significant compared with any already developed during the process of creep-damage under the design conditions.

2. The rate at which such a crack will propagate at typical service stresses and temperature, i.e. for the usable range of temperatures and the corresponding 100000 hour rupture stresses or alternatively, the maximum stress for zero crack propagation rate may be determined for comparison with the design stress.

3. The rate of increase of creep-crack propagation with increasing stress or temperature in the range of service conditions. The levels of critical crack acceleration will have to be studied experimentally before a 'safety criterion' can be established.

The first two factors are broadly analogous to those parts of fracture mechanics concerned respectively with initiation conditions and the propagation of a self-sustaining fracture. The third factor is an extension of the second factor that might permit a measurable, but small, rate of creep-crack propagation in practice, for a material in which the rate of increase of propagation rate is not highly stress or temperature dependent.

The determination of the level of local stress required to produce significant cracks will depend upon the availability of a means of detecting the onset of the cracks in the course of progressively loaded tests at elevated temperature. Possible techniques might include the 'crack opening displacement' method of WELLS [6] shown in Fig. 6, with the adaption for use at elevated temperatures, of an electrical resistance measurement method. The evaluation of the operative local stresses will depend upon the available elastic-plastic stress analysis method. It should be emphasised that near, or in, the 'creep range of temperatures', yielding is most likely to have occurred in the region being cracked. The extent of strain-hardening at elevated temperatures below the creep range is probably more comparable to that in tests near ambient conditions. The form of a suitable testpiece would be similar to that chosen for fracture toughness tests on ductile materials at low temperatures.

The determination of the second factor would require a time-dependent test, of sufficient duration to establish a reliable value of rate of propagation at approximately constant stress conditions, i.e. of the order of service stresses and temperatures. Alternatively, a time-dependent test, analogous to a stress-relaxation test, might be adopted in which the 'crack length' measurement would be used to control the load, and hence the minimum stress required to drive the crack would be determined. The form of testpiece required to establish these propagation rates will be dependent on the method of crack-length measurement available; also some provision will be required to regulate the stress in the slowly decreasing cross-section of the testpiece. When the creep

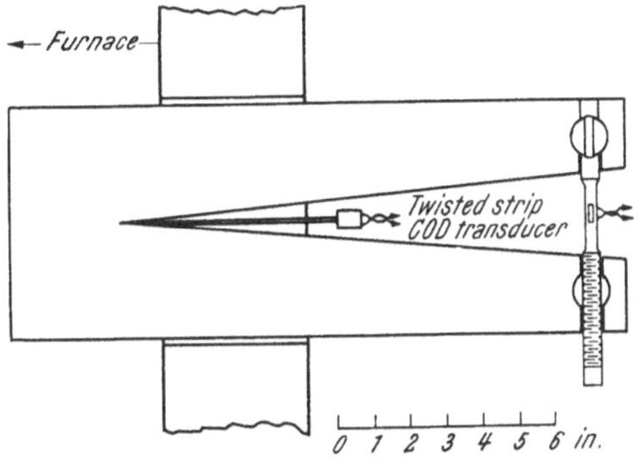

Fig. 6. Apparatus used by WELLS to study fracture at elevated temperatures [6]

crack rates are very slow, manual control of the creep machine would probably suffice, but more-sophisticated controls could be devised if electrical measurement of crack length were employed.

The determination of the third factor would require a form of test similar to that for the second but provision would have to be made for either a load increasing continuously and very slowly or for steps of load after the determination of specific rates of propagation — perhaps stepped while the rate remained very low, but changed to continuously increasing when it became possible to make 'rate', and 'change-of-rate' measurements from the crack-length/time plots.

5. Future Developments

Future experimental work at National Engineering Laboratory; East Kilbride, has been planned to make a more direct study of creep-crack propagation in an edge-cracked form of plate testpiece, to explore the possible development of a quantitative description of the liability of a material to brittleness under creep conditions. This will involve the systematic evaluation of the factors discussed above, which could form the basis of a 'High-Temperature Fracture Toughness Parameter'. If this extension of fracture mechanics to the analysis of high-temperature fracture is to be effective, an early comparison must be made with the results from notch-rupture testing, and the source of variations in these results caused by differences in testpiece geometry determined. Use of these experimental findings in specific design situations will depend on the application of the parameters describing crack behaviour to the conditions of complex stress and strain in components.

Acknowledgement

This paper is published by permission of the Director of the National Engineering Laboratory, Ministry of Technology. It is Crown copyright and is reproduced by permission of the Controller of H. M. Stationery Office.

References

[1] WEISS, V., and S. YUKAWA: Critical Appraisal of Fracture Mechanics. ASTM STP **381**, 1–29 (1965).
[2] PARIS, P. C., and C. M. SIH: Stress Analysis of Cracks. ASTM STP **381**, 30–83 (1965).
[3] CHEREPANOV, G. P.: Cracks in Solids. Int. J. Solids and Struct. **4** (8), 811–831 (1968).
[4] WILLIAMS, M. C.: The Kinetic Energy Contribution to Fracture Propagation in a Linearly Visco-elastic Material. Int. J. Fracture Mech. **4** (1), 69–78 (1968).
[5] MCCLINTOCK, F. A., and G. R. IRWIN: Plasticity Aspects of Fracture Mechanics. ASTM STP **381**, 84–132 (1965).
[6] WELLS, A. A., and F. H. MCBRIDE: Application of Fracture Mechanics to High-temperature Creep-Rupture. Can. Metall. Q. **6** (4), 347–368 (1967).
[7] KING, R. H., and A. I. SMITH: Thermal-Fatigue Processes and Testing Techniques. Symp. Thermal and High Strain Fatigue/London: Institute of Metals and Iron and Steel Institute, 1966.
[8] MANJOINE, M. J.: Size Effect on Notch Rupture Time. J. Bas. Engng. **84 D** (2), 220–221 (1962).
[9] ROBSON, K.: Unpublished results.
[10] GAROFALO, F.: Creep-Rupture Behaviour of Notched and Unnotched of Type 304, 316 and 321 Austenitic Stainless Steels. Proc. Amer. Soc. Test. Mater. **59**, 957–981 (1959).
[11] THUM, A., and K. RICHARD: Embrittlement and Damage of Heat Resisting Steels in Creep. Arch. Eisenhütt. Wes. **15**, 33–45 (1941–42). [In German]
[12] KNOTT, J. F., and A. H. COTTRELL: Notch Brittleness in Mild Steel. J. Iron Steel Inst. 249–260 (1963).
[13] DAY, M. F.: Unpublished results.

A Contribution to Creep Fracture under Combined Stress System

By

Shuji Taira and **Ryuichi Ohtani**

Kyoto (Japan)

Summary

An analytical investigation was conducted on the creep fracture of thin-walled cylinders subjected to combined tension and internal pressure, and the results were analyzed on the basis of the large strain theory. The experimental results on a 0.14% carbon steel at the test temperature of 500 °C proved the validity of the large strain theory combined with the von Mises criterion, and the time to rupture estimated by the large strain theory was in good agreement with the experimental results.

On the other hand, in the case of a material with less ductility, the large strain theory is not further applicable to its creep fracture. For instance, an 18-8 Nb austenitic stainless steel tested at 650 °C exhibited grain-boundary cracks, which were distributed uniformly on the surface of the specimen and were progressing perpendicular to the axis of the maximum tensile stress. The results imply that the criterion for creep fracture is closely related to the crack initiation and propagation.

Experimental study on creep fracture of cylinders under internal pressure were also conducted. The results were used to discuss on the criterion of the fracture. It was concluded that the mean diameter formula is valid for the design formula of pressure vessels and tubes irrespective of materials, wall-thickness and working conditions.

1. Introduction

Most of the common laboratory methods for studying creep and creep fracture under multiaxial stress conditions are to employ thin-walled cylinders under combined tension and torsion or thin- or thick-walled cylinders under combined internal pressure and axial load. A number of test results on this subject have hitherto been reported, and discussions were made on the relationship between stress and creep strain and on the stress criterion for creep fracture from the standpoint of multiaxial creep theory [1—4].

From the results of many investigations in this field, it can be said that the von Mises criterion is valid for the creep deformation of isotropic materials. On the other hand, for creep fracture, the von Mises criterion is not always the best criterion for the prediction of rupture life and the maximum principal stress criterion is superior in some cases.

An analysis of the conditions governing the multiaxial rupture was made by W. SIEGFRIED [5]. On the basis of his reason that the intergranular flow and subsequent fracture should be governed by the maximum effective shearing stress, he suggested that the hydrostatic stress might govern the rupture life for intergranular fracture, while for transgranular fracture the maximum stress deviator may be the leading criterion. Disagreement with the SIEGFRIED's prediction was shown in the tests carried out by A. E. JOHNSON and his colleagues [6, 7]. Few investigations, however, have been done to study the factors affecting the strength in creep fracture under multiaxial stresses.

From such a point of view, an analytical investigation was conducted on the creep instability of a thin-walled cylinder subjected to combined tension and internal pressure, and the rupture life was estimated by the large strain theory, which was proposed by N. J. HOFF for simple tension creep [8]. Creep tests under combined tension and internal pressure were carried out on a 0.14% carbon steel and an 18—8 stainless steel at the test temperatures of 500 and 650 °C, respectively, and the experimental results were used to determine the validity of the large strain theory. In addition, the stress criteria in the large strain theory were derived for creep of cylinders under internal pressure, and their relations to the design formulas of pressure vessels and boiler tubes were examined.

2. Creep Instability of a Thin-Walled Cylinder under Combined Tension and Internal Pressure

Let us now consider capped-end thin-walled cylinders subjected to combined constant tension and internal pressure. The wall thickness is small as compared with the outside diameter and the stress in the radial direction σ_r can be regarded as being zero. In conducting the analysis, the following assumptions are made:

a) The material is isotropic before and during creep.

b) The volume of the material remains constant so that hydrostatic pressure has no influence on creep deformation.

c) The true effective stress σ^* and the natural effective creep rate $\dot{\varepsilon}_c^*$ are related through the material constants b and α in the form

$$\dot{\varepsilon}_c^* = b\sigma^{*\alpha}, \tag{1}$$

where σ^* and $\dot{\varepsilon}_c^*$ are defined on the basis of the von Mises or the Tresca criterion [1]. According to the von Mises criterion, σ^* and $\dot{\varepsilon}_c^*$ are defined by

$$\left.\begin{aligned} \sigma^* &= \frac{1}{\sqrt{2}}\,[(\sigma_t - \sigma_r)^2 + (\sigma_r - \sigma_z)^2 + (\sigma_z - \sigma_t)^2]^{1/2} \\ \dot{\varepsilon}_c^* &= \frac{\sqrt{2}}{3}\,[(\dot{\varepsilon}_{tc} - \dot{\varepsilon}_{rc})^2 + (\dot{\varepsilon}_{rc} - \dot{\varepsilon}_{zc})^2 + (\dot{\varepsilon}_{zc} - \dot{\varepsilon}_{tc})^2]^{1/2}, \end{aligned}\right\} \tag{2}$$

and when the Tresca criterion is used as the stress criterion, these are represented by

$$\left.\begin{aligned} \sigma^* &= \sigma_t - \sigma_r \\ \dot{\varepsilon}_c^* &= \frac{2}{3}\,(\dot{\varepsilon}_{tc} - \dot{\varepsilon}_{rc}), \end{aligned}\right\} \tag{3}$$

in which the subscripts t, z and r mean tangential, axial and radial components, respectively, and the subscript c shows the creep component.

d) According to the theory proposed by F. K. G. Odqvist [3, 4], transient strain as well as initial elastic and plastic strains are substituted for total initial strain ε_0. The total strain ε is then expressed by the sum of total initial strain ε_0 and steady state creep strain ε_c as follows:

$$\varepsilon = \varepsilon_0 + \varepsilon_c = \varepsilon_0 + \int_0^t \dot{\varepsilon}_c dt. \tag{4}$$

e) By using the flow rule associated with the von Mises stress criterion of eq. (2), the components of creep rates are given by

$$\dot{\varepsilon}_{tc} = \frac{\dot{\varepsilon}_c^*}{\sigma^*}\left[\sigma_t - \frac{1}{2}\,(\sigma_r + \sigma_z)\right], \quad \text{etc.} \tag{5}$$

From these relations, the time dependent variation of stresses during creep can be obtained. Fig. 1 shows an example of calculated results [9], indicating how to change the stress state presented as a part of the ellipse of the von Mises effective stress for a 0.14% carbon steel used in the experiment. The ellipse at $t = 0$ hour represents the initial state of the effective stress of 18.0 kg/mm², and this initial state of stress changes in such a manner as inidicated by the full lines corresponding to 20, 40, 50, 60 and 80 hours. It is noted that the increase in true stress is the largest in the case of pure internal pressure $(\sigma_t/\sigma_z = 2)$ and the smallest for the stress ratio $\sigma_t/\sigma_z = 1/2$ in the entire range of the stress system studied. The thin dotted lines in the figure indicate the traces of the change in each true stress ratio. One can find out that, for the stress ratios of 2 (pure internal pressure), 1/2 (combined tension and internal pressure) and 0 (simple tension), the stress ratio remains unaltered during creep, but in the other case it becomes larger or smaller than the initial value with the increase in the creep deformation of tubes.

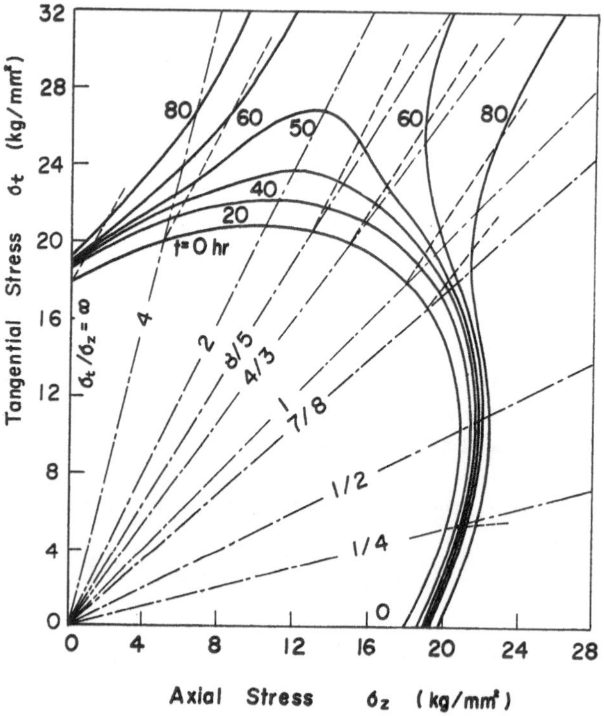

Fig. 1. Change with time in a part of the ellipse of the effective stress based on the von Mises criterion

Fig. 2 shows the analytical creep curves for the axially-loaded pressurized tubes obtained with consideration of the stress variation in the course of creep. The ordinate of the figure is the von Mises effective strain. The discrepancies among the creep curves are attributed to the difference of the time dependent increase in true stress in the respective case of stress ratio.

3. Comparison with the Experimental Results

In order to determine the validity of the concept of the creep instability under combined stresses mentioned above, the creep tests were conducted with thin-walled cylinders made of a 0.14% carbon steel under combined constant tension and internal pressure. Relevant details on the chemical compositions, the condition of heat treatment and the mechanical properties at room temperature of the material tested are listed in Table 1.

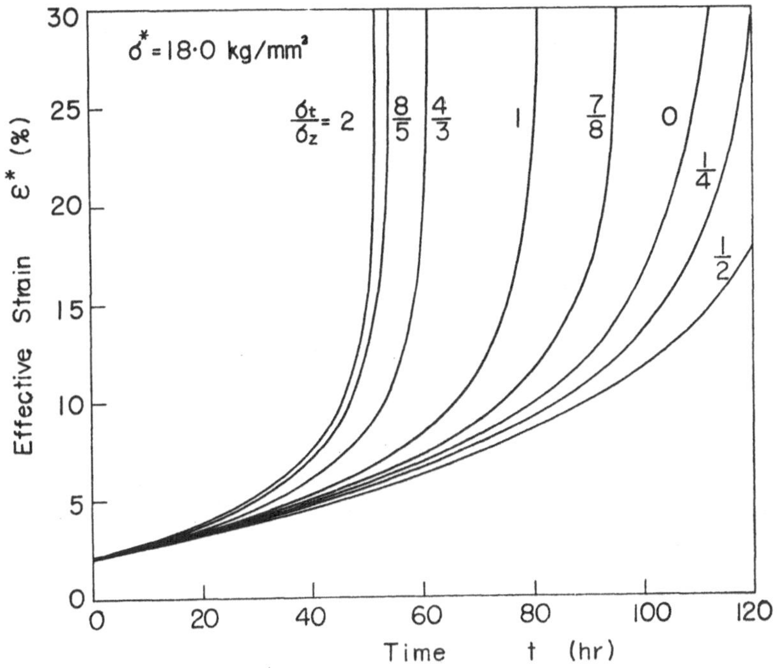

Fig. 2. Analytical creep curves for the axially-loaded pressurized tubes derived by the finite strain theory with the von Mises criterion

Table 1. *Chemical Composition, Condition of Heat Treatment and Mechanical Properties at Room Temperature*

Composition	C	Si	Mn	P	S	Fe
%	0.14	0.28	0.48	0.014	0.010	Bal.

920 °C × 1 hr, Cooling in Furnace

Yield Point kg/mm²	Tensile Strength kg/mm²	Elongation %	Reduction of Area %	Hardness H_{RB}
28.8	46.4	28.0	64.7	69.0

Fig. 3 shows a thin-walled tubular specimen with 25 mm inner diameter and 1 mm thick. Both ends of the test piece are closed by welding and a solid steel core is inserted into it in order to reduce the internal vapor volume.

The testing apparatus employed in this study is shown schematically in Fig. 4. The desired axial stress in the specimen *1* is generated through

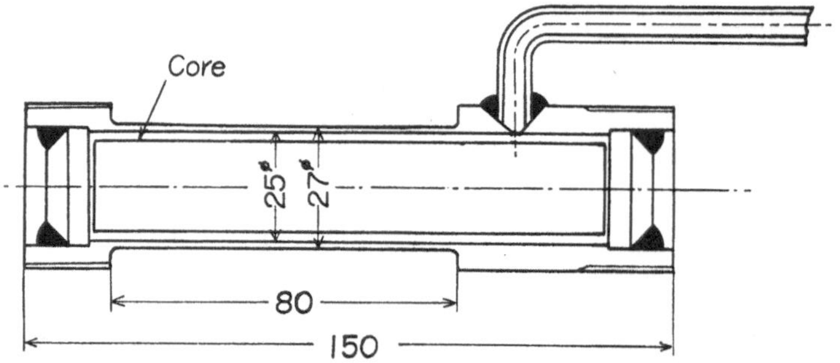

Fig. 3. Tubular specimen for creep tests under combined tension and internal pressure

a lever *5* by dead weights *6*. The internal pressure is supplied by the hand-operated water-pump *9* through a pipe *7* [10]. The torsional stress can also be produced through a disc *3* by dead weights *4*.

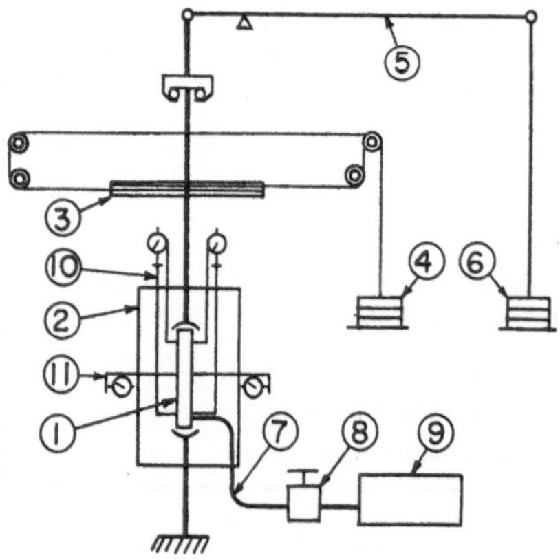

Fig. 4. Schematic diagram of the testing apparatus for combined stress creep

The axial strain of the specimen is measured by means of two longitudinal extensometers *10* set at the ends of the specimen gage length, and the tangential strain is determined by two lateral extensometers *11* set on the outer surface of the tubular specimen from the outside of the furnace *2*. These extensometers are composed of 1/100 mm scale dial gages.

In all creep tests, the initial effective stress was chosen to have the same value of 18.0 kg/mm^2, based on the von Mises stress criterion. The testing temperature is 500 °C. The stress conditions adopted in this experiment are indicated in Table 2.

Based on the large strain theory, a "creep failure time" is obtained by letting creep deformation approach infinity. With this line of approach, N. J. HOFF [8] has predicted theoretically the creep failure time for the case of simple tension creep and this has shown a good agreement with the experimental results. The creep failure times for a capped-end cylin-

Table 2. *Stress Conditions under Combined Tension and Internal Pressure*

Ratio of tangential stress to axial stress σ_t/σ_z	Tangential stress σ_t (kg/mm²)	Axial stress σ_z (kg/mm²)	Effective stress σ^* (kg/mm²)	Remarks
0	0	18.00	18.00	Simple tension
1/4	4.99	19.97	18.00	Combined tension and internal pressure
1/2	10.40	20.79	18.00	
7/8	16.69	19.07	18.00	
1	18.00	18.00	18.00	
4/3	19.94	14.96	18.00	
8/5	20.57	12.86	18.00	
2	20.79	10.40	18.00	Pure internal pressure

der under internal pressure [11—12], a hollow sphere under internal pressure [13] and a rotating cylinder [14] were discussed by F. P. J. Rimrott et al. It is found from these investigations that the creep failure time is helpful for the prediction of time at which specimen fractures.

We can predict, from Fig. 2, the rupture lives of the specimens tested, where the assumption is made that the creep fracture is occured when the effective strain approachs infinity. Fig. 5 is the diagram showing the

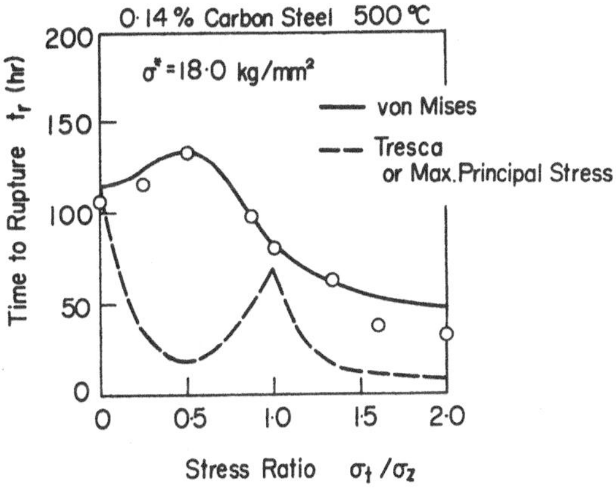

Fig. 5. Diagram showing the time to rupture as a function of the original stress ratio for combined stress creep fracture

time to rupture as a function of the original stress ratio. The full and the dotted lines indicate the analytical rupture lives derived by the large strain theory combined with the von Mises and the Tresca or the maximum principal stress criteria. The circles show the test results. It is found that the analytical curve based on the von Mises criterion agrees with the experimental data better than that of the Tresca criterion.

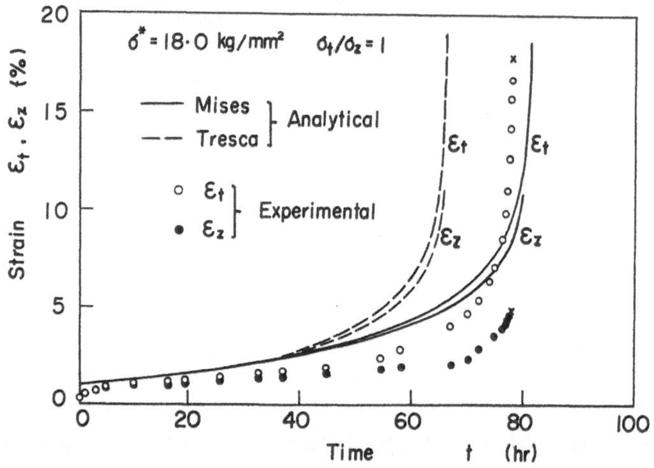

Fig. 6. Comparison of the analytical creep curves with the experimental results ($\sigma_t/\sigma_z = 1$)

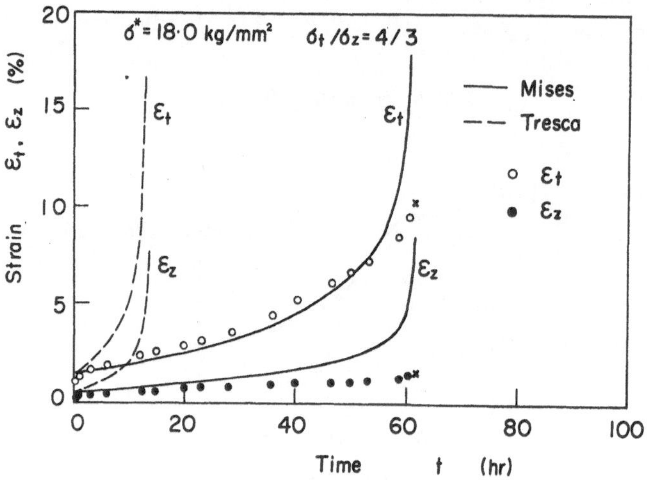

Fig. 7. Comparison of the analytical creep curves with the experimental results ($\sigma_t/\sigma_z = 4/3$)

The analytical creep curves for tangential and axial strain components are compared with the experimental results for the case of stress ratios of 1 and 4/3 in Figs. 6 and 7, respectively. The curves indicated by the full lines show the results calculated on the basis of the von Mises criterion, and the dotted line curves represent those of the Tresca criterion.

Of these two, the von Mises criterion gives results very close to experimental data up to fracture.

It is natural that the application of the concept of the large strain theory is justified only when the tertiary creep proceeds mainly due to the reduction of cross-sectional area of the specimen, and the rupture elongation of the material is large enough so that the difference between time to fracture and time to reach infinite deformation is insignificant. In the case of the material and the test conditions emplyed in the present study, it is appropriate to adopt the large strain theory based on the von Mises criterion for the predictions of rupture life as well as creep deformation. This implies that the material tested deforms by means of slip in the individual crystal grain according to the rule of critical shear stress, and that there are but few chances of void or crack formation up to near fracture and the tertiary creep appears owing to the increase in true stress in the specimen caused by the reduction of cross-sectional area.

4. Exception for the Application of the Large Strain Theory

As has been mentioned above, the large strain theory is not always applicable to the prediction of rupture life. For example, in the case of the material in which voids or cracks can be detected during creep, the tertiary creep will come into existence without noticeable reduction of the cross-sectional area of the specimen. As a typical example of the case, the test results for an 18—8 Nb austenitic stainless steel is cited, as the followings:

Table 3. *Chemical Composition. Condition of Heat Treatment and Mechanical Properties at Room Temperature*

Composition	C	Mn	Si	P	S	Cu	Ni	Cr	Nb + Ta	Fe
%	0.05	1.54	0.50	0.018	0.018	0.07	12.65	17.83	0.86	Bal.

1100°C × 1 hr, W. Q.

Yield Point kg/mm²	Tensile Strength kg/mm²	Elongation %	Reduction of Area %
28.8	61.4	64.5	69.2

Table 3 shows the chemical compositions, the condition of heat treatment and the mechanical properties at room temperature of the material tested. The specimen employed in the experiments is a thin-walled cylinder with 16 mm inner diameter, 1 mm wall thickness and

80 mm length of tubing. The creep tests were carried out under combined constant axial tension and internal pressure at the test temperature of 650 °C. The initial von Mises effective stress had the same value of 18.0 kg/mm² in all tests.

In Fig. 8 the creep curves are shown, which were obtained on the thin-walled tubular specimens in several stress ratios. There exists little discrepancies among minimum creep rates. When the experimental

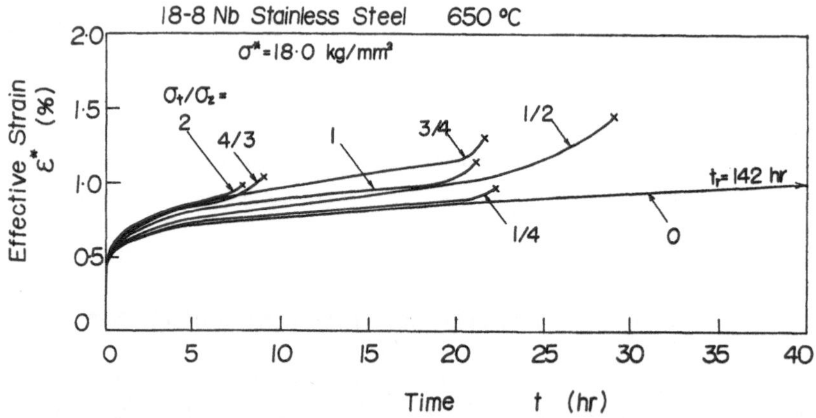

Fig. 8. Creep curves on an 18-8 Nb stainless steel under combined tension and internal pressure

results of the minimum creep rates are plotted in the μ versus ν diagram, in which μ and ν are the Lode's variables [15, 16] defined by

$$\left.\begin{aligned} \mu &= \frac{2\sigma_2 - (\sigma_1 + \sigma_3)}{\sigma_1 - \sigma_3}, \qquad \sigma_1 > \sigma_2 > \sigma_3 \\ \nu &= \frac{2\dot{\varepsilon}_2 - (\dot{\varepsilon}_1 + \dot{\varepsilon}_3)}{\dot{\varepsilon}_1 - \dot{\varepsilon}_3}, \end{aligned}\right\} \tag{6}$$

they agree fairly well with the straight line $\mu = \nu$. This suggests the validity of the von Mises flow rule and the hypothesis of isotropy.

On the other hand, a remarkable difference can be seen among rupture lives. The rupture life tends to shorten as the stress ratio becomes large. A similar tendency was found in the 0.14% carbon steel. In this case, however, the discrepancy among the rupture lives can not be wholly explained by the large strain theory, because strains are so small that the stress increase during creep has little or no effect on the deformation. Another characteristic feature is that the tertiary creep is little observed and the rupture elongation is quite small in every case of stress ratio. These will be closely related to the crack formation in the course of creep.

Figs. 9 and 10 are photographs, obtained by means of microscopic examination, showing the outer surface of tubular specimens at the secondary stage for the stress ratios of 2 and 1. It is found that the so-called wedge type cracks initiated at triple points of grains have developed to the typical grain-boundary cracks. The same type of cracks can be observed on the surfaces of the specimens for other stress ratios.

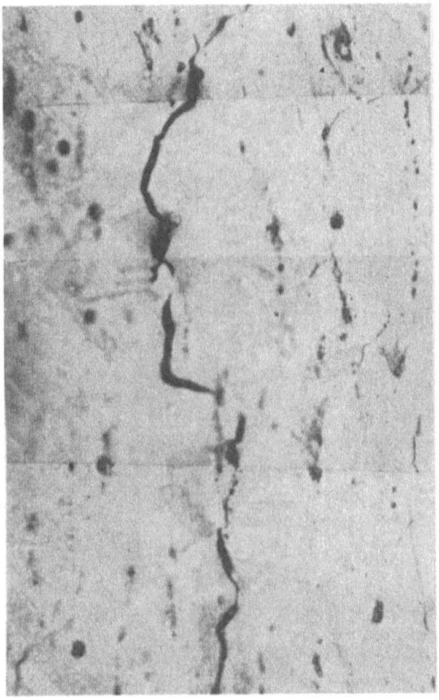

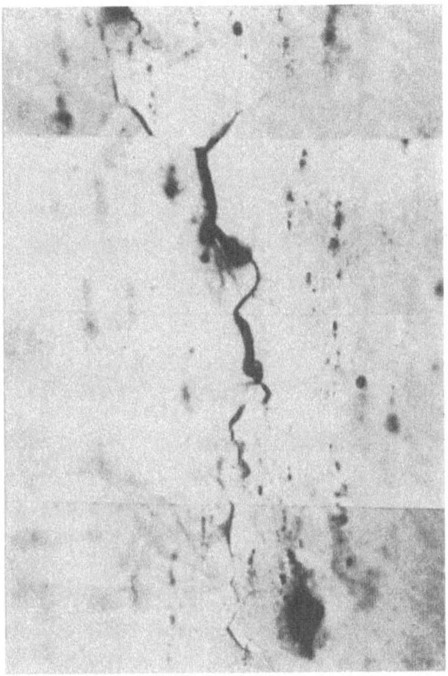

Fig. 9. $\sigma_t/\sigma_z = 2$ Fig. 10. $\sigma_t/\sigma_z = 1$

⟵ tangential direction ⟶

Typical cracks propagating along the grain-boundaries perpendicular to the maximum tensile stress

The cracks are seen to progress perpendicularly to the axis of the maximum tensile stress in all stress states, which demonstrates that the propagation of cracks is caused by the maximum tensile stress.

From these results, it is considered that the discrepancies among the rupture lives shown in Fig. 8 may be attributed to the difference of stress increase among various stress ratios. That is, the decrease in the effective cross-sectional area of specimens to hold the load, which is caused by the crack growth, and the magnitude of the maximum tensile stress to control the crack propagation are different in each case of the stress ratio.

5. Creep Instability of Pressurized Tubes

Based on the theoretical analysis derived by HOFF [8], eq. (1) is first taken up as the true stress-natural strain rate relation in simple tension. Integration of the equation yields

$$t = \frac{1}{\alpha b}\left(\frac{1}{\sigma_0}\right)^{\alpha}[1 - \exp(-\alpha\varepsilon)], \tag{7}$$

in which σ_0 is the initial tensile stress. The time to rupture is then obtained by letting the strain ε in eq. (7) approach infinity, that is,

$$t_r = \frac{1}{\alpha b}\left(\frac{1}{\sigma_0}\right)^{\alpha}. \tag{8}$$

For a tubular specimen of an isotropic material under internal pressure, the time t can be obtained as a function of the effective strain at outside diameter ε_0^*. That is,

$$\left.\begin{aligned} t &= K_1^{\alpha}\frac{(\alpha/2)^{\alpha}}{b}\left(\frac{1}{p}\right)^{\alpha} I \\ I &= \int_0^{\varepsilon_0^*}\left[\left\{\frac{(r_0/r_i)^2\exp 2K_2\varepsilon_0^*}{1+(r_0/r_i)^2(\exp 2K_2\varepsilon_0^* - 1)}\right\}^{1/\alpha} - 1\right]^{\alpha} d\varepsilon_0^*, \end{aligned}\right\} \tag{9}$$

where r_i and r_0 are the inner and outer radius, and p is the internal pressure. K_1 and K_2 are constants, which defines according to the von Mises or the Tresca criterion as follows:

$$K_1 = 2/\sqrt{3}, \quad K_2 = \sqrt{3}/2; \quad \text{von Mises}$$

$$K_1 = 1, \quad K_2 = 3/4; \quad \text{Tresca}$$

Thus, from eq. (9), the following relations can be obtained.

$$t_r = K_1^{\alpha}\frac{(\alpha/2)^{\alpha}}{b}\left(\frac{1}{p}\right)^{\alpha} I_{\infty}$$

$$I_{\infty} = \int_0^{\infty}\left[\left\{\frac{(r_0/r_i)^2\exp 2K_2\varepsilon_0^*}{1+(r_0/r_i)^2(\exp 2K_2\varepsilon_0^* - 1)}\right\}^{1/\alpha} - 1\right]^{\alpha} d\varepsilon_0^* = \lim_{\varepsilon_0^*\to\infty} I. \tag{10}$$

This expresses the time to rupture of the cylinder. The same relations were offered by RIMROTT [12].

From eqs. (7), (8) and eqs. (9), (10), one can get analytical creep curves for simple tension and for internal pressure derived on the basis of the large strain theory. Fig. 11 indicates an example. The solid lines show the creep curves for thin-walled ($D/d = 1.156$) and thick-walled ($D/d = 1.961$) tubular specimens, and the dotted line for simple tension. In the same figure the experimental results on a $2\frac{1}{4}$ Cr—1 Mo steel at 550°C [17] are shown together for comparison. The analytical creep

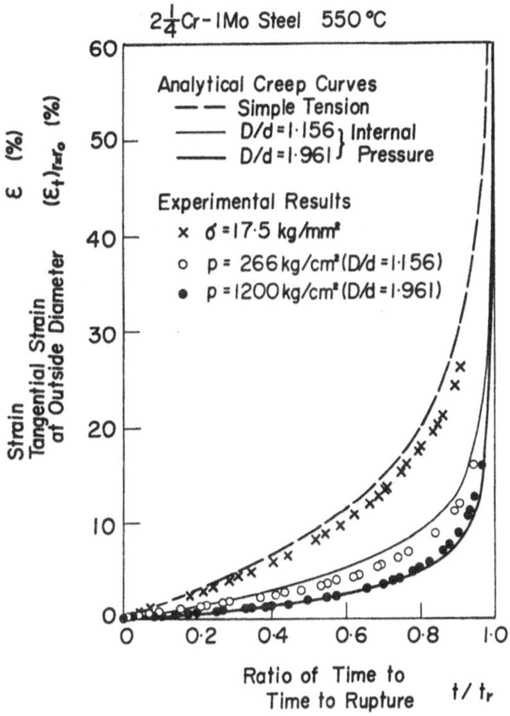

Fig. 11. Comparison of the analytical creep curves derived on the basis of the large strain theory with the experimental results in simple tension and internal pressure creeps

curves show a good agreement with the test results. Moreover, in this case, the rupture elongation in simple tension is more than 30% and those for tubular specimens under internal pressure are more than 10%, and thus, the magnitudes of t/t_r are nearly equal to 1. These lead to the conclusion that the large strain theory is applicable for the preciction of the rupture life.

When the time to rupture of a tube under an internal pressure of p is equal to that of a bar in simple tension under an initial tensile stress of σ_0, the relation between σ_0 and p is obtained from eqs. (8) and (10) as follows:

$$\sigma_0 = p\left(\frac{2}{K_1}\right)\alpha^{-(1+1/\alpha)}\,I_\infty^{-1/\alpha}. \tag{11}$$

On the other hand, the designe formulas of pressure vessels and tubes can be expressed in the form

$$\sigma = p\left[\frac{1}{2}\frac{D}{t} - y\right], \tag{12}$$

where D and t are the outer diameter and the wall thickness, and y is a constant inherent in the formula. The values of y in some typical formulas are as follows [18, 19]:

Outside diameter formula:	$y = 0$
Modified Lamé formula:	$y = 0.4$
Mean diameter formula:	$y = 0.5$
Creep common formula:	$y = 0.7$
Thin-walled formula:	$y = 1.0$

Comparison of eq. (11) with the design formulas leads to

$$\frac{y}{D/t} = \frac{1}{2} - \left(\frac{2}{K_1}\right)\alpha^{-(1+1/\alpha)}\,\frac{I_\infty^{-1/\alpha}}{D/t}. \tag{13}$$

Relations between D/d or D/t and $y/(D/t)$ obtained by the equation are shown in Fig. 12, being indicated by finite strains of $\varepsilon \to \infty$ and $\varepsilon_{to} \to \to \infty$. Similar relations derived from the infinitesimal strain theory are shown also in Fig. 12, represented by $\varepsilon = 0\%$ and $\varepsilon_{to} = 0\%$. The von Mises and the Tresca criteria for the large strain theory give smaller values of y than those for the infinitesimal strain theory. That is, the former is on the safe side as compared with the latter.

We now assume that the rupture elongation in simple tension is 10% and that in internal pressure creep is 5%. From eqs. (7) and (9), the same relations as eq. (13) can be obtained for these rupture elongations. Calculated results on the stress criteria for rupture in this case are also shown in Fig. 12. These curves show a fairly good agreement with the curves conducted from the large strain theory. This leads to the conclusion that it is sufficiently adequate to adopt the large strain theory, if the rupture elongation in simple tension creep is larger, from twice or three times, than that in internal pressure creep.

Furthermore, in such a case that a sufficiently small elongation is obtained at rupture in internal pressure creep in contrast to a large elongation in simple tension creep, the stress criterion is on the more critical side. An example for the case of $\varepsilon = 30\%$ and $\varepsilon_{to} = 5\%$ is indicated also in Fig. 12. It is seen that the rupture life of tubes is mainly concerned with the ductility of the material.

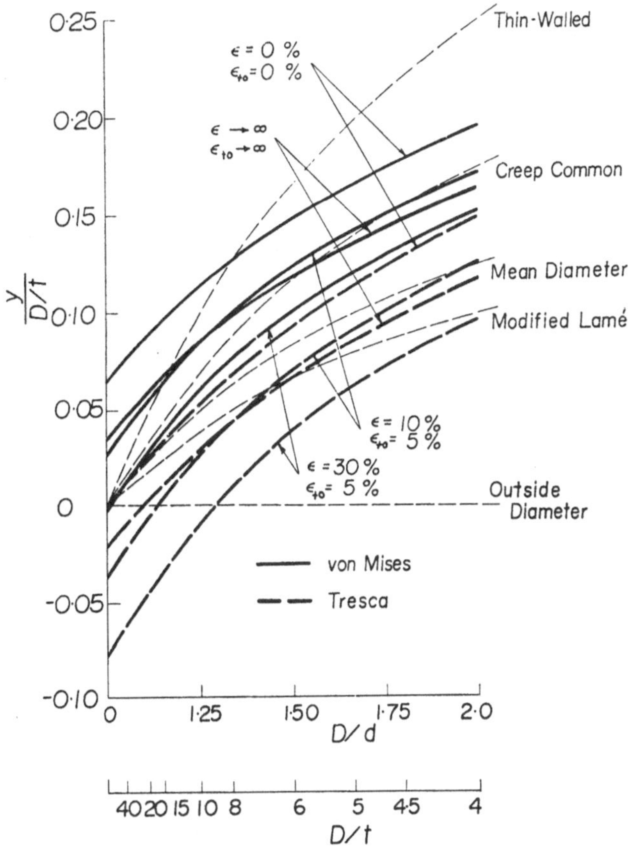

Fig. 12. Relationship between design formulas of pressure vessels and the von Mises and Tresca criteria in infinitesimal strain, finite strain and large strain theories

6. Comparison with the Experimental Results on Creep Rupture of Tubes

In Fig. 13, most of the existing data on creep rupture of tubes made of carbon steels, Cr–Mo steels and 18-8 austenitic stainless steels reported in the literatures [20—28] are plotted together with the design formulas. A number of date lie along the line indicating the mean diameter formula [28, 29], but they are scattered between creep common and modified Lamé formula.

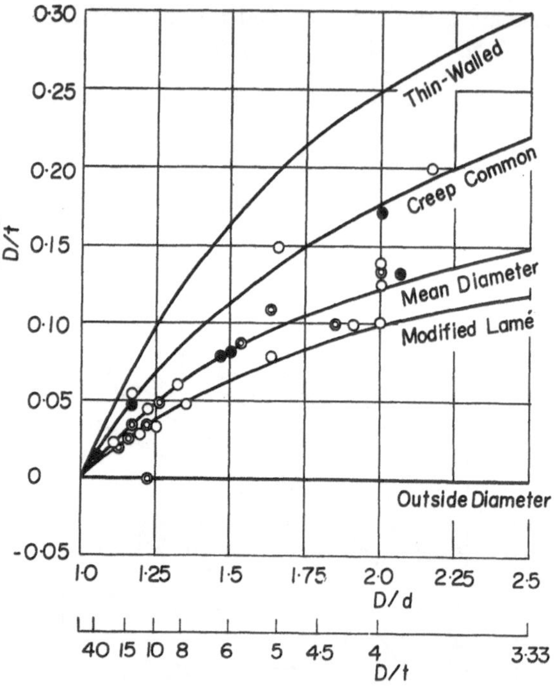

Fig. 13. Data on creep rupture of tubes reported in literatures

	Material	T °C	D/d	
○	Carbon steel	500	1.11, 1.22, 1.32	USSR [28]
	0.15% C	500	1.17, 1.67, 2.17	OHNAMI [23]
	0.23% C	454, 510	1.19, 1.35, 1.63, 1.91	KOOISTRA [22]
	Carbon	566	1.25, 2.0	VOORHEES [20]
	0.19% C	450, 500	2.0	TAIRA [27]
◎	$2^1/_4$ Cr—1 Mo	600	1.17, 1.53	SHINODA [26]
	$2^1/_4$ Cr—1 Mo	621	1.22	B & W [28]
	$2^1/_4$ Cr—1 Mo	593	1.26, 1.85	KOOISTRA [22]
	$2^1/_4$ Cr—1 Mo	566, 593	1.63	E. R. A. [28]
	$2^1/_4$ Cr—1 Mo	550	2.0	TAIRA [27]
	8 Cr—3 Mo—Ti	649	1.12, 1.16, 1.22	US Co. [28]
●	18 Cr—12 Ni—Mo	738, 816, 900	1.05	ROWE [25]
	18 Cr— 8 Ni—Mo	650	1.17	IKEJIMA [24]
	18 Cr—12 Ni—Nb	677	1.48	E. R. A. [28]
	18 Cr—10 Ni—Ti	650	1.5	USSR [28]
	18 Cr—12 Ni—Mo	649	2.0	DAVIS [21]
	18 Cr—12 Ni—Ti	649	2.06	KOOISTRA [22]

As was mentioned in the previous section, the von Mises criterion in the large strain theory is nearly equal to the creep common formula in the stress value. On the other hand, most of the experimental results shown in Fig. 13 give shorter rupture lives as compared with the rupture life predicted by the large strain theory. This will be attributed to the following factors:

1. In the case that the maximum principal stress criterion is applicable for the prediction of rupture life, it may not be helpful to adopt the large strain theory combined with the von Mises criterion. As the maximum principal stress is nearly equal to or slightly smaller than the Tresca effective stress, it gives the stress value between creep common and mean diameter formula.

2. In most of the cases that the maximum principal stress criterion is superior, a sufficiently small elongation is obtained at rupture in internal pressure creep in contrast to a large elongation in simple tension creep. Such phenomena were observed also in the creep on the 18-8 stainless steel used in our experiment. For example, if the rupture elongation at outside diameter of the tube is equal to 1/5 or 1/6 times of the rupture elongation in the bar specimen under simple tension, the maximum principal stress becomes nearly equal to the mean diameter or the modified Lamé formula.

The conceivable reasons why there is a fairly large discrepancies in the rupture elongations between the two cases are as follows [17]:

a) The rate of decrease in the effective cross-sectional area caused by the crack growth is larger in the tubular specimen under internal pressure than under simple tension.

b) The large stress concentration rises at the tip of the crack propagating in axial direction for internal pressure creep. Hence the strain concentration yields considerably large local elongations and the mean elongation over the diametrical gage length is small.

c) Tubular specimens under internal pressure have less ductility than solid bar specimens under simple tension. That is, the size effect of the specimen affects the rupture elongation.

d) The hydrostatic component of the stress is larger in the tubular specimen under internal pressure than under simple tension.

3. Many ordinary polystalline metallic materials with which the engineers must deal have approximately isotropic behaviour. This is due to the fact that strains are measured over volumes including a great many crystal grains, which are often random in orientation. However, anisotropy might be a property inherent in the material in its virgin state, or it could be developed during creep deformation from a virgin state of isotropy. If the anisotropy makes the creep resistance in tangen-

tial direction of the tube weaken, the rupture strength decreases, and the rupture stress will be smaller than that in the large strain theory.

Taking these factors as effective, we will be able to explain the justification of the mean diameter formula or the modified Lamé formula.

7. Conclusion

An analytical investigation was conducted on the creep instability of a thin-walled cylinder subjected to combined tension and internal pressure and thin- and thick-walled cylinders under internal pressure. The rupture life estimated by the large strain theory was compared with the test results and justified for a low carbon steel and a Cr–Mo alloy steel with large rupture elongations. In such a ductile material in which transgranular creep deformation is dominant and the time to crack initiation is nearly equal to the time to rupture, the large strain theory combined with the von Mises criterion is found to be helpful for the prediction of the rupture life.

On the other hand, in the material with little ductility such as an 18-8 austenitic stainless steel used in this study, the change of true stress due to creep is not so remarkable, and it is inadequate to apply the large strain theory. In such a case, the maximum tensile stress will be influencial on the rupture life.

With these in mind, an attempt was made to apply the large strain theory for the creep rupture of pressurized tubes. In the experiments, both simple tension bar specimens and pressurized tubular specimens fractured when the creep deformations reached some finite values, and some discrepancies were recognized in the rupture elongations between the two specimens. However, it was found that the large strain theory is applicable when the rupture elongations in simple tension and under internal pressure are larger than 30% and 10%, respectively. The magnitude of the von Mises effective stress in the large strain theory is nearly equal to that calculated by the creep common formula, which is unable to account for most of the existing data supporting the mean diameter formula. The further decrease in the rupture life of tubes is closely related to the reduction of the effective wall thickness due to gradually progressing cracks, the maximum tensile stress governing the fracture and the effect of anisotropy of the material.

References

[1] FINNIE, I., and W. R. HELLER: Creep of Engineering Materials. New York: McGraw-Hill. 1959.
[2] JOHNSON, A. E.: Metallurgical Rev. **5**, 447 (1960).

[3] Odqvist, F. K. G., and J. Hult: Creep Strength of Metallic Materials. Berlin-Göttingen-Heidelberg: Springer. 1961. [In German.]
[4] Odqvist, F. K. G.: Mathematical Theory of Creep and Creep Rupture. London: Oxford University Press. 1966.
[5] Siegfried, W.: J. Appl. Mech. **10**, 202 (1943).
[6] Johnson, A. E., and N. E. Frost: Engineer. **191**, 434 (1951).
[7] Johnson, A. E., J. Henderson, and V. D. Mathur: ibid. **202**, 261, 299 (1956).
[8] Hoff, N. J.: J. Appl. Mech. **20**, No. 1, 105 (1953).
[9] Taira, S., R. Ohtani, and A. Ishisaka: Proc. 11th Japan Congr. on Material Research, Soc. Mat. Sci., Kyoto, 76 (1968).
[10] Taira, S., R. Koterazawa, and R. Ohtani: Proc. 8th Japan Congr. on Testing Materials, Soc. Mat. Sci., Kyoto, 53 (1965).
[11] Rimrott, F. P. J.: Trans. ASME, Ser. E, **81**, 271 (1959).
[12] Rimrott, F. P. J., F. J. Mills, and J. Marin: ibid. **82**, 303 (1960).
[13] Rimrott, F. P. J., and J. R. Luke: ZAMM **41**, 485 (1961).
[14] Rimrott, F. P. J.: Ing.-Arch. **27**, 169 (1959).
[15] Lode, W.: Forsch. Gebiete Ingenieurw. **303** (1928).
[16] Lode, W.: Z. Physik **36**, 913 (1926).
[17] Taira, S., and R. Ohtani: Bulletin of JSME **11**, No. 46, 593 (1968).
[18] Buxton, W. J., and W. R. Burrows: Trans. ASME **73**, 575 (1951).
[19] Burrows, W. R., R. Michel, and A. W. Rankin: ibid. **76**, 427 (1954).
[20] Voorhees, H. R., C. M. Sliepcevich, and J. W. Freeman: Ind. Engng. Chem. **48**, 872 (1956).
[21] Davis, E. A.: Trans. ASME, Ser. D, **82**, No. 2, 453 (1960).
[22] Tucker, T. J., Jr., E. E. Coulter, and L. F. Kooistra: ibid., 465.
[23] Ohnami, M., and Y. Awaya: Proc. 6th Japan Congr. on Testing Materials, 61 (1962).
[24] Ikejima, T., et al.: J. Japan Soc. Mech. Test. **11**, No. 102, 165 (1962).
[25] Rowe, G. H., J. R. Stewart, and K. N. Burgess: Trans. ASME, Ser. D, **85**, No. 1, 71 (1963).
[26] Shinoda, N., et al.: J. Japan Soc. Mech. Test. **14**, No. 137, 78 (1965).
[27] Taira, S., and R. Ohtani: J. Japan Soc. Mech. Engr. **70**, No. 587 1737 (1967).
[28] Carlson, W. B., and D. Duval: Engineering **193**, 829 (1962).
[29] Chitty, A., and D. Duval: Joint Int. Conf. on Creep, New York, No. 4-1 (1963).

Thermomechanical Coupling in Viscoelastic Solids

By

T. R. Tauchert

Princeton, N.J. (U.S.A.)

Summary

Experimental data on the heat generated in polymeric specimens undergoing oscillatory deformations are presented. Various factors which affect the hysteresis, including strain amplitude, frequency and mode of loading, and ambient temperature are investigated. It is observed that not all of the energy added to a specimen during a cycle of deformation is dissipated as heat. A comparison of theory and experiment demonstrates that the remaining (or "stored") energy must be accounted for in predicting the thermomechanical behavior of such materials.

I. Introduction

When an inelastic solid is subjected to mechanical deformation, part of the work done by the external forces is dissipated as heat. If the deformation is oscillatory and a large number of cycles are involved, a significant rise in temperature may occur. A temperature increase will in turn affect the mechanical properties of the material. In a previous study [1] on dissipative heating in polyethylene rods it was observed that small cyclic shear strains produced a temperature rise which softened the material to such an extent that the stresses were reduced to 25% of their initial values.

In addition to the response of various inelastic materials to sustained oscillations, the detail of the response during a single cycle of deformation has been investigated. It is observed that not all of the mechanical work done per cycle is transformed into heat. The difference between the work done and the heat increase, or the "stored energy", varies considerably from one material to another. A summary of previous results in this regard is presented in Table 1. The same experimental technique and apparatus were used in each investigation [1—6]. It is

Table 1. *Conversion of Mechanical to Thermal Energy during Deformation of Inelastic Solids*

Material	% of Mechanical Energy Dissipated as Heat During Initial Cycles of Deformation	Source of Information
Aluminum, commercially pure (1100)	100 (approximately)	[2]
Lead, tellurium (99.95% pure)	75—100 (average of 96%)	[3]
Copper, OFHC	61—94 (average of 79%)	[4]
Polymethylmethacrylate	59—69 (average of 60%)	[5]
Phenolic resin, coarse canvas filler	73	[6]
Phenolic resin, saturated paper filler	56	[6]
Polyethylene, low-density, injection molded and machined	45—56	[1]
Polyethylene, low-density, commercially extruded		
Tensile loading	43—52 (average 45%)	Presented here
Shear loading	45—55 (average 47%)	

noted that the percentage of mechanical energy dissipated as heat is generally larger in metals than in polymers; or conversely, the percentage of stored energy is larger in polymers than in metals. Furthermore it is seen that a material with a multiphase molecular configuration (e.g., polyethylene) stores more energy than an amorphous material (e.g., polymethylmethacrylate) which is not susceptible to changes in crystallinity.

The internal heat generation problem is studied further in this paper. Experimental data for commercially extruded, low-density polyethylene tubes undergoing longitudinal and torsional oscillations are presented. The results point out some of the shortcomings of present continuum theories in predicting thermomechanical response.

II. Experimental Technique

The test apparatus used for longitudinal oscillatory loading is shown schematically in Fig. 1. The motor speed and the stroke of the loading arm can be varied to provide deformations of various frequencies (0 — 300 cpm) and amplitudes (0—0.25″). A load-cell is calibrated to measure the axial force exerted on the specimen, and a variable differential transformer is used as the displacement pick-up. Assuming that the oscillations are slow enough that inertia effects may be neglected and that the material obeys a linear stress-strain-time law, the average stress and strain in the specimen may then be determined using standard strength-of-materials formulae. The temperature at the midpoint of the specimen is monitored by an iron-constantan thermocouple (No. 30 wire) attached to the specimen with electrical tape. The lateral surfaces of the

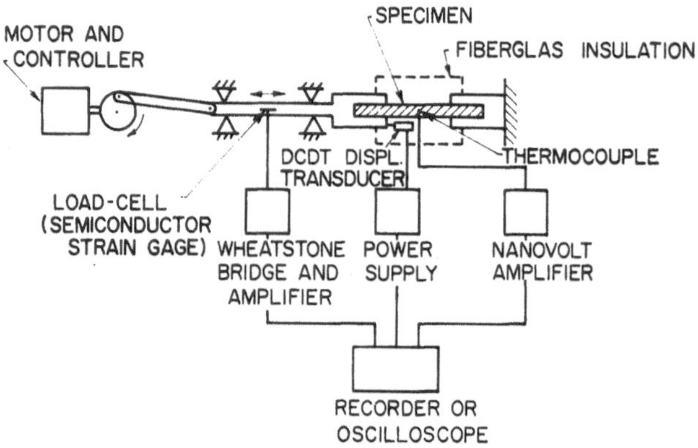

Fig. 1. Longitudinal loading apparatus

specimen are insulated with fiberglass, and a heating chamber is placed over the entire apparatus for testing at elevated ambient temperatures. The stress (force), strain (displacement), and temperature histories are recorded on an oscilloscope.

A similar apparatus was used for torsional loading. The details of this apparatus and of the temperature measuring technique have been described previously [1, 7].

III. Experimental Results

Typical stress, strain and temperature records corresponding to the initial cycles of torsional deformation of a polyethylene tube (3″ long, 0.5″ O. D., 0.375″ I. D.) are shown in Fig. 2. The strain (narrow trace) is seen to lag the stress (wide trace). The temperature change is seen to be greatest during those portions of a cycle when the strain *rate* is maximum. When the strain rate is zero there is no heat generation and the temperature remains nearly constant. The shape of the temperature curve is the same during each half-cycle, i.e., the temperature generation is independent of the direction of the applied loading.

For comparison purposes, strain and temperature histories for a tube subjected to longitudinal deformation are shown in Fig. 3. In the upper record the loading was initially tensile (+), and in the lower record it was initially compressive (−). During tensile loading the temperature decreases and during compression it increases (compare, for example, the

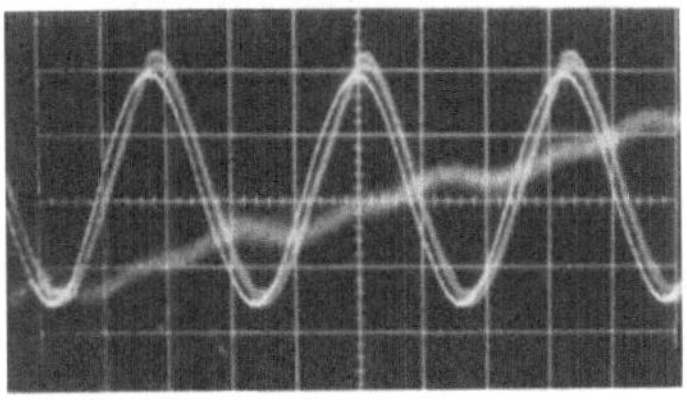

Fig. 2.
Typical stress, strain and temperature histories during torsional oscillations

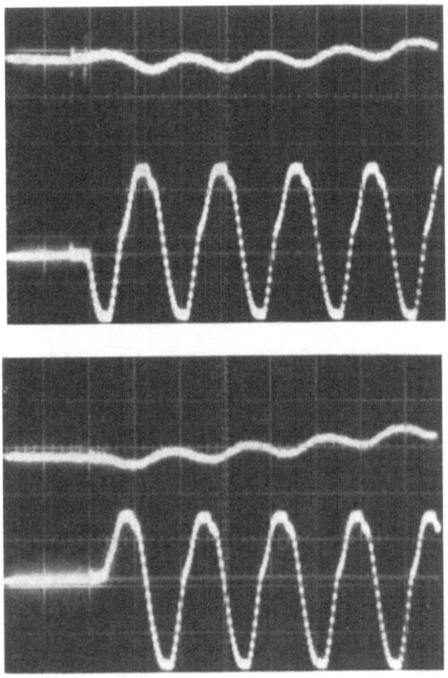

Fig. 3.
Typical strain and temperature histories during longitudinal oscillations

temperature change during the first half cycle in each record). In addition to these temperature fluctuations which result from volume expansion and contraction, there is a gradual rise in temperature associated with the dissipative heating.

Correlation of Mechanical and Thermal Energies

The hysteresis loops that are developed during the oscillations can be displayed directly on the oscilloscope and recorded photographically. If one scale (horizontal) measures the average strain and the other scale (vertical) measures the average stress, the area inside the loop is a measure of the mechanical energy per unit volume U^* expended in twisting the rod through one cycle. If we assume that the process is adiabatic (which is nearly the case for the initial cycles), the temperature rise per cycle $\Delta\theta$ is converted into heat increase per cycle Q by

$$Q = \varrho c \, \Delta\theta$$

where ϱc is the heat capacity of the material[1]. We now compare the mechanical and thermal energies in the polyethylene tubes for various ambient temperatures, amplitudes and frequencies of oscillation, and for both longitudinal and torsional loading conditions. Nearly one hundred specimens were tested, but only a few typical results are reported.

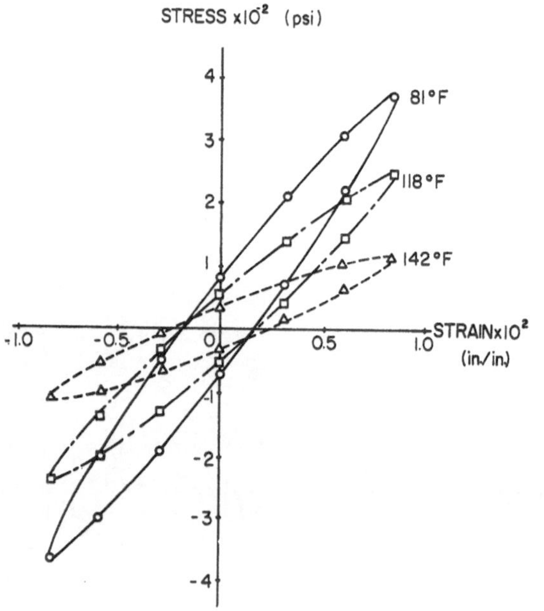

Fig. 4. Hysteresis loops during longitudinal loading at various ambient temperatures (Strain amplitude ε_0 = 0.008 in./in., frequency ω = 100 cpm)

Typical hysteresis loops corresponding to longitudinal loadings at three different ambient temperatures are shown in Fig. 4. The work done and the heat generated per cycle for each case are tabulated in Table 2. An increase in the ambient temperature evidently softens the material so that less energy is required to deform the specimen. The heat generated is correspondingly less, and the ratio of the thermal to the mechanical energy is not significantly affected.

The effect of the strain amplitude on the work done and the heat generated for two specimens, one loaded axially and the other loaded in torsion, is shown in Table 3. In both cases an increase in the strain amplitude is seen to result in an increase in both the mechanical and

[1] The values of the density ϱ and the specific heat c for the polyethylene used here were supplied by the manufacturer.

Table 2. *Effect of Ambient Temperature on the Mechanical Work Done and the Heat Generated in a Polyethylene Tube*

(Strain amplitude $\varepsilon_0 = \frac{\Delta L}{L} = 0.008$ in./in., frequency $\omega = 100$ cpm)

	Temperature (°F)	Mechanical work per cycle, U^* (in. lb/in.^{3})	Heat generated per cycle, Q (in. lb/in.^{3})	Percent of work that appears as heat $[(Q/U^*) \times 100]$
Specimen No. L—S 25				
Longitudinal loading	81	3.10	1.35	43
	118	2.43	1.05	43
	142	2.03	0.97	48

Table 3. *Effect of Strain Amplitude on the Mechanical Work Done and the Heat Generated in a Polyethylene Tube*

(Frequency $\omega = 100$ cpm)

	Strain amplitude, $\varepsilon_0 = \frac{\Delta L}{L}$ (longitudinal) or $\gamma_0 = \frac{\Phi R}{L}$ (torsional) (in./in.)	Mechanical work per cycle, U^* (in. lb/in.^{3})	Heat generated per cycle, Q (in. lb/in.^{3})	Percent of work that appears as heat $[(Q/U^*) \times 100]$
Specimen No. L—S1				
Longitudinal loading	.007	2.60	1.26	48
	.010	3.63	1.60	44
Specimen No. T—T6				
Torsional loading	.015	3.58	1.63	46
	.020	5.40	2.52	48
	.030	11.6	5.42	47

Table 4. *Effect of Frequency on the Mechanical Work Done and the Heat Generated in a Polyethylene Tube*

	Frequency, ω (cpm)	Mechanical work per cycle, (in. lb/in.^{3})	Heat generated per cycle, Q (in. lb/in.^{3})	Percent of work that appears as heat $[(Q/U^*)\times 100]$
Specimen No. L—S4				
Longitudinal loading				
$\left(\varepsilon_0 = \frac{\Delta L}{L} = 0.008 \text{ in./in.}\right)$	100	3.02	1.35	45
	150	3.03	1.39	46
	200	3.10	1.41	46
Specimen No. T—T5				
Torsional loading				
$\left(\gamma_0 = \frac{\Phi R}{L} = 0.02 \text{ in./in.}\right)$	10	4.34	1.95	45
	100	5.20	2.64	51
	200	5.25	2.60	50

thermal energy, but has little effect on the ratio between the two. The ratio is approximately the same for each mode of loading.

Table 4 presents results for specimens which were oscillated at various frequencies ω. Although the heat generation and the work increase, in general, with an increase in frequency, the percentage of work which appears as heat is not significantly affected.

IV. Conclusions

The data cited above indicate that only approximately 50 percent of the work done during the deformation of a polyethylene specimen is dissipated as heat. Although the magnitude of both the work and the heat is affected by such factors as the ambient temperature, strain amplitude and frequency of oscillation, the ratio of the two is apparantly not affected. It is possible that the heat increase is actually slightly higher than the values calculated, since heat losses have not been accounted for. However all observations were made within the period of time when conduction affects were minimal, i.e., during the first ten to twenty cycles; no measurable changes in the hysteresis were observed during this interval of time. Furthermore, a near perfect balance of mechanical and thermal energy was obtained for certain metallic specimens (Table 1) using the same experimental technique as used here. If heat losses were found to be negligible in the case of metals, they should be even less important for polymers, which have a much lower coefficient of thermal conductivity.

The fact that not all the mechanical work done on an inelastic material is transformed into heat suggests that changes in the molecular configuration of the material occur. This point was discussed previously [1, 5]. These changes, which in turn affect the heat generation, obviously can not be predicted strictly by continuum theories. The concept that some energy is stored must, however, be accounted for if one wishes to predict temperature changes resulting from mechanical deformation of such materials. A simple correction to the theory of coupled thermoviscoelasticity is examined in the Appendix. The "corrected" theoretical solution is compared with experimental results for a polyethylene specimen.

Appendix

A Comparison of Theoretical Solutions with Experimental Data

A theoretical solution was presented earlier [8] for the temperature distribution in a linear viscoelastic thin-walled tube subject to torsional oscillations. The relaxation modulus for the material was assumed to be of the form

$$G_1(t) = \sum_{i=1}^{N} G_{1i} e^{-t/\tau_{1i}} \tag{A.1}$$

where G_{1i} and τ_{1i} are material constants. The shear strain e_{23} was defined by the expression

$$e_{23}^2 = \left(\frac{\Phi R}{2L}\right)^2 \sin^2 \omega t \sum_{n=1}^{\infty} D_n \sin \frac{n\pi x_3}{L} \tag{A.2}$$

Here R, L, Φ denote the mean radius and length of the tube and the angle-of twist, respectively. A Fourier sine series having Fourier coefficients D_n was taken in order to permit specification of various strain distributions in the axial (x_3) direction. In this case, the coupled heat equation, which governs the temperature field θ, has the form

$$k \nabla^2 \theta = \varrho c \frac{\partial \theta}{\partial t} - D \tag{A.3}$$

where k denotes the thermal conductivity of the material, and the rate of energy dissipation D is given by

$$D = 2 \left(\frac{\Phi R}{2L}\right)^2 \sum_{i=1}^{N} (a_i \cos \omega t + b_i \sin \omega t)^2 \sum_{n=1}^{\infty} D_n \sin \frac{n\pi x_3}{L} \tag{A.4}$$

where

$$\begin{aligned} a_i &= \frac{\sqrt{\dfrac{G_{1i}}{\tau_{1i}}}\, \omega \tau_{1i}}{1 + (\omega \tau_{1i})^2} \\ b_i &= \frac{\sqrt{\dfrac{G_{1i}}{\tau_{1i}}}\, (\omega \tau_{1i})^2}{1 + (\omega \tau_{1i})^2}. \end{aligned} \tag{A.5}$$

The temperature rise in the tube for an adiabatic process involving uniform strain (which holds for small time) was found to be

$$\begin{aligned} \theta = \left(\frac{\Phi R}{2L}\right)^2 \frac{1}{\varrho c \omega} \sum_{n=1}^{\infty} \sum_{i=1}^{N} D_n \Big\{ a_i^2 \left[\omega t + \frac{1}{2} \sin 2\omega t\right] + \\ + 2 a_i b_i [\sin^2 \omega t] + b_i^2 \left[\omega t - \frac{1}{2} \sin 2\omega t\right]\Big\} \sin \frac{n\pi x_3}{L}. \end{aligned} \tag{A.6}$$

In order to compare (A.6) with the measured temperature response, the following values of the material constants in (A.1) are assumed:

$$\begin{aligned} G_{11} &= 27{,}100 \text{ psi} \\ G_{12} &= 54{,}800 \text{ psi} \\ \tau_{12} &= 0.0155 \text{ sec.} \\ G_{1i} &= \tau_{1i} = 0 \text{ otherwise}. \end{aligned} \tag{A.7}$$

These values were obtained by curve-fitting the complex shear modulus[1] $G^*(\omega) = G'(\omega) + iG''(\omega)$ for polyethylene. The components $G'(\omega)$ and $G''(\omega)$ corresponding to the assumed three-parameter model (A.7) are compared with the experimentally determined moduli in Fig. 5. It is seen that the model adequately characterizes the actual material's behavior over a very limited frequency range. A larger number of parameters could be used in order to provide a better curve-fit; however the present model suffices for this example.

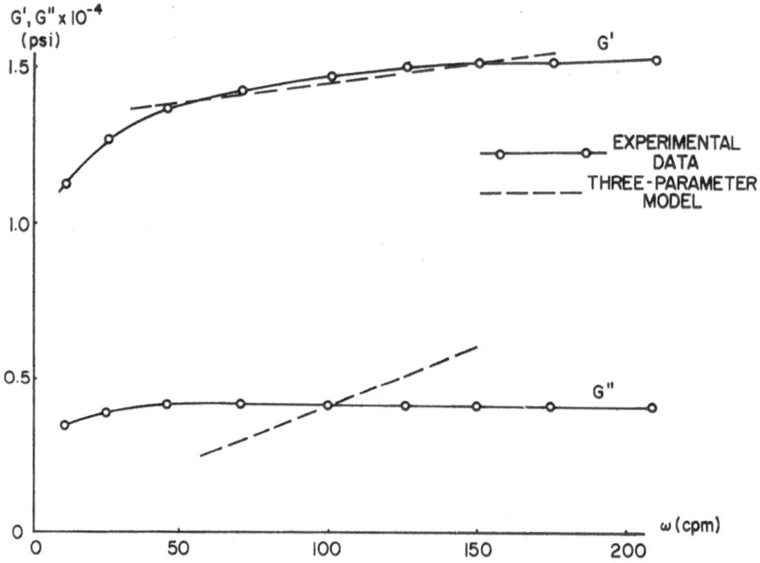

Fig. 5. Real (G') and imaginary (G'') components of the complex shear modulus for polyethylene

The temperature history predicted by (A.6) and (A.7) is compared with experimental data for a typical test ($\omega = 100$ cpm, $\Phi R/2L = 0.01$ in./in.) in Fig. 6. The agreement is seen to be relatively poor as might be expected in view of the fact that the theoretical solution ignores the phenomenon of stored energy. To account for the condition that only approximately 50% of the residual energy in a polyethylene specimen is transformed into heat, a "corrected" solution was calculated based upon a reduced rate of heat generation D^*; i.e., the rate of energy dissipation D [eq. (A.4)] was reduced by 50%. The corrected solution

[1] The mathematical relationship between the complex modulus $G^*(\omega)$ and the relaxation function $G_1(t)$ [eq. (A.1)] may be found in FERRY [9], p. 54. The real and imaginary components of $G^*(\omega)$ were obtained for polyethylene from the measured stress amplitude, strain amplitude and phase angle as described on p. 12 of [9].

is also plotted in Fig. 6 and it is seen to agree extremely well with the experimental curve. Naturally the agreement becomes less good for large time since the effects of heat conduction and temperature dependence of the material properties have been neglected in the theoretical solutions.

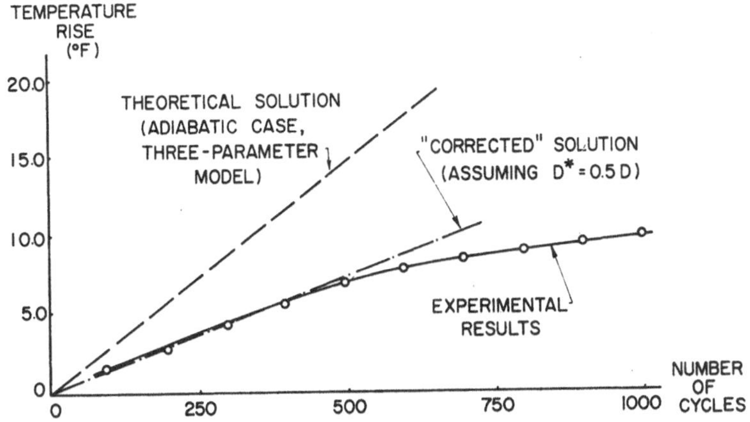

Fig. 6. A comparison of theoretical and experimental temperature histories during torsional oscillations

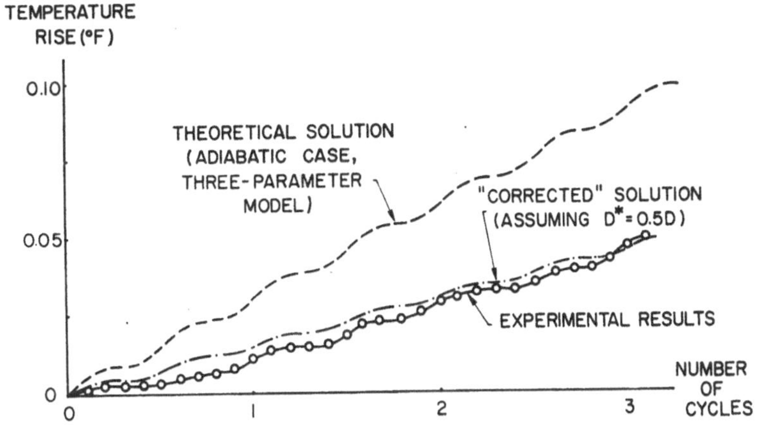

Fig. 7. A comparison of theoretical and experimental short-time temperature histories

The details of the predicted and measured temperature response during the first few cycles of oscillation are shown in Fig. 7. Again good agreement between the corrected solution and the experimental data is observed.

In conclusion it must be admitted that this rather simple correction is not *the final word.* A full explanation of the heat generation process requires an understanding of the energy-absorbing mechanisms which take place within the material. The example does illustrate, however, that the stored energy phenomenon can play an important role and must be accounted for in predicting the thermomechanical behavior of inelastic solids.

References

[1] Tauchert, T. R.: The Temperature Generated During Torsional Oscillations of Polyethylene Rods. Int. J. Engng. Sci. **5**, 353—365 (1967).
[2] Dillon, O. W., Jr.: Coupled Thermoplasticity. J. Mech. Phys. Solids **11**, 21—33 (1963).
[3] Watson, J. P.: A Demonstration of Thermoviscoelastic Phenomena in the Torsional Loading of Tellurium Lead Tubing. M. S. E. Thesis. Princeton University 1967.
[4] Dillon, O. W., Jr.: The Heat Generated During the Torsional Oscillations of Copper Tubes. Int. J. Solids Struct. **2**, 181—204 (1966).
[5] Tauchert, T. R., and S. M. Afzal: Heat Generated During Torsional Oscillations of Polymethylmethacrylate Tubes. J. Appl. Phys. **38**, 4568—4572 (1967).
[6] Afzal, S. M.: The Heat Generated During the Torsional Oscillations of Polymethylmethacrylate Tubes and Rods. M. S. E. Thesis. Princeton University 1966.
[7] Dillon, O. W., Jr., and T. R. Tauchert: The Experimental Technique for Observing the Temperature Due to the Coupled Thermoelastic Effect. Int. J. Solids Struct. **2**, 385—391 (1966).
[8] Tauchert, T. R.: Transient Temperature Distributions in Viscoelastic Solids Subject to Cyclic Deformations. Acta Mechanica **6**, 239—252 (1968).
[9] Ferry, J. D.: Viscoelastic Properties of Polymers. New York: John Wiley & Sons. 1961.

Influence de la température sur les réactions viscoélastiques de vulcanisats de caoutchouc susceptibles de cristalliser au cours d'essais de traction monoaxiale

Par

P. Thirion et **R. Chasset**

Paris (France)

Summary

Though the basically entropic phenomena mainly conditioning the elasticity of rubber vulcanisates are now well understood, several other processes usually intervene also in their mechanical behaviour, the exact analysis of which is therefore still a controversial matter. In particular if the structure of the molecular chains presents a sufficient regularity, as for most usual elastomers, a strain induced crystallisation may combine with the delayed elastic effects to make the achievement of an equilibrium state of the deformed materiel extremely difficult to realize experimentally. In practice, rather large hysteresis loops are always observed during cyclic tensile tests at room temperature, whatever the range of elongation considered.

By means of a specially designed recording dynamometer, it has been found that a sufficient raising of the temperature gradually reduces both crystallisation and the viscoelastic component of the stress, so that hysteresis is practically absent from successive stress-strain curves, at a temperature of about 120 °C, even when the maximum elongation approaches the conditions of rupture. This technique provides a quick and precise mean of checking, in almost completely reversible conditions, the phenomenological or molecular theories of rubber elasticity, and of deriving unambiguously the corresponding mechanical and structural parameters.

A condition que la température excède notablement celle de leur transition vitreuse, les vulcanisats de caoutchouc semblent revenir presque instantanément à leurs dimensions initiales, même lorsqu'on les libère après leur avoir fait subir des déformations considérables.

Toutefois, ce type d'élasticité apparemment réversible est en réalité généralement associé à des effets secondaires d'hystérésis [1], qui se sont montrés jusqu'ici très gênants pour l'analyse quantitative du phénomène, car les moyens assez empiriques utilisés pour tenter de les éliminer laissaient beaucoup à désirer.

C'est pourquoi, on s'est proposé dans cette étude de la thermoélasticité des vulcanisats du type „pure-gomme“ de chercher si l'élévation de la température d'essai pourrait, grâce à la réduction des frottements intermoléculaires et de la cristallisation, rendre davantage caractéristiques de l'état d'équilibre les données fournies par un essai de traction monoaxiale classique.

I. Défauts des méthodes courantes de réduction de l'hystérésis

La méthode permettant d'approcher le plus près possible d'états d'équilibre véritablement réversibles est sans doute celle de Gee, qui consiste à gonfler une éprouvette étirée au contact des vapeurs d'un solvant, qu'on élimine ensuite par évaporation sous vide [2]. Cependant, comme il est nécessaire de répéter l'opération pour chaque valeur de l'allongement, un vieillissement notable risque de se produire en cours d'essai, de sorte que cette technique, par ailleurs extrêmement laborieuse, ne s'est pas répandue. Faute de mieux, on se contente donc d'effectuer des conditionnements empiriques, soit en amenant au préalable le caoutchouc à une température et à un allongement supérieurs à ceux de mesure [3—7], soit en augmentant la charge par incréments successifs, les allongements étant alors mesurés à l'issue d'une période de stabilisation arbitrairement choisie, d'environ 5 à 15 minutes [8—11]. Ce procédé, qui se prête mal à l'automatisation, est cependant lui aussi long et délicat à mettre en oeuvre. De plus, une étude de la relaxation viscoélastique des vulcanisats a montré que, sauf pour les caoutchoucs fortement rétifiés à haut module d'élasticité, il entraîne normalement des erreurs importantes, de l'ordre de 10 à 20% au moins. En effet, la relaxation des réseaux macromoléculaires peut s'étaler sur des durées extrêmement longues, particulièrement dans le cas des caoutchoucs peu vulcanisés, pour lesquels le retour à des états d'équilibre, évidemment virtuels en raison du vieillissement, nécessite une dizaine d'années, à la température ambiante, après une perturbation mécanique quelconque [12, 13].

Cependant, au moins jusqu'à des allongements de l'ordre de 100%. les connaissances ainsi recueillies sur la cinétique de relaxation viscoélastique permettent actuellement d'envisager une détermination précise de relations contrainte — déformation à l'équilibre, dans ce domaine de température; il suffirait d'opérer encore par courts paliers de déformation, l'extrapolation à temps infini des contraintes transitoires mesurées pouvant maintenant être effectuée avec une précision de l'ordre du 1/1000e [14]. Mais cette méthode est d'un faible rendement et elle devient en outre inapplicable aux très grands allongements en raison de la rupture prématurée des éprouvettes ou des modifications de la cinétique de relaxation introduites par la cristallisation de polymères présentant, comme les polyènes stéréospécifiques, des structures de chaîne régulières.

Ces considérations nous ont amenés à examiner les possibilités offertes à ce point de vue par un enregistrement continu de la tension au cours d'un allongement ou d'une rétraction isothermes effectués à chaud at à vitesse constante, à l'aide des dispositifs dynamométriques semblables à ceux couramment employés pour le contrôle industriel.

Comme des méthodes optiques très sensibles l'ont récemment confirmé [15], on savait en effet qu'au dessus de 80° à 90°C environ la cristallisation ne devrait plus être appréciable, même pour du caoutchouc naturel fortement étiré [16]. En outre, il avait été établi à l'occasion de l'étude précédente de la relaxation des vulcanisats que, pour ces systèmes, les effets viscoélastiques transitoires peuvent être aussi efficacement accélérés par une élévation de température que par le gonflement des polymères dans un solvant [12]. Ainsi, se dessinait-il un moyen à la fois commode et rapide de déterminer les réactions des vulcanisats à l'équilibre, à condition de préciser au préalable les limites opératoires imposées, soit par l'appareillage, soit par les altérations possibles des réseaux macromoléculaires en cours d'essais à chaud et sous grandes déformations [17].

II. Description du dispositif dynamométrique d'extension

Lorsqu'on sollicite le caoutchouc en compression, ou en cisaillement, il est difficile de réaliser des déformations homogènes importantes dans une grande partie des éprouvettes, en raison des perturbations exercées par les supports, et la précision des mesures s'en trouve fortement affectée.

Au contraire, dans le cas de l'extension, il suffit que la longueur des éprouvettes soit grande par rapport à leurs dimensions transversales pour éviter cet inconvénient, ce qui explique que l'essai de traction soit de loin le plus utilisé pour caractériser les propriétés mécaniques du caoutchouc.

En pratique, deux sortes d'éprouvettes sont concurremment employées, les unes en forme d'anneaux qu'on étire entre deux galets ou pitons, les autres en forme d'haltères dont les têtes sont serrées entre des mâchoires de traction. Dans les deux cas, il est possible d'enregistrer simultanément la tension en fonction de l'allongement; mais les haltères se prêtent moins facilement à cette détermination car il est alors nécessaire de suivre à l'aide de dispositifs optiques ou électrique l'écartement de marques médianes, toujours plus ou moin bien localisées, en raison de la déformabilité de ces matériaux. Par contre, dans la mesure où la déformation dans les sections droites, ainsi que le long des parties épousant les galets, est homogène, l'allongement d'un anneau se trouve en rapport géométrique direct avec la distance séparant les galets, laquelle peut être aisément enregistrée avec une grande précision. C'est pourquoi on a opté ici pour ce type d'éprouvettes, les diamètres intérieur et extérieur

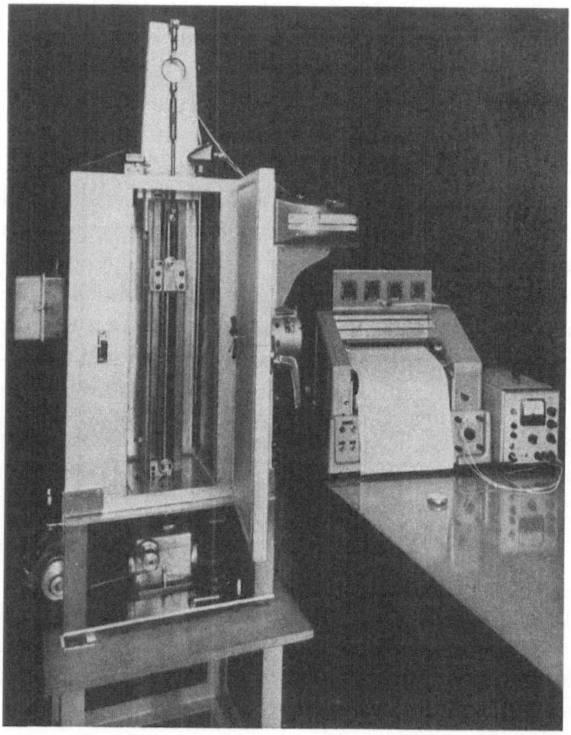

Fig. 1. Dynamomètre enregistreur

des anneaux étant 23 et 27 mm respectivement, l'épaisseur 2 mm et le diamètre des gorges des galets 12 mm.

Le dispositif dynamométrique est lui-même représenté Fig. 1. La vitesse uniforme d'écartement ou de rapprochement des galets est comprise entre 0,12 cm/min. et 72 cm/min. et, comme de coutume, afin d'égaliser les tensions dans chacun des brins de l'éprouvette étirée, l'un des galets est fou sur on axe. De plus, la rotation imposée à l'autre galet entraîne l'anneau à la façon d'une courroie. Toutefois, à la différence d'appareils similaires, où la rotation des galets est commandée par leur écartement, un moteur auxiliaire assure une vitesse angulaire réglable de façon entièrement indépendante de la vitesse d'allongement de l'éprouvette.

Ce détail de conception, qui pourrait paraître superflu à première vue, s'est révélè indispensable à l'usage pour assurer un rapport rigoureusement direct entre l'espacement des galets et l'extension de la fibre neutre de l'anneau, de façon à bénéficier pleinement du principal avantage de

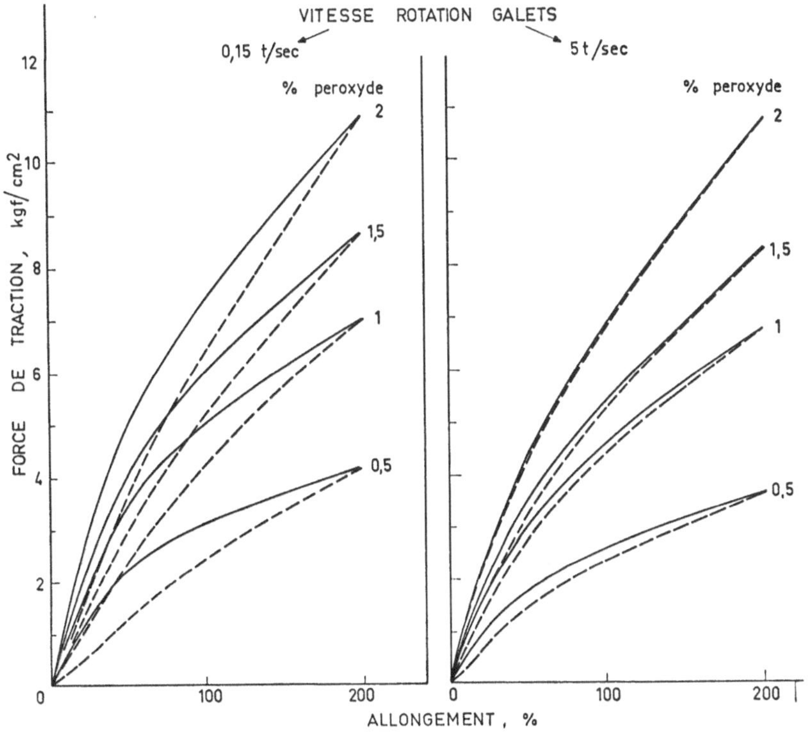

Fig. 2. Cycles d'hystérésis de vulcanisats de cis polyisoprène pour diverses densités de réctification par un peroxyde. Température 120 °C — Vitesse d'étirement 60 cm/min

cette méthode. La Fig. 2 montre en effet à quels résultats aberrants on aboutit en opérant avec une vitesse de rotation des galets trop faible: 0,15 tours/min. Alors que ce facteur ne devrait manifestement exercer aucune influence sur les boucles d'hystérésis observées au cours de cycles d'extension et de rétraction, on constate des pertes d'énergie nettement plus grandes qu'en tournant à 5 tours/seconde, pour la même vitesse d'allongement.

Cette anomalie, due comme on va le voir, à l'adhérence du caoutchouc sur les gorges des galets, constitue un «artefact».

III. Distribution des allongements subis par une éprouvette anneau le long de sa fibre neutre

Si le frottement pouvait être totalement supprimé entre les galets et la surface de l'anneau, les contraintes seraient idéalement uniformes tout le long de celui-ci, abstraction faite des tensions secondaires créées par les actions de contact normales aux surfaces et par la diminution du rayon de courbure de l'éprouvette au voisinage des galets. Comme il n'est pas possible de réduire suffisamment le coefficient élevé de frottement du caoutchouc, la distribution réelle des allongements le long de la fibre neutre de l'anneau se rapproche toujours plus ou moins de celle représentée sur le schéma de la Fig. 3.

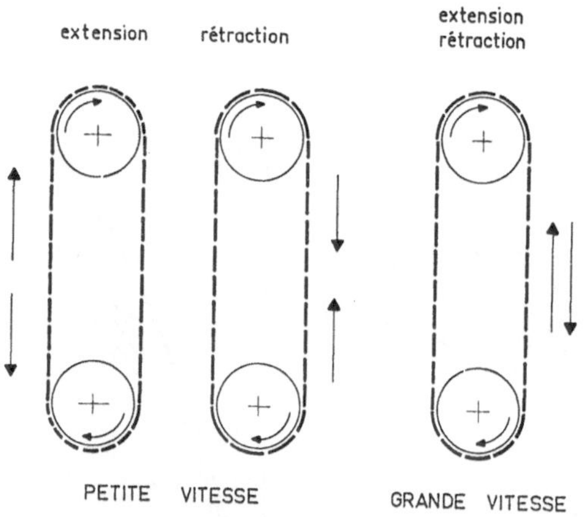

Fig. 3. Distribution schématique de l'allongement d'une éprouvette anneau selon la vitesse de rotation des galets

En raison des forces d'adhérence précédentes, tout se passe en effet comme si des éléments venant juste au contact d'un galet étaient figés dans la configuration où ils se trouvaient à cet instant et conservaient celle-ci jusqu'à ce qu'ils quittent le galet. Lorsque la vitesse linéaire de rotation de l'anneau est relativement lente par rapport à la vitesse de déformation des brins, deux cas peuvent se produire selon que les galets sont en train de s'écarter ou de se rapprocher (partie gauche du schéma). Pendant la phase d'extension, les éléments qui viennent successivement au contact sont évidemment de plus en plus étirés et il s'en suit, à un instant donné, un gradient négatif d'allongement le long des parties curvilignes, les éléments situés vers l'amont du galet étant plus déformés que vers l'aval. L'allongement moyen, calculé à partir de la longueur totale de la fibre neutre est, dans ces conditions, inférieur à celui des brins, dont dépend directement en définitive la force de traction exercée par l'éprouvette.

Inversement, pendant la phase de rétraction, les éléments situés vers l'amont sont alors au contraire moins étirés que vers l'aval et l'allongement moyen supérieur à celui des brins.

Ainsi, l'erreur commise par rapport à la valeur de la déformation dans la région utile de l'éprouvette change de signe pour la branche aller et la branche retour d'un cycle de traction complet et il en résulte un apparent effet d'hystérésis sans aucun rapport avec les propriétés mécaniques intrinsèques de l'éprouvette.

L'importance de ces gradients devant évidemment diminuer lorsqu'on fait croître, toutes choses égales, la vitesse de rotation des galets, on a cherché la vitesse d'entraînement de ces derniers à partir de laquelle cet effet devient négligeable en supposant, afin de simplifier le problème, qu'aucun glissement ne se produit entre l'anneau et les galets.

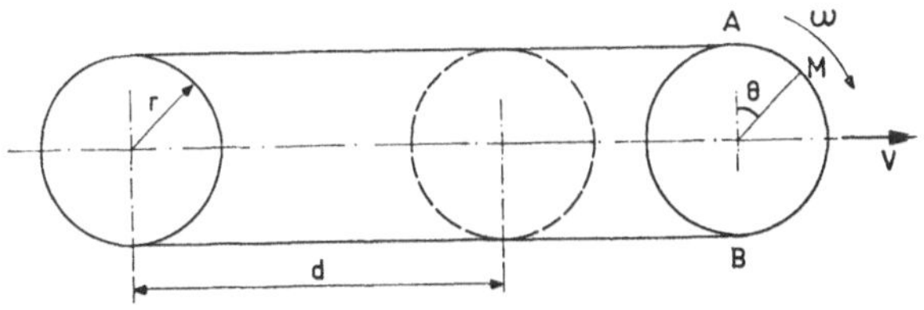

Fig. 4. Représentation de la fibre neutre de l'anneau au repos ($t = 0$) et étiré à l'instant t

Soient alors $2y_1(t)$ et $2y_2(t)$ les longueurs qu'auraient au repos les parties de la fibre neutre qui se trouvent respectivement à l'instant t, compté à partir du début de l'extension, dans des configurations respectivement rectilignes et circulaires; soient en outrer le rayon de courbure de la fibre neutre autour des galets, d la distance entre les axes des galets au temps $t = \theta$, v leur vitesse relative de translation, ω leur vitesse angulaire et θ une variable d'intégration permettant de balayer tous les éléments d'une demi-circonférence de contact AB (Fig. 4).

Le probleme revient à trouver la différence entre l'extension $\lambda(t)$ des brins et l'extension moyenne $\lambda_m(t)$ de la totalité de la fibre neutre respectivement égales à:

$$\lambda_m = \frac{d + \pi r + vt}{d + \pi r}$$

$$\lambda = \frac{d + vt}{y_1(t)} .$$

D'apres ce qui précède, l'extension $\lambda_{\theta, t}$ de l'élément M, situé au temps t à la distance $r\theta$ du point d'attaque A, est égale à la valeur de λ au temps $t - \frac{\theta}{\omega}$.

A partir de l'instant $t = \frac{\pi}{\omega}$ où s'installe un régime quasi-permanent de déformation le long de l'anneau, on a donc, lorsque les galets s'écartent[1]:

$$y_2(t) = r \int_0^\pi \frac{d\theta}{\lambda_{\theta, t}} = r \int_0^\pi \frac{y_1(t - \theta/\omega)}{d + \left(t - \frac{\theta}{\omega}\right) v} \, d\theta.$$

D'autre part, puisque $y_1 + y_2 = d + \pi r$, on est conduit à une équation intégrale dont la solution ne nous est pas apparue:

$$y_1(t) = d + \pi r - r \int_0^\pi \frac{y_1(t - \theta/\omega)}{d + \left(t - \frac{\theta}{\omega}\right) v} \cdot d\theta.$$

Cependant, l'hypothèse d'une adhérence totale aux galets n'est probablement pas tout à fait exacte et la complexité des propriétés de frottement du caoutchouc ne permet guère de tenir compte plus explicitement de ce facteur [18, 19]. Dans ces conditions, comme la fonction $y_1(t)$, pour des raisons physiques, doit être continuellement croissante, on peut se contenter de calculer les bornes de l'intégrale en remplaçant $y_1\left(t - \frac{\theta}{\omega}\right)$ par $y_1(t)$ et $y_1\left(t - \frac{\pi}{\omega}\right)$ qui sont alors indépendants de θ.

On aboutit finalement à une limite supérieure et une limite inférieure de l'erreur absolue commise sur l'extension:

$$\lambda - \lambda_m \leq - \varrho \left[\frac{1}{x} \log (1 - x) + 1\right]$$

et

$$\lambda - \lambda_m \geq - \varrho \left[\frac{\frac{1}{x} \log (1 - x)}{1 + \gamma \log \frac{(1 - x)^2}{1 - 2x}} + 1\right].$$

Dans ces expressions, ne figurent que les quantités sans dimension suivantes:

— le rapport ϱ de r au rayon de la fibre neutre de l'anneau non déformé.

[1] Dans le cas où les galets se rapprochent, un calcul analogue peut être effectué avec des conditions initiales différentes.

— le rapport $\gamma = \frac{r\omega}{v}$ de la vitesse linéaire de rotation de l'anneau à la vitesse d'écartement des galets.

— la variable $x = \frac{\varrho}{\gamma(\lambda - \varrho)}$, liée indirectement au temps par l'extension λ.

Toutefois, les conditions de validité sont plus strictes pour la borne inférieure que pour l'autre, car on est amené à poser dans le premier cas $t > \frac{2\pi}{\omega}$, soit $\lambda > 1 + \frac{2\varrho}{\gamma}$, au lieu de $t > \frac{\pi}{\omega}$, soit $\lambda > 1 + \frac{\varrho}{\gamma}$ dans le second.

Lorsque γ est suffisamment grand, on peut utiliser en première approximation le premier terme des développements limités, ce qui conduit à:

$$\frac{1}{2}\,\varrho x\left(1 - \frac{2\varrho}{\lambda - \varrho}\right) \leq \lambda - \lambda_m \leq \frac{1}{2}\,\varrho x$$

ou encore:

$$\frac{\varrho^2}{2\gamma(\lambda - \varrho)}\left(1 - \frac{2\varrho}{\lambda - \varrho}\right) \leq \lambda - \lambda_m \leq \frac{\varrho^2}{2\gamma(\lambda - \varrho)}\,.$$

Sous cette forme, on voit que l'erreur $\lambda - \lambda_m$ doit être d'autant plus faible que le rapport ϱ des rayons est plus petit et le rapport γ des vitesses plus grand. En outre, l'erreur doit diminuer à mesure que l'allongement augmente, ainsi qu'on devait s'y attendre.

A titre d'exemple, on a calculé les limites supérieures d'erreur d'allongement dans le cas des conditions expérimentales de la Fig. 2. Pour une rotation des galets de 0,15 t/sec., on trouve à 100% d'allongement une erreur absolue de 25% d'allongement, alors qu'elle est seulement, au même allongement, respectivement de 2,3% et 0,4% pour des vitesses de rotation de 1 t/sec. et de 5 t/sec. Ces chiffres prouvent qu'en se plaçant dans ces dernières conditions, on est certain que cette source d'erreur reste largement inférieure à celles commises sur la détermination de la force de traction.

IV. Evolution de la forme des courbes de traction sous l'effet de la température

L'importance de l'hystérésis de traction des vulcanisats à la température ambiante, dont la Fig. 5 donne un exemple pour un caoutchouc synthétique courant: le cis polyisoprene „Natsyn“ et une vitesse de déformation moyenne pour l'appareillage précédent, est déjà bien connue. On voit qu'elle se traduit au cours de deux cycles de déformation successifs par un écart massif entre chacune des courbes aller retour, la branche aller du deuxième cycle étant cependant située nettement au-dessous de celle du premier cycle.

Comme la tension observée à un instant donné dépend alors de l'ensemble des déformations antérieurement imposées, on a intérêt à considérer, au moins en première analyse, plutôt l'extension initiale que les phases suivantes de ces cycles, ce qui permet en outre de suivre le phénomène jusqu'à la rupture. Les Figs. 6, 7 et 8 montrent les relations tension — allongement ainsi obtenues à différentes températures avec un caoutchouc Natsyn peu (N 20), ou au contraire très rétifié (N 90), ainsi

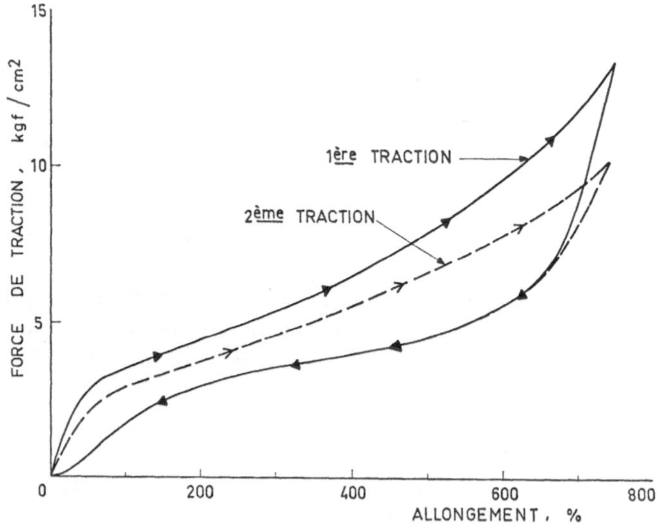

Fig. 5. Premier et deuxième cycle sur le vulcanisat Natsyn-Peroxyde $N_{0,5}$. Température 30 °C — Vitesse d'étirement 60 cm/min.

qu'avec un cis polybutadiène „BR 11" peu rétifié (B 40)[1]. On a évidemment tenu compte sur ces graphiques de la dilatation thermique pour calculer l'allongement réel des anneaux à chaque température T. En outre, on a porté au lieu des forces F expérimentalement observées, rapportées à la section au repos, les valeurs de $F \frac{\varrho_0 T_0}{\varrho T}$ qui correspondent à des forces réduites à une même température de référence $T_0 = 303$ °K, ϱ et ϱ_0 représentant respectivement les masses volumiques des vulcanisats aux températures T et T_0. Ce mode de représentation permet en effet de mettre en évidence l'influence de la température sur la compo-

[1] La composition des mélanges et les conditions de vulcanisation sont indiquées en annexe.

sante transitoire de la tension, puisque des courbes de traction idéalement réversibles devraient alors toutes se superposer, lorsque la température varie, dans le cas d'une élasticité d'origine principalement entropique comme celle du caoutchouc.

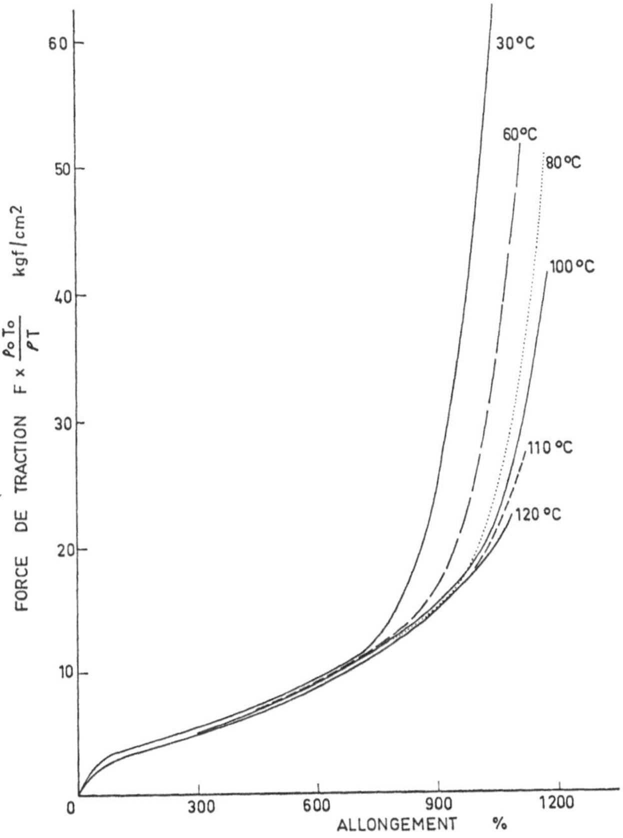

Fig. 6. Influence de la température sur les courbes tension — allongement normées à 30 °C. Cas du vulcanisat Natsyn-peroxyde N_{30}. Vitesse d'étirement 60 cm/min.

De fait, tant que l'allongement n'excède pas 500% environ, les écarts séparant les courbes de traction réduites restent relativement minimes et indépendants de la déformation pour chacun des trois vulcanisats. La cristallisation ne pouvant, même à 30 °C, guère entrer en jeu aux faibles allongements, on est conduit à attribuer dans cette zone la légère diminution de tension constatée à température élevée à la réduction des frottements internes provoqués par les réarrangements de la structure des réseaux macromoléculaires sous l'effet de la déformation.

Dans le domaine des très grands allongements, on ne peut tirer aucune conclusion des résultats fournis par les vulcanisats N 90 et B 40, car ils se rompent à des allongements trop faibles dès que la température s'élève. Pour le vulcanisat N 20, qui permet d'atteindre des allongements de l'ordre de 1100% à 110 °C, on constante une divergence rapide des courbes de traction à partir de 700% d'allongement. De plus, les forces

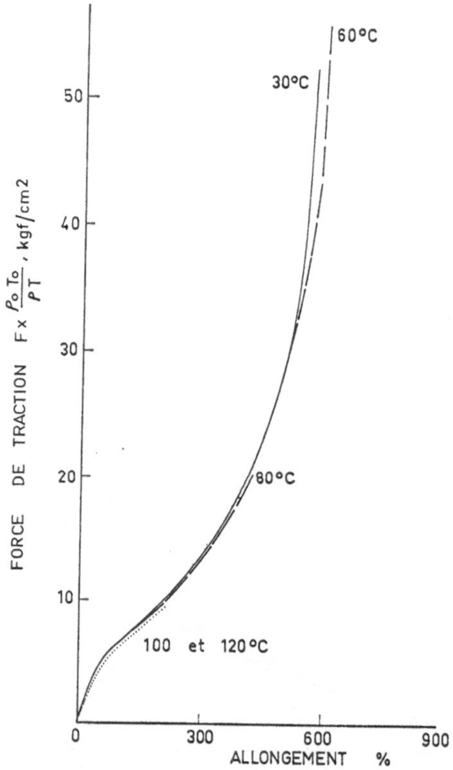

Fig. 7. Influence de la température sur les courbes tension — allongement normées à 30 °C. Cas du vulcanisat Natsyn-peroxyde N_{90}. Vitesse d'étirement 60 cm/min.

de traction correspondant dans cette zone à un même allongement diminuent considérablement entre 30 ° et 80 °C, mais varient relativement peu ensuite lorsque la température s'élève à 120 °C.

Ces faits indiquent que c'est alors la cristallisation, et non plus la viscoélasticité, qui provoque le relèvement beaucoup plus rapide des courbes de traction lorsque la température tombe au-dessous de 80 °C environ. En accord avec d'autres observations [15, 16], la cristallisation ne semble, par contre, guère intervenir à plus haute température, même pour des allongements aussi grands, car les différences de tension restent dans ce cas de l'ordre de celles causées aux allongements modérés par la viscoélasticité.

Au total, il apparaît donc à ce stade qu'une élévation suffisamment importante de la température d'essai permet de déformer les vulcanisats dans des conditions très voisines de l'équilibre. Par suite, les courbes réduites de traction de la Fig. 9, relatives à une série de vulcanisats

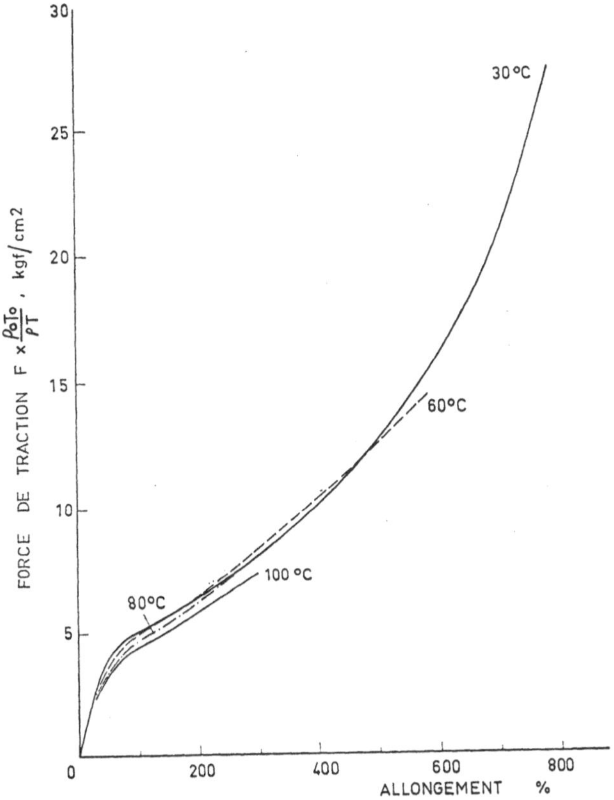

Fig. 8. Influence de la température sur les courbes tension — allongement normées à 30 °C. Cas du vulcanisat Br 11-peroxyde B_{40}. Vitesse d'étirement 60 cm/min.

Natsyn N_2, $N_{1,5}$, N_1, $N_{0,5}$ étirés à 120 °C, donnent une idée plus objective que les essais de traction ordinaires à température ambiante de l'influence du degré de rétification sur les propriétés purement élastiques de ces matériaux. L'allure de ces courbes, caractérisée par une zone d'inflexion, reste d'ailleurs classique.

Cependant, sur un plan quantitatif, les données précédentes constituent une confirmation directe de l'existence, souvent contestée jusqu'à présent, de fortes déviations par rapport aux théories de l'élasticité des réseaux macromoléculaires basées sur une statistique de chaîne gaussienne aux faibles déformations [1, 20]. Si ces théories étaient vérifiées par l'expérience, on ne devrait en effet observer, en fonction de l'extension λ, aucune variation de $F/(\lambda - \lambda^{-2})$, alors qu'on constate encore dans les conditions précédentes une pente positive notable en fonction $1/\lambda$,

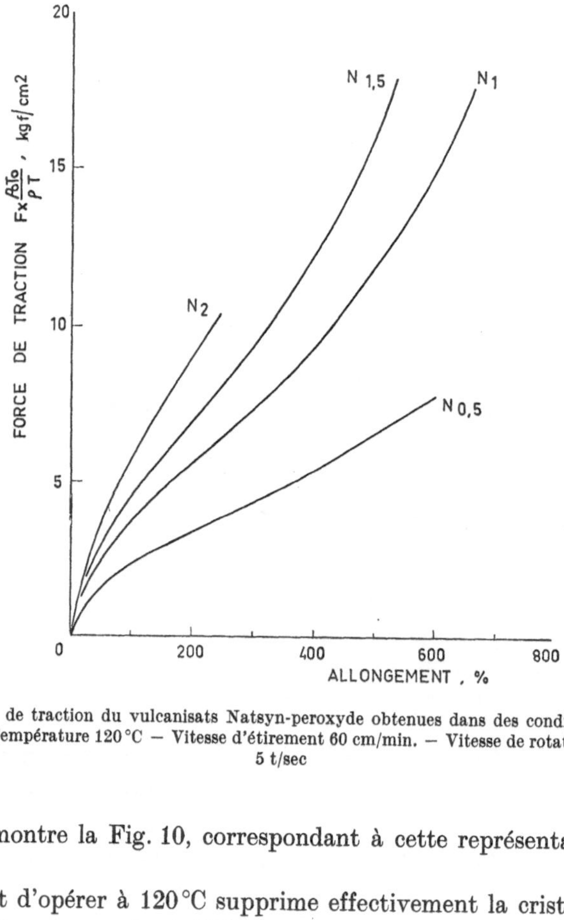

Fig. 9. Courbes de traction du vulcanisats Natsyn-peroxyde obtenues dans des conditions proches de l'équilibre. Température 120 °C – Vitesse d'étirement 60 cm/min. – Vitesse de rotation des galets 5 t/sec

comme le montre la Fig. 10, correspondant à cette représentation conventionelle.

Si le fait d'opérer à 120 °C supprime effectivement la cristallisation, il ne permet pas cependant d'éliminer à coup sûr les effets transitoires d'origine viscoélastique, notamment dans le cas de vulcanisats peu rétifiés, et par suite dotés d'une forte hystérésis [12]. A cet égard, la preuve la plus directe d'un comportement absolument réversible est la coïncidence des branches aller – retour des courbes de traction obtenues au cours d'un cycle d'extension et de rétraction complet, c'est-à-dire la disparition de la boucle d'hystérésis. On s'est donc proposé de recueillir par ce dernier moyen des indications plus précises sur le résidu d'hystérésis viscoélastique à 120 °C.

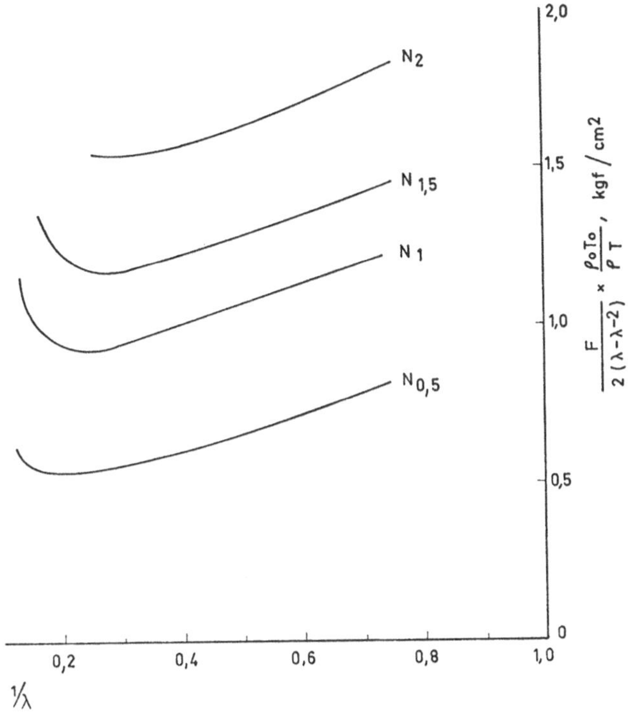

Fig. 10. Représentation conventionelle des courbes tension — allongement obtenues dans des conditions proches de l'équilibre. Vulcanisats Natsyn-peroxyde. Température 120 °C — Vitesse d'étirement 60 cm/min. — Vitesse de rotation des galets 5 t/sec

V. Influence de la viscoélasticité sur les boucles d'hystérésis, en l'absence de cristallisation

En se reportant à la partie droite de la Fig. 2, relative à des conditions de fonctionnement correctes du dynamomètre, on constate que les branches aller — retour d'un cycle effectué entre 0 et 200 % d'allongement sont pratiquement confondues, dans le cas d'un vulcanisat très rétifié tel que N_2. A mesure que décroît le degré de vulcanisation, et par suite le «module», une hystérésis résiduelle croissante se manifeste encore à cette température par des pertes d'énergie, d'ailleurs minimes, au cours de chaque cycle (Tableau 1). Cette conclusion, qui se dégageait aussi de l'étude de la relaxation de tension [13], reste d'ailleurs valable même lorsque l'allongement maximum atteint au cours d'un cycle approche les conditions de la rupture (Fig. 11). Il semble donc que les modifications irréversibles de structure des vulcanisats, souvent notées lors d'essais de traction effectués à la température ambiante [21], ne puissent se produire que lorsque le polymère cristallise sous l'effet de l'étirement.

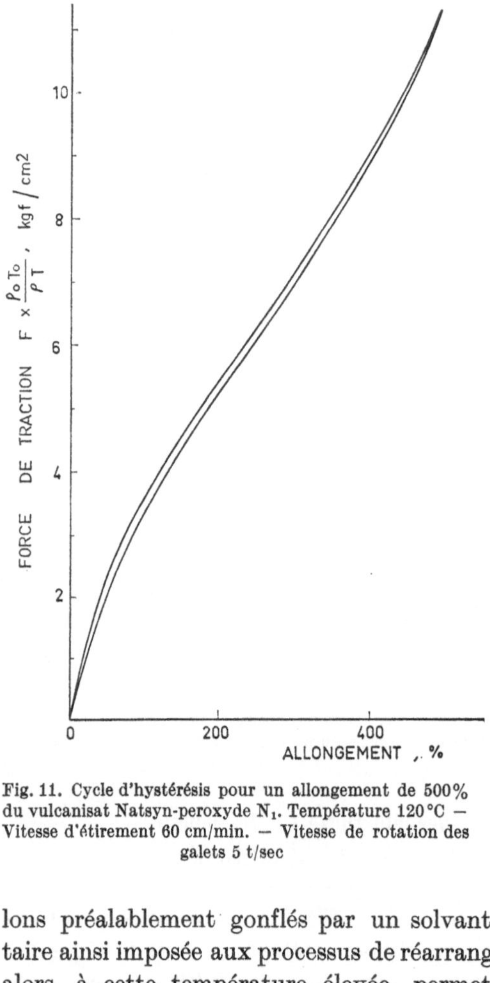

Fig. 11. Cycle d'hystérésis pour un allongement de 500% du vulcanisat Natsyn-peroxyde N_1. Température 120 °C — Vitesse d'étirement 60 cm/min. — Vitesse de rotation des galets 5 t/sec

Dans l'ensemble, la technique proposée peut fournir des courbes de traction réversibles dans le cas de vulcanisats pure-gomme assez rétifiés, et permettrait donc de caractériser leur structure à l'aide d'une théorie statistique de l'élasticité adaptée au comportement mécanique réel du caoutchouc [22, 23]. Toutefois, ceci requiert une grande précision des résultats et, malgré l'amélioration réalisée par rapport aux moyens de conditionnement courants, on ne saurait directement déduire les réactions à l'équilibre de ces données lorsque l'hystérésis résiduelle est encore notable à 120 °C comme pour les caoutchoucs peu rétifiés.

Cet aspect du problème sera d'abord envisagé à notre laboratoire sous l'angle expérimental, en opérant cette fois sur des échantillons préalablement gonflés par un solvant. L'accélération supplémentaire ainsi imposée aux processus de réarrangements moléculaires devrait alors, à cette température élevée, permettre d'éviter dans tous les cas tout effet d'hystérésis. Une fois ce résultat atteint, on pourrait vérifier s'il est possible de calculer les réactions à l'équilibre, à partir des données de traction obtenues, à l'état sec, dans des conditions non absolument réversibles.

Les faits déjà connus sur la cinétique de relaxation viscoélastique des vulcanisats [12—14] rendent en principe accessible une approche rhéologique de ces phénomènes héréditaires, malgré les sérieuses difficultés d'analyse résultant de leur caractère non linéaire [24].

Tableau 1. *Hystérésis mesurée à 120 °C pour un cycle d'allongement de 200%*

Vulcanisat	*Perte d'énergie* %
$N_{0,5}$	5,7
N_1	3,8
$N_{1,5}$	2,2
N_2	1

Conclusions

L'étude de la thermoélasticité a été gênée jusqu'à présent par l'absence d'une méthode commode et précise permettant de distinguer les effets transitoires des propriétés purement élastiques.

Il ressort de ce travail qu'un essai d'allongement effectué dans un intervalle de températures allant de 30° à 120 °C constitue à cet égard un moyen d'investigation efficace, à condition d'utiliser des éprouvettes en forme d'anneau étirées entre deux galets tournant à une vitesse suffisante. Comme l'indique le calcul, on supprime ainsi des gradiants d'allongement le long des galets, ce qui évite des erreurs de mesure importantes et une source d'hystérésis purement artificielle au cours de cycles de déformation de grande amplitude.

Grâce à cette précaution, on a pu enregistrer des courbes tension — allongement reflétant objectivement les propriétés intrinsèques de deux élastomères stéréospécifiques (cis polyisoprène et cis polybutadiène) plus ou moins rétifiés par un peroxyde organique. La forme de ces courbes montre qu'en accord avec les conclusions d'autres études optiques ou thermodynamiques on ne décèle plus de cristallisation par étirement à des températures supérieures à 80° ou 90 °C environ. Toutefois, une hystérésis viscoélastique, d'ailleurs minime, subsiste encore à cette température élevée, du moins pour les caoutchoucs peu rétifiés, qui donne lieu dans ces conditions à des pertes d'énergie de l'ordre de 5%.

On espère neanmoins aboutir encore en ce cas à des courbes de traction totalement réversibles jusqu'au voisinage des conditions de rupture, en étendant ultérieurement cette étude aux vulcanisats gonflés par un solvant et en s'appuyant, pour les caoutchoucs présentant une certaine hystérésis à l'état sec, sur une analyse rhéologique de ces phénomènes héréditaires.

Remerciements

Les auteurs expriment leur gratitude à l'International Institute of Synthetic Rubber Producers pour le soutien financier accordé à ce travail.

Annexe

Composition des vulcanisats

Référence	N_{20}	N_{90}	$N_{0,5}$	N_1	$N_{1,5}$	N_2	B_{40}
Caoutchouc Natsyn	100	100	100	100	100	100	—
Br—11	—	—	—	—	—	—	100
Peroxyde de dicumyle	3,5	3,5	0,5	1,0	1,5	2,0	0,4
Vulcanisation température °C	135	135	147	147	147	147	135
Durée, min.	20	90	40	40	40	40	40

Références

[1] Treloar, L. R. G.: The Physics of Rubber Elasticity, 2ème édit., p. 30 et 91. Oxford: Clarendon Press. 1958.
[2] Gee, G.: Trans. Faraday Soc. **42**, 585 (1946).
[3] Anthony, R. L., R. H. Caston et E. Guth: J. Phys. Chem. **46**, 826 (1942).
[4] Wood, L. A., et F. L. Roth: J. Appl. Phys. **15**, 781 (1944).
[5] Thirion, P., et R. Chasset: Rev. Gén. Caoutchouc **36**, 688 (1959).
[6] Ciferri, A.: J. Polymer Sci. **54**, 149 (1961).
[7] Roe, R. J., et W. R. Krigbaum: J. Polymer Sci. **61**, 167 (1962).
[8] Mullins, L.: J. Polymer Sci. **19**, 225 (1956).
[9] Gumbrell, S. M., L. Mullins et R. S. Rivlin: Trans. Faraday Soc. **49**, 1495 (1953).
[10] Kraus, G., et G. A. Mocvzgemba: J. Polymer Sci. **2 A**, 277 (1964).
[11] Hoff, B. M. E. van der, et E. J. Buckler: J. Macromol. Sci. (Chem.) **A1** (4), 747 (1967).
[12] Chasset, R., et P. Thirion: Proceed. Int. Conf. on Physics of Non Crystalline Solids, p. 346. Delft: North Holland Publ. Comp. 1964. — Rev. Gén. Caoutchouc **44**, 1041 (1967).
[13] Thirion, P., et R. Chasset: Chimie et Ind. (Génie Chimique) **97**, 617 (1967).
[14] Chasset, R., et P. Thirion: Communication à la Natural Rubber Conference, Kuala Lumpur (1968), [à paraitre].
[15] Yan, W., et R. S. Stein: J. Polymer Sci. **26 A**, 1 (1968).
[16] Smith, K. J., Jr., A. Greene et A. Ciferri: Rubber Chem. Tech. **39**, 685 (1966).
[17] Thirion, P., et R. Chasset: Proc. 4th Rubber Tech. Conf., p. 338. Cambridge: Heffer. 1962. — Rev. Gén. Caoutchouc **41**, 271 (1964).
[18] Thirion, P., et R. Chasset: Rev. Gén. Caoutchouc **43**, 211 (1966).
[19] Schallamach, A.: Rubber Chem. Tech. **41**, 209 (1968).
[20] Krigbaum, W. R., et R. J. Roe: Rubber Chem. Tech. **38**, 1039 (1965).
[21] Harwood, J. A. C., et A. R. Payne: J. Appl. Polymer Sci. **10**, 1203 (1966).
[22] Thirion, P., et R. Chasset: C. R. Acad. Sci. Paris **264**, 958 (1967).
[23] Thirion, P., et R. Chasset: Rev. Gén. Caoutchouc **45**, 859 (1968).
[24] Zapas, L. J., et T. Craft: J. Res. NBS **69 A**, 541 (1965).